Embedded Hardware

Newnes Know It All Series

PIC Microcontrollers: Know It All
Lucio Di Jasio, Tim Wilmshurst, Dogan Ibrahim, John Morton,
Martin Bates, Jack Smith, D.W. Smith, and Chuck Hellebuyck
ISBN: 978-0-7506-8615-0

Embedded Software: Know It All
Jean Labrosse, Jack Ganssle, Tammy Noergaard, Robert Oshana, Colin Walls, Keith Curtis,
Jason Andrews, David J. Katz, Rick Gentile, Kamal Hyder, and Bob Perrin
ISBN: 978-0-7506-8583-2

Embedded Hardware: Know It All
Jack Ganssle, Tammy Noergaard, Fred Eady, Creed Huddleston, Lewin Edwards,
David J. Katz, Rick Gentile, Ken Arnold, Kamal Hyder, and Bob Perrin
ISBN: 978-0-7506-8584-9

Wireless Networking: Know It All
Praphul Chandra, Daniel M. Dobkin, Alan Bensky, Ron Olexa,
David A. Lide, and Farid Dowla
ISBN: 978-0-7506-8582-5

RF & Wireless Technologies: Know It All
Bruce Fette, Roberto Aiello, Praphul Chandra, Daniel Dobkin,
Alan Bensky, Douglas Miron, David A. Lide, Farid Dowla, and Ron Olexa
ISBN: 978-0-7506-8581-8

For more information on these and other Newnes titles visit: www.newnespress.com

Embedded Hardware

Jack Ganssle
Tammy Noergaard
Fred Eady
Lewin Edwards
David J. Katz
Rick Gentile
Ken Arnold
Kamal Hyder
Bob Perrin
Creed Huddleston

AMSTERDAM • BOSTON • HEIDELBERG • LONDON
NEW YORK • OXFORD • PARIS • SAN DIEGO
SAN FRANCISCO • SINGAPORE • SYDNEY • TOKYO

Newnes is an imprint of Elsevier

Cover image by iStockphoto
Newnes is an imprint of Elsevier
30 Corporate Drive, Suite 400, Burlington, MA 01803, USA
Linacre House, Jordan Hill, Oxford OX2 8DP, UK

Copyright © 2008 by Elsevier Inc. All rights reserved.

No part of this publication may be reproduced, stored in a retrieval system, or transmitted in any form or by any means, electronic, mechanical, photocopying, recording, or otherwise, without the prior written permission of the publisher.

Permissions may be sought directly from Elsevier's Science & Technology Rights Department in Oxford, UK: phone: (+44) 1865 843830, fax: (+44) 1865 853333, E-mail: permissions@elsevier.com. You may also complete your request online via the Elsevier homepage (http://elsevier.com), by selecting "Support & Contact" then "Copyright and Permission" and then "Obtaining Permissions."

 Recognizing the importance of preserving what has been written, Elsevier prints its books on acid-free paper whenever possible.

Library of Congress Cataloging-in-Publication Data
Ganssle, Jack G.
  Embedded hardware / Jack Ganssle ... [et al.].
    p. cm.
  Includes index.
  ISBN 978-0-7506-8584-9 (alk. paper)
  1. Embedded computer systems. I. Title.
  TK7895.E42G37 2007
  004.16—dc22

                                                              2007027559

British Library Cataloguing-in-Publication Data
A catalogue record for this book is available from the British Library.

For information on all Newnes publications
visit our Web site at www.books.elsevier.com

07 08 09 10   10 9 8 7 6 5 4 3 2 1

Typeset by Charon Tec Ltd (A Macmillan Company), Chennai, India
www.charontec.com
Printed in the United States of America

Working together to grow
libraries in developing countries

www.elsevier.com | www.bookaid.org | www.sabre.org

ELSEVIER   BOOK AID International   Sabre Foundation

Contents

About the Authors xiii

Chapter 1: Embedded Hardware Basics 1
  1.1 Lesson One on Hardware: Reading Schematics 1
  1.2 The Embedded Board and the von Neumann Model 5
  1.3 Powering the Hardware 9
      1.3.1 A Quick Comment on Analog Vs. Digital Signals 10
  1.4 Basic Electronics 12
      1.4.1 DC Circuits 12
      1.4.2 AC Circuits 21
      1.4.3 Active Devices 28
  1.5 Putting It Together: A Power Supply 32
      1.5.1 The Scope 35
      1.5.2 Controls 35
      1.5.3 Probes 38
  Endnotes 41

Chapter 2: Logic Circuits 43
  2.1 Coding 43
      2.1.1 BCD 46
  2.2 Combinatorial Logic 47
      2.2.1 NOT Gate 47
      2.2.2 AND and NAND Gates 48
      2.2.3 OR and NOR Gates 49
      2.2.4 XOR 50
      2.2.5 Circuits 50
      2.2.6 Tristate Devices 53
  2.3 Sequential Logic 53
      2.3.1 Logic Wrap-Up 57
  2.4 Putting It All Together: The Integrated Circuit 58
  Endnotes 61

Chapter 3: Embedded Processors ... 63
3.1 Introduction ... 63
3.2 ISA Architecture Models ... 65
- 3.2.1 Operations ... 65
- 3.2.2 Operands ... 68
- 3.2.3 Storage ... 69
- 3.2.4 Addressing Modes ... 71
- 3.2.5 Interrupts and Exception Handling ... 72
- 3.2.6 Application-Specific ISA Models ... 72
- 3.2.7 General-Purpose ISA Models ... 74
- 3.2.8 Instruction-Level Parallelism ISA Models ... 76

3.3 Internal Processor Design ... 78
- 3.3.1 Central Processing Unit (CPU) ... 82
- 3.3.2 On-Chip Memory ... 99
- 3.3.3 Processor Input/Output (I/O) ... 113
- 3.3.4 Processor Buses ... 130

3.4 Processor Performance ... 131
- 3.4.1 Benchmarks ... 133

Endnotes ... 133

Chapter 4: Embedded Board Buses and I/O ... 137
4.1 Board I/O ... 137
4.2 Managing Data: Serial vs. Parallel I/O ... 140
- 4.2.1 Serial I/O Example 1: Networking and Communications: RS-232 ... 144
- 4.2.2 Example: Motorola/Freescale MPC823 FADS Board RS-232 System Model ... 146
- 4.2.3 Serial I/O Example 2: Networking and Communications: IEEE 802.11 Wireless LAN ... 148
- 4.2.4 Parallel I/O ... 153
- 4.2.5 Parallel I/O Example 3: "Parallel" Output and Graphics I/O ... 153
- 4.2.6 Parallel and Serial I/O Example 4: Networking and Communications—Ethernet ... 156
- 4.2.7 Example 1: Motorola/Freescale MPC823 FADS Board Ethernet System Model ... 158
- 4.2.8 Example 2: Net Silicon ARM7 (6127001) Development Board Ethernet System Model ... 160
- 4.2.9 Example 3: Adastra Neptune x86 Board Ethernet System Model ... 161

4.3 Interfacing the I/O Components ... 161
- 4.3.1 Interfacing the I/O Device with the Embedded Board ... 162
- 4.3.2 Interfacing an I/O Controller and the Master CPU ... 164

4.4 I/O and Performance 165
4.5 Board Buses.. 166
4.6 Bus Arbitration and Timing.. 168
    4.6.1 Nonexpandable Bus: I^2C Bus Example 174
    4.6.2 PCI (Peripheral Component Interconnect)
          Bus Example: Expandable 175
4.7 Integrating the Bus with Other Board Components 179
4.8 Bus Performance 180

Chapter 5: Memory Systems 183
5.1 Introduction 183
5.2 Memory Spaces 183
    5.2.1 L1 Instruction Memory 186
    5.2.2 Using L1 Instruction Memory for Data Placement.................................... 186
    5.2.3 L1 Data Memory 187
5.3 Cache Overview 187
    5.3.1 What Is Cache? 188
    5.3.2 Direct-Mapped Cache 190
    5.3.3 Fully Associative Cache.. 190
    5.3.4 N-Way Set-Associative Cache 191
    5.3.5 More Cache Details... 191
    5.3.6 Write-Through and Write-Back Data Cache... 193
5.4 External Memory.. 195
    5.4.1 Synchronous Memory 195
    5.4.2 Asynchronous Memory... 203
    5.4.3 Nonvolatile Memories 206
5.5 Direct Memory Access... 214
    5.5.1 DMA Controller Overview 215
    5.5.2 More on the DMA Controller 216
    5.5.3 Programming the DMA Controller 218
    5.5.4 DMA Classifications 228
    5.5.5 Register-Based DMA 228
    5.5.6 Descriptor-Based DMA 231
    5.5.7 Advanced DMA Features.. 234
Endnotes 236

Chapter 6: Timing Analysis in Embedded Systems.. 239
6.1 Introduction 239
6.2 Timing Diagram Notation Conventions 239
    6.2.1 Rise and Fall Times... 241

    6.2.2 Propagation Delays 241
    6.2.3 Setup and Hold Time..241
    6.2.4 Tri-State Bus Interfacing 243
    6.2.5 Pulse Width and Clock Frequency 244
6.3 Fan-Out and Loading Analysis: DC and AC 244
    6.3.1 Calculating Wiring Capacitance..247
    6.3.2 Fan-Out When CMOS Drives LSTTL 249
    6.3.3 Transmission-Line Effects 251
    6.3.4 Ground Bounce 253
6.4 Logic Family IC Characteristics and Interfacing 255
    6.4.1 Interfacing TTL Compatible Signals to 5 V CMOS 258
6.5 Design Example: Noise Margin Analysis Spreadsheet..261
6.6 Worst-Case Timing Analysis Example..270
Endnotes 272

Chapter 7: Choosing a Microcontroller and Other Design Decisions 273
7.1 Introduction 273
7.2 Choosing the Right Core 276
7.3 Building Custom Peripherals with FPGAs..281
7.4 Whose Development Hardware to Use—Chicken or Egg?...282
7.5 Recommended Laboratory Equipment..285
7.6 Development Toolchains 286
7.7 Free Embedded Operating Systems 289
7.8 GNU and You: How Using "Free" Software Affects Your Product............................295

Chapter 8: The Essence of Microcontroller Networking: RS-232.........................301
8.1 Introduction 301
8.2 Some History...303
8.3 RS-232 Standard Operating Procedure 305
8.4 RS-232 Voltage Conversion Considerations 308
8.5 Implementing RS-232 with a Microcontroller 310
    8.5.1 Basic RS-232 Hardware..310
    8.5.2 Building a Simple Microcontroller RS-232 Transceiver 313
8.6 Writing RS-232 Microcontroller Routines in BASIC 333
8.7 Building Some RS-232 Communications Hardware...339
    8.7.1 A Few More BASIC RS-232 Instructions..339
8.8 I^2C: The Other Serial Protocol 342
    8.8.1 Why Use I^2C?...343
    8.8.2 The I^2C Bus 344
    8.8.3 I^2C ACKS and NAKS 347

8.8.4 More on Arbitration and Clock Synchronization 347
8.8.5 I^2C Addressing 351
8.8.6 Some I^2C Firmware 352
8.8.7 The AVR Master I^2C Code 352
8.8.8 The AVR I^2C Master-Receiver Mode Code 358
8.8.9 The PIC I^2C Slave-Transmitter Mode Code 359
8.8.10 The AVR-to-PIC I^2C Communications Ball 365
8.9 Communication Options 378
  8.9.1 The Serial Peripheral Interface Port 378
  8.9.2 The Controller Area Network 380
  8.9.3 Acceptance Filters 386
Endnote 387

Chapter 9: Interfacing to Sensors and Actuators 389
9.1 Introduction 389
9.2 Digital Interfacing 389
  9.2.1 Mixing 3.3 and 5 V Devices 389
  9.2.2 Protecting Digital Inputs 392
  9.2.3 Expanding Digital Inputs 398
  9.2.4 Expanding Digital Outputs 402
9.3 High-Current Outputs 404
  9.3.1 BJT-Based Drivers 405
  9.3.2 MOSFETs 409
  9.3.3 Electromechanical Relays 411
  9.3.4 Solid-State Relays 417
9.4 CPLDs and FPGAs 418
9.5 Analog Interfacing: An Overview 420
  9.5.1 ADCs 420
  9.5.2 Project 1: Characterizing an Analog Channel 421
9.6 Conclusion 434
Endnote 435

Chapter 10: Other Useful Hardware Design Tips and Techniques 437
10.1 Introduction 437
10.2 Diagnostics 437
10.3 Connecting Tools 438
10.4 Other Thoughts 439
10.5 Construction Methods 440
  10.5.1 Power and Ground Planes 441
  10.5.2 Ground Problems 441

- 10.6 Electromagnetic Compatibility ... 442
- 10.7 Electrostatic Discharge Effects ... 442
  - 10.7.1 Fault Tolerance ... 443
- 10.8 Hardware Development Tools ... 444
  - 10.8.1 Instrumentation Issues ... 445
- 10.9 Software Development Tools ... 445
- 10.10 Other Specialized Design Considerations ... 446
  - 10.10.1 Thermal Analysis and Design ... 446
  - 10.10.2 Battery-Powered System Design Considerations ... 447
- 10.11 Processor Performance Metrics ... 448
  - 10.11.1 IPS ... 448
  - 10.11.2 OPS ... 448
  - 10.11.3 Benchmarks ... 449

Appendix A: Schematic Symbols ... 451

Appendix B: Acronyms and Abbreviations ... 459

Appendix C: PC Board Design Issues ... 469
- C.1 Introduction ... 469
- C.2 Resistance of Conductors ... 470
- C.3 Voltage Drop in Signal Leads—"Kelvin" Feedback ... 471
- C.4 Signal Return Currents ... 472
- C.5 Grounding in Mixed Analog/Digital Systems ... 474
- C.6 Ground and Power Planes ... 475
- C.7 Double-Sided versus Multilayer Printed Circuit Boards ... 477
- C.8 Multicard Mixed-Signal Systems ... 478
- C.9 Separating Analog and Digital Grounds ... 479
- C.10 Grounding and Decoupling Mixed-Signal ICs with Low Digital Currents ... 480
- C.11 Treat the ADC Digital Outputs with Care ... 481
- C.12 Sampling Clock Considerations ... 483
- C.13 The Origins of the Confusion About Mixed-Signal Grounding: Applying Single-Card Grounding Concepts to Multicard Systems ... 485
- C.14 Summary: Grounding Mixed-Signal Devices with Low Digital Currents in a Multicard System ... 486
- C.15 Summary: Grounding Mixed-Signal Devices with High Digital Currents in a Multicard System ... 487
- C.16 Grounding DSPs with Internal Phase-Locked Loops ... 487
- C.17 Grounding Summary ... 488
- C.18 Some General PC Board Layout Guidelines for Mixed-Signal Systems ... 489

C.19	Skin Effect 491
C.20	Transmission Lines 493
C.21	Be Careful with Ground Plane Breaks.. 494
C.22	Ground Isolation Techniques 495
C.23	Static PCB Effects 497
C.24	Sample MINIDIP and SOIC Op Amp PCB Guard Layouts............... 500
C.25	Dynamic PCB Effects 502
C.26	Stray Capacitance 503
C.27	Capacitive Noise and Faraday Shields... 504
C.28	The Floating Shield Problem 506
C.29	Buffering ADCs Against Logic Noise.. 506

Endnotes 509
Acknowledgments 509

Index 511

About the Authors

Ken Arnold (Chapters 6 and 10) is the author of *Embedded Controller Hardware Design*. He is the Embedded Computer Engineering Program Coordinator and an instructor at UCSD Extension, as well as founding director of the On-Line University of California, where he manages, develops and teaches courses in engineering and embedded systems design. Ken has been developing commercial embedded systems and teaching others how for more than two decades. As the champion of the embedded program at UCSD, he lead the inception and growth of the program as well as introducing the world's first on-line embedded course well over a decade ago. Ken was also the founder and CEO of HiTech Equipment Corp., CTO of Wireless Innovation, and engineering chief at General Dynamics.

Fred Eady (Chapter 8) is the author of Networking and Internetworking with Microcontrollers. As an engineering consultant, he has implemented communications networks for the space program and designed hardware and firmware for the medical, retail and public utility industries. He currently writes a monthly embedded design column for a popular electronics enthusiast magazine. Fred also composes monthly articles for a popular robotics magazine. Fred has been dabbling in electronics for over 30 years. His embedded design expertise spans the spectrum and includes Intel's 8748 and 8051 microcontrollers, the entire Microchip PIC microcontroller family and the Atmel AVR microcontrollers. Fred recently retired from his consulting work and is focused on writing magazine columns and embedded design books.

Lewin Edwards (Chapter 7) is the author of *Embedded System Design on a Shoestring*. He hails from Adelaide, Australia. His career began with five years of security and encryption software at PC-Plus Systems. The next five years were spent developing networkable multimedia appliances at Digi-Frame in Port Chester, NY. Since 2004 he has been developing security and fire safety devices at a Fortune 100 company in New York. He has written numerous technical articles and three embedded systems books, with a fourth due in early 2008.

Jack Ganssle (Chapters 1, 2, and 10) is the author of *The Firmware Handbook*. He has written over 500 articles and six books about embedded systems, as well as a book about his sailing fiascos. He started developing embedded systems in the early 70s using the 8008. He's started and sold three electronics companies, including one of the bigger embedded tool businesses. He's developed or managed over 100 embedded products, from deep-sea navigation gear to the White House security system... and one instrument that analyzed cow poop! He's currently a member of NASA's Super Problem Resolution Team, a group of outside experts formed to advise NASA in the wake of Columbia's demise, and serves on the boards of several high-tech companies. Jack now gives seminars to companies world-wide about better ways to develop embedded systems.

Rick Gentile (Chapter 5) is the author of Embedded Media Processing. Rick joined ADI in 2000 as a Senior DSP Applications Engineer, and he currently leads the Processor Applications Group, which is responsible for Blackfin, SHARC and TigerSHARC processors. Prior to joining ADI, Rick was a Member of the Technical Staff at MIT Lincoln Laboratory, where he designed several signal processors used in a wide range of radar sensors. He has authored dozens of articles and presented at multiple technical conferences. He received a B.S. in 1987 from the University of Massachusetts at Amherst and an M.S. in 1994 from Northeastern University, both in Electrical and Computer Engineering.

Creed Huddleston (Chapter 8) is the author of *Intelligent Sensor Design Using the Microchip dsPIC*. With over twenty years of experience designing real-time embedded systems, he is President and founder of Real-Time by Design, LLC, a certified Microchip Design Partner based in Raleigh, NC that specializes in the creation of hard real-time intelligent sensing systems. In addition to his duties with Real-Time by Design, Creed also serves on the Advisory Board of Quickfilter Technologies Inc., a Texas-based company producing mixed-signal integrated circuits that provide high-speed analog signal conditioning and digital signal processing in a single package. A graduate of Rice University in Houston, TX with a BSEE degree, Creed performed extensive graduate work in digital signal processing at the University of Texas at Arlington before heading east to start Omnisys. To her great credit and his great fortune, Creed and his wife Lisa have been married for 23 years and have three wonderful children: Kate, Beth, and Dan.

Kamal Hyder (Chapter 9) is the author of *Embedded Systems Design Using the Rabbit 3000 Microprocessor*. He started his career with an embedded microcontroller manufacturer. He then wrote CPU microcode for Tandem Computers for a number of years, and was a Product Manager at Cisco Systems, working on next-generation switching platforms. He is currently with Brocade Communications as Senior Group Product Manager. Kamal's BS is in EE/CS from the University of Massachusetts, Amherst, and he has an MBA in finance/marketing from Santa Clara University.

David Katz (Chapter 5) is the author of Embedded Media Processing. He has over 15 years of experience in circuit and system design. Currently, he is the Blackfin Applications Manager at Analog Devices, Inc., where he focuses on specifying new convergent processors. He has published over 100 embedded processing articles domestically and internationally, and he has presented several conference papers in the field. Previously, he worked at Motorola, Inc., as a senior design engineer in cable modem and automation groups. David holds both a B.S. and M. Eng. in Electrical Engineering from Cornell University.

Walt Kester (Appendix C) is the editor of *Data Conversion Handbook*. He is a corporate staff applications engineer at Analog Devices. For more than 35 years at Analog Devices, he has designed, developed, and given applications support for high-speed ADCs, DACs, SHAs, op amps, and analog multiplexers. Besides writing many papers and articles, he prepared and edited eleven major applications books which form the basis for the Analog Devices world-wide technical seminar series including the topics of op amps, data conversion, power management, sensor signal conditioning, mixed-signal, and practical analog design techniques. Walt has a BSEE from NC State University and MSEE from Duke University.

Tammy Noergaard (Chapters 1, 2, 3, 4, Appendices B and C) is the author of *Embedded Systems Architecture*. Since beginning her embedded systems career in 1995, she has had wide experience in product development, system design and integration, operations, sales, marketing, and training. Noergaard worked for Sony as a lead software engineer developing and testing embedded software for analog TVs. At Wind River she was the liaison engineer between developmental engineers and customers to provide design expertise, systems configuration, systems integration, and training for Wind River embedded software (OS, Java, device drivers, etc.) and all associated hardware for a variety of embedded systems in the Consumer Electronic market. Most recently she was a Field Engineering Specialist and Consultant with Esmertec North America, providing project management, system design, system integration, system configuration, support and expertise for various embedded Java systems using Jbed in everything from control systems to medical devices to digital TVs. Noergaard has lectured to engineering classes at the University of California at Berkeley and Stanford, the Embedded Internet Conference, and the Java User's Group in San Jose, among others.

Bob Perrin (Chapter 9) is the author of *Embedded Systems Design Using the Rabbit 3000 Microprocessor*. He got his start in electronics at the age of nine when his mother gave him a "150-in-one Projects" kit from Radio Shack for Christmas. He grew up programming a Commodore PET. In 1990, Bob graduated with a BSEE from Washington State University. Since then Bob has been working as an engineer designing digital and analog electronics. He has published about twenty technical articles, most with *Circuit Cellar*.

CHAPTER 1
Embedded Hardware Basics

Jack Ganssle
Tammy Noergaard

1.1 Lesson One on Hardware: Reading Schematics

This section is equally important for embedded hardware and software engineers. Before diving into the details, note that it is important for all embedded designers to be able to understand the diagrams and symbols that hardware engineers create and use to describe their hardware designs to the outside world. These diagrams and symbols are the keys to quickly and efficiently understanding even the most complex hardware design, regardless of how much or little practical experience one has in designing hardware. They also contain the information an embedded programmer needs to design any software that requires compatibility with the hardware, and they provide insight to a programmer as to how to successfully communicate the hardware requirements of the software to a hardware engineer.

There are several different types of engineering hardware drawings, including:

- *Block diagrams*, which typically depict the major components of a board (processors, buses, I/O, memory) or a single component (a processor, for example) at a systems architecture or higher level. In short, a block diagram is a basic overview of the hardware, with implementation details abstracted out. While a block diagram can reflect the actual physical layout of a board containing these major components, it mainly depicts how different components or units within a component function together at a systems architecture level. Block diagrams are used extensively throughout this book (in fact, Figures 1.5a–e later in this chapter are examples of block diagrams) because they are the simplest method by which to depict and describe the components within a system. The symbols used within a block diagram are simple, such as squares or rectangles for chips and straight lines for buses. Block diagrams are typically not detailed enough for a software designer to be able to write all the low-level software accurately enough to control the hardware (without a lot of headaches, trial and error, and even some burned-out hardware!). However, they are very useful in communicating a basic overview of the hardware, as well as providing a basis for creating more detailed hardware diagrams.

- *Schematics*. Schematics are electronic circuit diagrams that provide a more detailed view of all the devices within a circuit or within a single component—everything from processors down to resistors. A schematic diagram is not meant to depict the physical layout of the board or component, but provides information on the flow of data in the system, defining what signals are assigned where—which signals travel on the various lines of a bus, appear on the pins of a processor, and so on. In schematic diagrams, *schematic symbols* are used to depict all the components within the system. They typically do not look anything like the physical components they represent but are a type of "shorthand" representation based on some type of schematic symbol standard. A schematic diagram is the most useful diagram to both hardware and software designers trying to determine how a system actually operates, to debug hardware, or to write and debug the software managing the hardware. See Appendix A for a list of commonly used schematic symbols.

- *Wiring diagrams*. These diagrams represent the bus connections between the major and minor components on a board or within a chip. In wiring diagrams, vertical and horizontal lines are used to represent the lines of a bus, and either schematic symbols or more simplified symbols (that physically resemble the other components on the board or elements within a component) are used. These diagrams may represent an approximate depiction of the physical layout of a component or board.

- *Logic diagrams/prints*. Logic diagrams/prints are used to show a wide variety of circuit information using logical symbols (AND, OR, NOT, XOR, and so on) and logical inputs and outputs (the 1's and 0's). These diagrams do not replace schematics, but they can be useful in simplifying certain types of circuits in order to understand how they function.

- *Timing diagrams*. Timing diagrams display timing graphs of various input and output signals of a circuit, as well as the relationships between the various signals. They are the most common diagrams (after block diagrams) in hardware user manuals and data sheets.

Regardless of the type, to understand how to read and interpret these diagrams, it is important to first learn the standard *symbols, conventions*, and *rules* used. Examples of the symbols used in timing diagrams are shown in Table 1.1, along with the conventions for input/output signals associated with each of the symbols.

An example of a timing diagram is shown in Figure 1.1. In this figure, each row represents a different signal. In the case of the signal rising and falling symbols within the diagram, the *rise time* or *fall time* is indicated by the time it takes for the signal to move from LOW to HIGH or vice versa (the entire length of the diagonal line of the symbol). In comparing two signals, a delay is measured at the center of the rising or falling symbols of each signal being compared. In Figure 1.1, there is a fall time delay between signals B and C and signals A and C in the first falling symbol. In comparing the first falling symbol of signals A and B in the figure, no delay is indicated by the timing diagram.

Table 1.1: Timing diagrams symbol table.[1.1]

Symbol	Input Signals	Output Signals
———	Input signal must be valid	Output signal will be valid
✕✕✕	Input signal doesn't affect system, will work regardless	Indeterminate output signal
⟩———	Garbage signal (nonsense)	Output signal not driven (floating), tristate, HiZ, high impedance
╱———	If the input signal rises:	Output signal will rise
╲———	If the input signal falls:	Output signal will fall

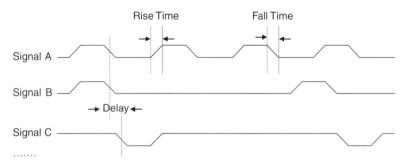

Figure 1.1: Timing diagram example.

Schematic diagrams are much more complex than their timing diagram counterparts. As introduced earlier this chapter, schematics provide a more detailed view of all the devices within a circuit or within a single component. Figure 1.2 shows an example of a schematic diagram.

In the case of schematic diagrams, some of the conventions and rules include:

- A *title section* is located at the bottom of each schematic page, listing information that includes, but is not limited to, the name of the circuit, the name of the hardware engineer responsible for the design, the date, and a list of revisions made to the design since its conception.

- The use of *schematic symbols* indicating the various components of a circuit (see Appendix A).

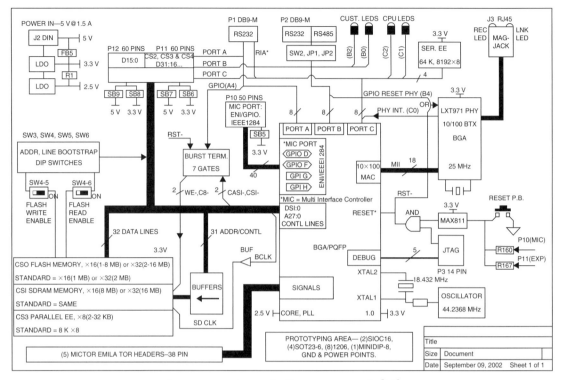

Figure 1.2: Schematic diagram example.[1.2]

- Along with the assigned symbol comes a *label* that details information about the component (i.e., size, type, power ratings, etc.). Labels for components of a symbol, such as the pin numbers of an IC, signal names associated with wires, and so forth are usually located outside of the schematic symbol.

- *Abbreviations* and *prefixes* are used for common units of measurement (i.e., *k* for kilo or 10^3, *M* for mega or 10^6) and these prefixes replace writing out the units and larger numbers.

- *Functional groups* and subgroups of components are typically separated onto different pages.

- *I/O* and *voltage source/ground terminals*. In general, positive voltage supply terminals are located at the top of the page, and negative supply/ground at the bottom. Input components are usually on the left, and output components are on the right.

At the very least, the block and schematic diagrams should contain nothing unfamiliar to anyone working on the embedded project, whether they are coding software or prototyping the hardware. This means becoming familiar with everything from where the name of the diagram is located to how the states of the components shown within the diagrams are represented.

One of the most efficient ways of learning how to learn to read and/or create a hardware diagram is via the *Traister* and *Lisk* method[1.3], which involves:

Step 1. Learning the basic symbols that can make up the type of diagram, such as timing or schematic symbols. To aid in the learning of these symbols, rotate between this step and steps 2 and/or 3.

Step 2. Reading as many diagrams as possible until reading them becomes boring (in that case, rotate between this step and steps 1 and/or 3) or comfortable (so there is no longer the need to look up every other symbol while reading).

Step 3. Writing a diagram to practice simulating what has been read, again until it becomes either boring (which means rotating back through steps 1 and/or 2) or comfortable.

1.2 The Embedded Board and the von Neumann Model

In embedded devices, all the electronics hardware resides on a board, also referred to as a *printed wiring board (PW)* or *printed circuit board (PCB)*. PCBs are often made of thin sheets of fiberglass. The electrical path of the circuit is printed in copper, which carries the electrical signals between the various components connected on the board. All electronic components that make up the circuit are connected to this board, either by soldering, plugging into a socket, or some other connection mechanism. All the hardware on an embedded board is located in the hardware layer of the Embedded Systems Model (see Figure 1.3).

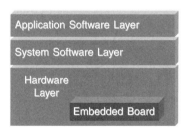

Figure 1.3: Embedded board and the Embedded Systems Model.

At the highest level, the major hardware components of most boards can be classified into five major categories:

- *Central processing unit (CPU)*. The master processor.
- *Memory*. Where the system's software is stored.
- *Input device(s)*. Input slave processors and relative electrical components.

- *Output device(s)*. Output slave processors and relative electrical components.

- *Data pathway(s)/bus(es)*. Interconnects the other components, providing a "highway" for data to travel on from one component to another, including any wires, bus bridges, and/or bus controllers.

These five categories are based on the major elements defined by the *von Neumann model* (see Figure 1.4), a tool that can be used to understand any electronic device's hardware architecture. The von Neumann model is a result of the published work of John von Neumann in 1945, which defined the requirements of a general-purpose electronic computer. Because embedded systems are a type of computer system, this model can be applied as a means of understanding embedded systems hardware.

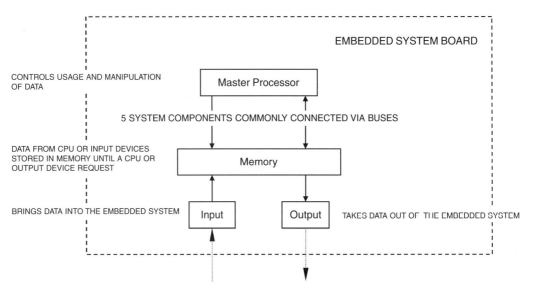

Figure 1.4: Embedded system board organization.[1.4]

Based on the von Neumann architecture model (also referred to as the Princeton architecture).

While board designs can vary widely, as demonstrated in the examples of Figures 1.5a–d, all the major elements on these embedded boards—*and on just about any embedded board*—can be classified as either the master CPU(s), memory, input/output, or bus components.

To understand how the major components on an embedded board function, it is useful to first understand what these components consist of and why. All the components on an embedded board, including the major components introduced in the von Neumann model, are made up of one or some combination of interconnected *basic electronic devices*, such as wires, resistors,

Embedded Hardware Basics

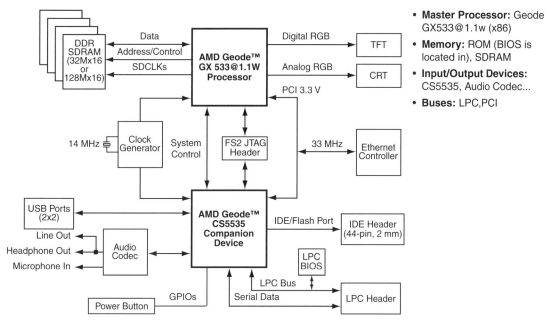

Figure 1.5a: AMD/National Semiconductor x86 reference board.[1.5]

© 2004 Advanced Micro Devices, Inc. Reprinted with permission.

- **Master Processor:** Geode GX533@1.1w (x86)
- **Memory:** ROM (BIOS is located in), SDRAM
- **Input/Output Devices:** CS5535, Audio Codec...
- **Buses:** LPC, PCI

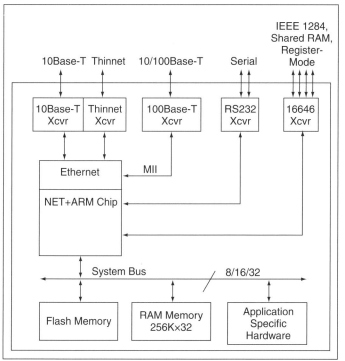

- **Master Processor:** Net+ARM ARM7
- **Memory:** Flash, RAM
- **Input/Output Devices:** 10Base-T transceiver, Thinnet transceiver, 100Base-T transceiver, RS-232 transceiver, 16646 transceiver, ...
- **Buses:** System Bus, MII, ...

Figure 1.5b: Net Silicon ARM7 reference board.[1.6]

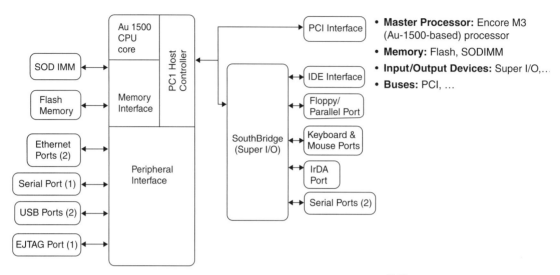

Figure 1.5c: Ampro MIPS reference board.[1.7]

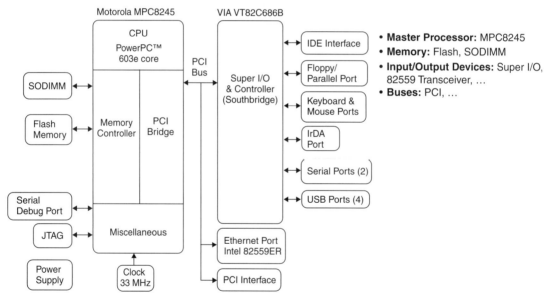

Figure 1.5d: Ampro PowerPC reference board.[1.8]
Copyright Freescale Semiconductor, Inc., 2004. Used by permission.

capacitors, inductors, and diodes. These devices also can act to connect the major components of a board together. At the highest level, these devices are typically classified as either *passive* or *active* components. In short, passive components include devices such as wires, resistors, capacitors and inductors that can only receive or store power. Active components, on the other hand, include devices such as transistors, diodes, and integrated circuits (ICs) that are capable

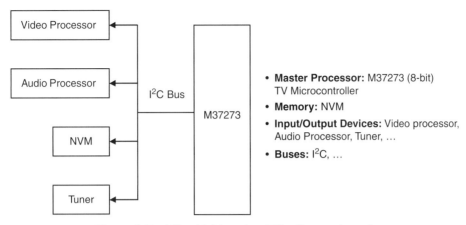

Figure 1.5e: Mitsubishi analog TV reference board.

of delivering as well as receiving and storing power. In some cases, active components themselves can be made up of passive components. Within the passive and active families of components, these circuit devices essentially differ according to how they respond to *voltage* and *current*.

1.3 Powering the Hardware

Power is the rate that energy is expended or work is performed. This means that in alternating current (AC) and direct current (DC) circuits, the power associated with each element on the board equals the current through the element multiplied by the voltage across the element (P = VI). Accurate power and energy calculations must be done for all elements on an embedded board to determine the power consumption requirements of that particular board. This is because each element can only handle a certain type of power, so AC-DC converters, DC-AC converters, direct AC-AC converters, and so on may be required. Also, each element has a limited amount of power that it requires to function, that it can handle, or that it dissipates. These calculations determine the type of voltage source that can be used on a board and how powerful the voltage source needs to be.

In embedded systems, both AC and DC voltage sources are used because each current generation technique has its pros and cons. AC is easier to generate in large amounts using generators driven by turbines turned by everything from wind to water. Producing large amounts of DC from electrochemical cells (batteries) is not as practical. Also, because transmitting current over long transmission lines results in a significant loss of energy due to the resistance of the wire, most modern electric company facilities transmit electricity to outlets in AC current, since AC can be transformed to lower or higher voltages much more easily than DC. With AC, a device called a *transformer*, located at the service provider, is used to efficiently transmit current over long distances with lower losses. The transformer is a device that transfers electrical energy from one circuit to another and can make changes to the current and voltage during the transfer. The service provider transmits lower levels of current at a higher voltage rate from the power plant, and then a transformer at the customer site decreases the voltage to the value required. On the flip side, at very high voltages, wires offer less resistance to DC than AC, thus making DC more efficient to transmit than AC over very long distances.

Some embedded boards integrate or plug into *power supplies*. Power supplies can be either AC or DC. To use an AC power supply to supply power to components using only DC, an AC-to-DC converter can be used to convert AC to the lower DC voltages required by the various components on an embedded board, which typically require 3.3, 5, or 12 volts.

> **Note:** Other types of converters, such as DC-to-DC, DC-to-AC, or direct AC-to-AC can be used to handle the required power conversions for devices that have other requirements.

Other embedded boards or components on a board (such as nonvolatile memory, discussed in more detail in Chapter 5) rely on *batteries* as voltage sources, which can be more practical for providing power because of their size. Battery-powered boards don't rely on a power plant for energy, and they allow portability of embedded devices that don't need to be plugged into an outlet. Also, because batteries supply DC current, no mechanism is needed to convert AC to DC for components that require DC, as is needed with boards that rely on a power supply and outlet supplying AC. Batteries, however, have a limited life and must be either recharged or replaced.

1.3.1 A Quick Comment on Analog vs. Digital Signals

A digital system processes only digital data, which is data represented by only 0's and 1's. On most boards, two voltages represent "0" and "1," since all data is represented as some combination of 1's and 0's. No voltage (0 volts) is referred to as ground, VSS, or low, and 3, 5, or 12 volts are commonly referred to as VCC, VDD, or high. All signals within the system are one of the two voltages or are transitioning to one of the two voltages. Systems can define "0" as low and "1" as high, or some range of 0–1 volts as LOW and 4–5 volts as HIGH, for instance. Other signals can base the definition of a "1" or "0" on edges (low to high) or (high to low).

Because most major components on an embedded board, such as processors, inherently process the 1's and 0's of digital signals, a lot of embedded hardware is digital by nature. However, an embedded system can still process analog signals, which are continuous—that is, not only

1's and 0's but values in between as well. Obviously, a mechanism is needed on the board to convert analog signals to digital signals. An analog signal is digitized by a sampling process, and the resulting digital data can be translated back into a voltage "wave" that mirrors the original analog waveform.

Real-World Advice

> **Inaccurate Signals: Problems with Noise in Analog and Digital Signals**
>
> One of the most serious problems in both the analog and digital signal realm involves noise distorting incoming signals, thus corrupting and affecting the accuracy of data. Noise is generally any unwanted signal alteration from an input source, any part of the input signal generated from something other than a sensor, or even noise generated from the sensor itself. Noise is a common problem with analog signals. Digital signals, on the other hand, are at greater risk if the signals are not generated locally to the embedded processor, so any digital signals coming across a longer transmission medium are the most susceptible to noise problems.
>
> Analog noise can come from a wide variety of sources—radio signals, lightning, power lines, the microprocessor, or the analog sensing electronics themselves. The same is true for digital noise, which can come from mechanical contacts used as computer inputs, dirty slip rings that transmit power/data, limits in accuracy/dependability of input source, and so forth.
>
> The key to reducing either analog or digital noise is: (1) to follow basic design guidelines to avoid problems with noise. In the case of analog noise, this includes not mixing analog and digital grounds, keeping sensitive electronic elements on the board a sufficient distance from elements switching current, limiting length of wires with low signal levels/high impedance, etc. With digital signals, this means routing signal wires away from noise-inducing high current cables, shielding wires, transmitting signals using correct techniques, etc. (2) to clearly identify the root cause of the problem, which means exactly what is causing the noise. With point (2), once the root cause of the noise has been identified, a hardware or software fix can be implemented. Techniques for reducing analog noise include filtering out frequencies not needed and averaging the signal inputs, whereas digital noise is commonly addressed via transmitting correction codes/parity bits and/or adding additional hardware to the board to correct any problems with received data.
>
> —Based on the articles "Minimizing Analog Noise" (May 1997), "Taming Analog Noise" (November 1992), and "Smoothing Digital Inputs" (October 1992), by Jack Ganssle, in Embedded Systems Programming Magazine.

1.4 Basic Electronics

In this section, we will review some electronics fundamentals.

1.4.1 DC Circuits

DC means *direct current*, a fancy term for signals that don't change. They're flatlined, like a corpse's EEG or the output from a battery (Figure 1.6). Your PC's power supply makes DC out of the building's AC (alternating current) mains. All digital circuits require DC power supplies.

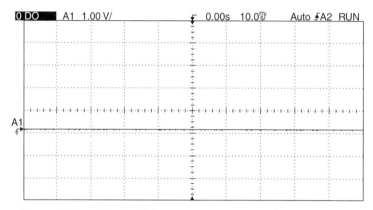

Figure 1.6: A DC signal has a constant, unvarying amplitude.

1.4.1.1 Voltage and Current

We measure the quantity of electricity using voltage and amperage, but both arise from more fundamental physics. Atoms that have a shortage or surplus of electrons are called *ions*. An ion has a positive or negative charge. Two ions of opposite polarity (one plus, meaning it's missing electrons, and the other negative, with one or more extra electrons) attract each other. This attractive force is called the *electromotive force*, commonly known as EMF.

Charge is measured in *coulombs*, where one coulomb is 6.25×10^{18} electrons (for negative charges) or protons for positive ones.

An *ampere* is one coulomb flowing past a point for one second. *Voltage* is the force between two points for which one ampere of current will do one *joule* of work, a joule per second being one watt.

Embedded Hardware Basics 13

Figure 1.7: A VOM, even an old-fashioned analog model like this $10 Radio Shack model, measures DC voltage as well or better than a scope.

But few electrical engineers remember these definitions, and none actually use them.

An old but still apt analogy uses water flow through a pipe: Current would be the amount of water flowing through a pipe per unit of time, whereas voltage is the pressure of the water.

The unit of current is the ampere (amp), though in computers an amp is an awful lot of current. Most digital and analog circuits require much less. Table 1.2 shows the most common nomenclatures.

Table 1.2: Common nomenclatures.

Name	Abbreviation	Number of Amps	Where Likely Found
amp	A	1	Power supplies; very high-performance processors may draw many tens of amps
milliamp	mA	.001 amp	Logic circuits, processors (tens or hundreds of mA), generic analog circuits
microamp	μA	10^{-6} amp	Low-power logic, low-power analog, battery-backed RAM
picoamp	pA	10^{-12} amp	Very sensitive analog inputs
femtoamp	fA	10^{-15} amp	The cutting edge of low-power analog measurements

Most embedded systems have a far less extreme range of voltages. Typical logic and microprocessor power supplies range from a volt or 2–5 volts. Analog power supplies rarely exceed plus and minus 15 volts. Some analog signals from sensors might go down to the millivolt (.001 volt) range. Radio receivers can detect microvolt-level signals, but they do this using quite sophisticated noise-rejection techniques.

1.4.1.2 Resistors

As electrons travel through wires, components, or accidentally through a poor soul's body, they encounter *resistance*, which is the tendency of the conductor to limit electron flow.

A vacuum is a perfect resistor: no current flows through it. Air's pretty close, but since water is a decent conductor, humidity does allow some electricity to flow in air.

Superconductors are the only materials with zero resistance, a feat achieved through the magic of quantum mechanics at extremely low temperatures, on the order of that of liquid nitrogen and colder. Everything else exhibits some resistance, even the very best wires. Feel the power cord of your 1500 watt ceramic heater—it's warm, indicating some power is lost in the cord due to the wire's resistance.

We measure resistance in ohms, the more ohms, the poorer the conductor. The Greek capital omega (Ω) is the symbol denoting ohms.

Resistance, voltage, and amperage are all related by the most important of all formulas in electrical engineering. Ohm's Law states:

$$E = I \times R$$

where E is voltage in volts, I is current in amps, and R is resistance in ohms. (EEs like to use E for volts because it indicates electromotive force.)

What does this mean in practice? Feed one amp of current through a one-ohm load and there will be one volt developed across the load. Double the voltage and, if resistance stays the same, the current doubles.

Though all electronic components have resistance, a *resistor* is a device specifically made to reduce conductivity (Figure 1.8 and Table 1.3). We use them everywhere. The volume control on a stereo (at least, the nondigital ones) is a resistor whose value changes as you rotate the knob; more resistance reduces the signal and hence the speaker output.

What happens when you connect resistors together? For resistors in series, the total effective resistance is the sum of the values:

$$R_{\text{eff}} = R_1 + R_2$$

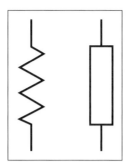

Figure 1.8: The squiggly thing on the left is the standard symbol used by engineers to denote a resistor on their schematics. On the right is the symbol used by engineers in the United Kingdom. As Churchill said, we are two peoples divided by a common language.

Table 1.3: Range of values for real-world resistors.

Name	Abbreviation	Ohms	Where Likely Found
milliohm	m Ω	.001 ohm	Resistance of wires and other good conductors
ohm	Ω	1 ohm	Power supplies may have big dropping resistors in the few to tens of ohms range
hundreds of ohms			In embedded systems, it's common to find resistors in the few hundred ohm range used to terminate high-speed signals
kiloohm	k Ω or just k	1000 ohms	Resistors from a half-k to a hundred or more k are found all over every sort of electronic device; "pullups" are typically a few k to tens of k
megaohm	M Ω	10^6 ohms	Low signal-level analog circuits
hundreds of M Ω		10^8++ ohms	Geiger counters and other extremely sensitive apps; rarely seen since resistors of this size are close to the resistance of air

For two resistors in parallel, the effective resistance is:

$$R_{\text{eff}} = \frac{R_1 \times R_2}{R_1 + R_2}$$

(Thus, two identical resistors in parallel are effectively half the resistance of either of them: two 1 ks is 500 ohms. Now add a third: that's essentially a 500-ohm resistor in parallel with a 1k, for an effective total of 333 ohms.)

The general formula for more than two resistors in parallel (Figure 1.9) is:

$$R_{\text{eff}} = \frac{1}{\dfrac{1}{R_1} + \dfrac{1}{R_2} + \dfrac{1}{R_3} + \dfrac{1}{R_4} + \cdots}$$

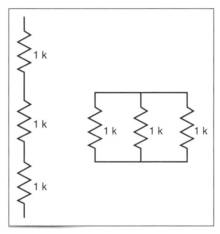

Figure 1.9: The three series resistors on the left are equivalent to a single 3000-ohm part. The three paralleled on the right work out to one 333-ohm device.

Manufacturers use color codes to denote the value of a particular resistor. Although at first this may seem unnecessarily arcane, in practice it makes quite a bit of sense. Regardless of orientation, no matter how it is installed on a circuit board, the part's color bands are always visible (Figure 1.10 and Table 1.4).

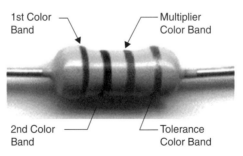

Figure 1.10: This black-and-white photo masks the resistor's color bands. However, we read them from left to right, the first two designating the integer part of the value, the third band giving the multiplier. A fourth gold (5%) or silver (10%) band indicates the part's tolerance.

Table 1.4: The resistor color code. Various mnemonic devices designed to help one remember these are no longer politically correct; one acceptable but less memorable alternative is Big Brown Rabbits Often Yield Great Big Vocal Groans When Gingerly Slapped.

Color band	Value	Multiplier
Black	0	1
Brown	1	10
Red	2	100
Orange	3	1000
Yellow	4	10,000
Green	5	100,000
Blue	6	1,000,000
Violet	7	Not used
Gray	8	Not used
White	9	Not used
Gold (third band)		÷10
Silver (third band)		÷100

The first two bands, reading from the left, give the integer part of the resistor's value. The third is the multiplier. Read the first two bands' numerical values and multiply by the scale designated by the third band. For instance: brown black red = 1 (brown) 0 (black) times 100 (red), or 1000 ohms, more commonly referred to as 1 k. Table 1.5 has more examples.

Table 1.5: Examples showing how to read color bands and compute resistance.

First Band	Second Band	Third Band	Calculation	Value (Ohms)	Commonly Called
Brown	Red	Orange	12 × 1000	12,000	12 k
Red	Red	Red	22 × 100	2,200	2.2 k
Orange	Orange	Yellow	33 × 10,000	330,000	330 k
Green	Blue	Red	56 × 100	5,600	5.6 k
Green	Blue	Green	56 × 100,000	5,600,000	5.6 M
Red	Red	Black	22 × 1	22	22
Brown	Black	Gold	10 ÷ 10	1	1
Blue	Gray	Red	68 × 100	6,800	6.8 k

Resistors come in standard values. Novice designers specify parts that do not exist; the experienced engineer knows that, for instance, there's no such thing as a 1.9 k resistor. Engineering is a very practical art; one important trait of the good designer is using standard and easily available parts.

1.4.1.3 Circuits

Electricity always flows in a loop. A battery left disconnected discharges only very slowly because there's no loop, no connection of any sort (other than the nonzero resistance of humid air) between the two terminals. To make a lamp light, connect one lead to each battery terminal; electrons can now run in a loop from the battery's negative terminal, through the lamp, and back into the battery.

There are only two types of circuits: series and parallel. All real designs use combinations of these. A *series circuit* connects loads in a circular string; current flows around through each load in sequence (Figure 1.11). In a series circuit, the current is the same in every load.

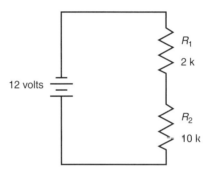

Figure 1.11: In a series circuit, the electrons flow through one load and then into another. The current in each resistor is the same; the voltage dropped across each depends on the resistor's value.

It's easy to calculate any parameter of a series circuit. In Figure 1.11, a 12-volt battery powers two series resistors. Ohm's Law tells us that the current flowing through the circuit is the voltage (12 in this case) divided by the resistance (the sum of the two resistors, or 12 k).

Total current is thus:

$$I = V \div R = (12\ volts) \div (2000 + 10{,}000\ ohms) = 12 \div 12000 = 0.001\ amp = 1\ mA$$

(Remember that *mA* is the abbreviation for milliamps.)

So what's the voltage across either of the resistors? In a series circuit, the current is identical in all loads, but the voltage developed across each load is a function of the load's resistance and the current. Again, Ohm's Law holds the secret. The voltage across R_1 is the current in the resistor times its resistance, or:

$$V_{R_1} = I_{R_1} = 0.001\ amps \times 2000\ ohms = 2\ volts$$

Since the battery places 12 volts across the entire resistor string, the voltage dropped on R_2 must be 12 – 2, or 10 volts. Don't believe that? Use Mr. Ohm's wonderful equation on R_2 to find:

$$V_{R_2} = I_{R_2} = 0.001 \text{ amps} \times 10{,}000 \text{ ohms} = 10 \text{ volts}$$

It's easy to extend this to any number of parts wired in series.

Parallel circuits have components wired so both pins connect (Figure 1.12). Current flows through both parts, though the amount of current depends on the resistance of each leg of the circuit. The voltage on each component, though, is identical.

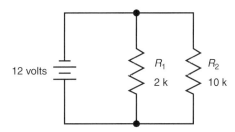

Figure 1.12: R_1 and R_2 are in parallel, both driven by the 12-volt battery.

We can compute the current in each leg much as we did for the series circuit. In the preceding case, the battery applies 12 volts to both resistors. The current through R_1 is:

$$I_{R_1} = 12 \text{ volts} \div 2{,}000 \text{ ohms} = 12 \div 2000 = 0.006 \text{ amps} = 6 \text{ mA}$$

Through R_2:

$$I_{R_2} = 12 \text{ volts} \div 10{,}000 \text{ ohms} = 0.0012 \text{ amps} = 1.2 \text{ mA}$$

Real circuits are usually a combination of series and parallel elements (Figure 1.13). Even in these more complex, more realistic cases, it's still very simple to compute anything one wants to know.

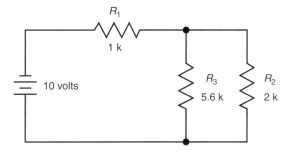

Figure 1.13: A series/parallel circuit.

Let's analyze the circuit shown in Figure 1.13. There's only one trick: cleverly combine complicated elements into simpler ones. Let's start by figuring the current flowing out of the battery. It's much too hard to do this calculation until we remember that two resistors in parallel look like a single resistor with a lower value.

Start by figuring the current flowing out of the battery and through R_1. We can turn this into a series circuit (in which the current flowing is the same through all the components) by replacing R_3 and R_2 with a single resistor with the same effective value as these two paralleled components. That's:

$$R_{EFF} = \frac{R_2 \times R_3}{R_1 + R_3} = \frac{5600 \times 2000}{5600 + 2000} = 1474 \ ohms$$

So the circuit is identical to one with two series resistors: R_1, still 1k, and R_{EFF} at 1474 ohms. Ohm's Law gives the current flowing out of the battery and through these two resistors:

$$i = \frac{V}{R_1 + R_{EFF}} = \frac{10}{1000 + 1474} = 0.004 \ amps = 4 \ mA$$

Ohm's Law remains the font of all wisdom in basic circuit analysis and readily tells us the voltage dropped across R_1:

$$V = iR_1 = 0.004 \ amps \times 1000 \ ohms = 4 \ volts$$

Clearly, since the battery provides 10 volts, the voltage across the paralleled pair R_2 and R_3 is 6 volts.

1.4.1.4 Power

Power is the product of voltage and current and is expressed in *watts*. One watt is one volt times one amp. A *milliwatt* is a thousandth of a watt; a *microwatt* is a millionth.

You can think of power as the total amount of electricity present. A thousand volts sounds like a lot of electricity, but if there's only a microamp available, that's a paltry milliwatt—not much power at all.

Power is also current^2 times resistance:

$$P = I^2 \times R$$

Electronic components like resistors and ICs consume a certain amount of volts and amps. An IC doesn't move, make noise, or otherwise release energy (other than exerting a minimal amount of energy in sending signals to other connected devices), so almost all the energy consumed gets converted to heat. All components have maximum power dissipation ratings; exceed these at your peril.

If a part feels warm it's dissipating a reasonable fraction of a watt. If it's hot but you can keep your finger on it, then it's probably operating within specs, though many analog components want to run cooler. If you pull back, not burned, but the heat is too much for your finger, then in most cases (be wary of the wimp factor; some people are more heat sensitive than others) the device is too hot and either needs external cooling (heat sink, fan, etc.), has failed, or your circuit exceeds heat parameters. A burn or near burn or discoloration of the device means there's trouble brewing in all but exceptional conditions (e.g., high-energy parts like power resistors).

A PC's processor has so many transistors, each losing a bit of heat, that the entire part might consume and eliminate 100+ watts. That's far more than the power required to destroy the chip. Designers expend a huge effort in building heat sinks and fans to transfer the energy in the part to the air.

Figure 1.14: This 10-ohm resistor, with 12 volts applied, draws 833 mA. $P = I^2R$, so it's sucking about 7 watts. Unfortunately, this particular part is rated for $1/4$ watt max, so it is on fire. Few recent college grads have a visceral feel for current, power, and heat, so this demo makes their eyes go like saucers.

The role of heat sinks and fans is to remove the heat from the circuits and dump it into the air before the devices burn up. The fact that a part dissipates a lot of energy and wants to run hot is not bad as long as proper thermal design removes the energy from the device before it exceeds its max temp rating (Figure 1.14).

1.4.2 AC Circuits

AC is short for *alternating current*, which is any signal that's not DC. AC signals vary with time. The mains in your house supply AC electricity in the shape of a sine wave: the voltage varies from a large negative to a large positive voltage 60 times per second (in the United States and Japan) or 50 times per second (in most of the rest of the world).

AC signals can be either *periodic*, which means they endlessly and boringly repeat forever, or *aperiodic*, the opposite. Static from your FM radio is largely aperiodic since it's quite random. The bit stream on any address or data line from a micro is mostly aperiodic, at least over short times, as it's a complex changing pattern driven by the software.

The rate at which a periodic AC signal varies is called its *frequency*, which is measured in *hertz* (Hz for short). One Hz means the waveform repeats once per second. A thousand Hz is a kHz (kilohertz), a million Hz is the famous MHz by which so many microprocessor clock rates are defined, and a billion Hz is a GHz.

The reciprocal of Hz is *period*. That is, where the frequency in hertz defines the signal's repetition rate, the period is the time it takes for the signal to go through a cycle.

Mathematically:

$$\text{Period in seconds} = 1 \div \text{frequency in Hz}$$

Thus, a processor running at 1 GHz has a clock period of 1 nanosecond—one billionth of a second. No kidding. In that brief flash of time, even light goes but a bare foot. Though your 1.8 GHz PC may seem slow loading Word, it's cranking instructions at a mind-boggling rate.

Wavelength relates a signal's period—and thus its frequency—to a physical "size." It's the distance between repeating elements and is given by:

$$\text{Wavelength in meters} = \frac{c}{frequency} = \frac{300,000,000 \; meters/second}{frequency \; in \; Hz}$$

where c is the speed of light.

An FM radio station at about 100 MHz has a wavelength of 3 meters. AM signals, on the other hand, are around 1 MHz, so each sine wave is 300 meters long. A 2.4-GHz cordless phone runs at a wavelength a bit over 10 cm.

As the frequency of an AC signal increases, things get weird. The basic ideas of DC circuits still hold but need to be extended considerably. Just as relativity builds on Newtonian mechanics to describe fast-moving systems, electronics needs new concepts to properly describe fast AC circuits.

Resistance, in particular, is really a subset of the real nature of electronic circuits. It turns out that there are three basic kinds of resistive components; each behaves somewhat differently. We've already looked at resistors; the other two components are capacitors and inductors.

Both of these parts exhibit a kind of resistance that varies depending on the frequency of the applied signal; the amount of this "AC resistance" is called *reactance*.

1.4.2.1 Capacitors

A *capacitor*, colloquially called the "cap," is essentially two metal plates separated from each other by a thin insulating material. This insulation, of course, means that a DC signal cannot flow through the cap. It's like an open circuit.

But in the AC world, strange things happen. It turns out that AC signals can make it across the gap between the two plates; as the frequency increases, the effective resistance of this gap decreases. This resistive effect is called *reactance*; for a capacitor it's termed *capacitive reactance* (Figure 1.15). There's a formula for everything in electronics; for capacitive reactance it's:

$$X_c = \frac{1}{2\pi f c}$$

where:

X_c = capacitive reactance
f = frequency in Hz
c = capacitance in farads

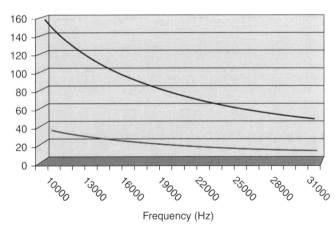

Figure 1.15: Capacitive reactance of a 0.1 μF cap (top) and a 0.5 μF cap (bottom curve). The vertical axis is reactance in ohms. See how larger caps have lower reactances, and as the frequency increases reactance decreases. In other words, a bigger cap passes AC better than a smaller one, and at higher frequencies all caps pass more AC current. Not shown: at 0 Hz (DC), reactance of all caps is essentially infinite.

Capacitors thus pass only *changing* signals (Table 1.6). The current flowing through a cap is:

$$I = \frac{dV}{dt}$$

(If your calculus is rusty or nonexistent, this simply means that the current flow is proportional to the change in voltage over time.)

In other words, the faster the signal changes, the more current flows.

Table 1.6: Range of values for real-world capacitors.

Name	Abbreviation	Farads	Where Likely Found
picofarad	pF	10^{-12} farad	Padding caps on microprocessor crystals, oscillators, analog feedback loops.
microfarad	μF	10^{-6} farad	Decoupling caps on chips are about .01 to .1 μF; low-freq decoupling runs about 10 μF, big power supply caps might be 1000 μF.
farad	F	1 farad	One farad is a huge capacitor and generally does not exist. A few vendors sell "supercaps" that have values up to a few farads, but these are unusual. Sometimes used to supply backup power to RAM when the system is turned off.

In real life there's no such thing as a perfect capacitor. All leak a certain amount of DC and exhibit other more complex behavior. For that reason, there's quite a range of different types of parts.

In most embedded systems you'll see one of two types of capacitors (Figure 1.16). The first are the polarized ones, devices which have a plus and a minus terminal. Connect one backward and the part will likely explode!

Polarized devices have large capacitance values: tens to thousands of microfarads. They're most often used in power supplies to remove the AC component from filtered signals. Consider the equation of capacitive reactance: large cap values pass lower-frequency signals efficiently. Typical construction today is from a material called *tantalum*; seasoned EEs often call these devices *tantalums*. You'll see tantalum caps on PC boards to provide a bit of bulk storage of the power supply.

Smaller caps are made from a variety of materials. These have values from a few picofarads to a fraction of a microfarad. They're often used to "decouple" the power supply on a PCB (i.e., to short high-frequency switching from power to ground, so the logic signals don't get coupled into the power supply). Most PCBs have dozens or hundreds of these parts scattered around.

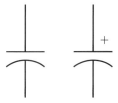

Figure 1.16: Schematic symbols for capacitors. The one on the left is a generic, generally low-valued (under 1 μF) part. On the right the plus sign shows that the cap is polarized. Installed backward, it's likely to explode.

We can wire capacitors in series and in parallel; compute the total effective capacitance using the rules opposite those for resistors. So, for two caps in parallel, sum their values to get the effective capacitance. In a series configuration the total effective capacitance is:

$$C_{eff} = \frac{1}{\frac{1}{C_1} + \frac{1}{C_2} + \frac{1}{C_3} + \cdots}$$

Note that this rule is for figuring the total capacitance of the circuit, *not* for computing the total reactance. More on that shortly.

One useful characteristic of a capacitor is that it can store a charge. Connect one to a battery or power supply and it will store that voltage. Remove the battery and (for a perfect, lossless part) the capacitor will still hold that voltage. Real parts leak a bit; ones rated at under 1 μF or so discharge rapidly. Larger parts store the charge longer.

Interesting things happen when you wire a cap and a resistor in series. The resistor limits current to the capacitor, causing it to charge slowly. Suppose the circuit shown in Figure 1.17 is dead, no voltage at all applied. Now turn on the switch. Though we've applied a DC signal, the sudden transition from 0 to 5 volts is AC.

Current flows due to the $I = \frac{dV}{dt}$ rule; dV is the sudden edge from flipping the switch. But the input goes from an AC-edge to steady-state DC, so current stops flowing pretty quickly. How fast? That's defined by the circuit's *time constant*.

A resistor and capacitor in series is colloquially called an *RC circuit*. The graph shows how the voltage across the capacitor increases over time. The time constant of any circuit is pretty well approximated by:

$$t = RC$$

for *R* in ohms, *C* in farads, and *t* in seconds.

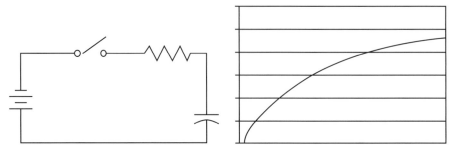

Figure 1.17: Close the switch and the voltage applied to the RC circuit looks like the top curve. The lower graph shows how the capacitor's voltage builds slowly with time, headed asymptotically toward the upper curve.

This formula tells us that after RC seconds the capacitor will be charged to 63.2% of the battery's voltage. After another RC seconds, another 63.2%, for a total of 86.5%.

Analog circuits use a lot of RC circuits; in a microprocessor it's still common to see them controlling the CPU's reset input. Apply power to the system and all the logic comes up, but the RC's time constant keeps reset asserted low for a while, giving the processor time to initialize itself.

The most common use of capacitors in the digital portion of an embedded system is to *decouple* the logic chips' power pins. A medium value part (0.01 to 0.1 μF) is tied between power and ground very close to the power leads on nearly every digital chip. The goal is to keep power supplied to the chips as clean as possible—close to a perfect DC signal.

Why would this be an issue? After all, the system's power supply provides a nearly perfect DC level. It turns out that as a fast logic chip switches between zero and one it can draw immense amounts of power for a short, subnanosecond, time. The power supply cannot respond quickly enough to regulate that, and since there's some resistance and reactance between the supply and the chip's pins, what the supply provides and what the chip sees are somewhat different. The decoupling capacitor shorts this very high-frequency (i.e., short transient) signal on Vcc to ground. It also provides a tiny bit of localized power storage that helps overcome the instantaneous voltage drop between the power supply and the chip.

Most designs also include a few tantalum bulk storage devices scattered around the PC board, also connected between Vcc and ground. Typically these are 10 to 50 μF each. They are even more effective bulk storage parts to help minimize the voltage drop chips would otherwise see.

You'll often see very small caps (on the order of 20 pF) connected to microprocessor drive crystals. These help the device oscillate reliably.

Analog circuits make many wonderful and complex uses of caps. It's easy to build integrators and differentiators from these parts, as well as analog hold circuits that memorize a signal for a short period of time. Common values in these sorts of applications range from 100 pF to fractions of a microfarad.

1.4.2.2 Inductors

An *inductor* is, in a sense, the opposite of a capacitor. Caps block DC but offer diminishing resistance (really, reactance) to AC signals as the frequency increases. An inductor, on the other hand, passes DC with zero resistance (for an idealized part), but the resistance (reactance) increases proportionately to the frequency.

Physically an inductor is a coil of wire and is often referred to as a *coil*. A simple straight wire exhibits essentially no inductance. Wrap a wire in a loop and it's less friendly to AC signals. Add more loops, or make them smaller, or put a bit of ferrous metal in the loop, and inductance increases. Electromagnets are inductors, as is the field winding in an alternator or motor.

An *iron core* inductor is wound around a slug of metal, which increases the device's inductance substantially (Figure 1.18).

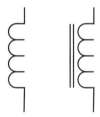

Figure 1.18: Schematic symbols of two inductors. The one on the left is an "air core"; the one on the right is an "iron core."

Inductance is measured in henries (H). *Inductive reactance* is the tendency of an inductor to block AC and is given by:

$$X_L = 2\pi L f$$

where:
X_L = Inductive reactance
f = frequency in Hz
L = inductance in henries

Clearly, as the frequency goes to zero (DC), reactance does as well.

Inductors follow the resistor rules for parallel and series combinations: add the value (in henries) when in series, and use the division rule when in parallel.

Inductors are much less common in embedded systems than are capacitors, yet they are occasionally important. The most common use is in switching power supplies. Many datacomm circuits use small inductors (generally millihenries) to match the network being driven.

Power supplies usually have a *transformer* which reduces the AC mains (from the wall) to a lower voltage more appropriate for embedded systems (Figure 1.19).

Figure 1.19: The schematic symbol for a transformer.

Transformers are two inductors wrapped around each other, with an iron core. The input AC generates a changing magnetic field, which induces a voltage in the output ("secondary") inductor.

If both inductors have the same number of wire loops, the output voltage is the same as the input. If the secondary has fewer loops, the voltage is less.

Sometimes signals, especially those flowing off a PC board, will have a *ferrite bead* wrapped around the wire. These beads are small cylinders (a few mm long) made of a ferromagnetic material. Like all inductors, they help block AC so are used to minimize noise of signal wires.

1.4.3 Active Devices

Resistors, capacitors and inductors are the basic *passive* components, passive meaning "dumb." The parts can't amplify or dramatically change applied signals. By contrast, *active* parts can clip, amplify, distort, and otherwise change an applied signal. The earliest active parts were vacuum tubes, called "valves" in the UK.

Consider the schematic in Figure 1.20, which is a single tube that contains two identical active elements, each called a *triode*, as each has three terminals. Tubes are easy to understand; let's see how one works.

A *filament* heats the cathode, which emits a stream of electrons. They flow through the grid, a wire mesh, and are attracted to the plate. Electrons are negatively charged, so applying a very small amount of positive voltage to the grid greatly reduces their flow. This is the basis of amplification: a small control signal greatly affects the device's output.

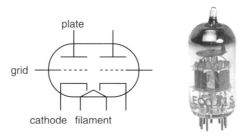

Figure 1.20: On the left, a schematic of a dual triode vacuum tube. The part itself is shown on the right.

Of course, in the real world tubes are almost unheard of today. When Bardeen, Brattain, and Shockley invented the *transistor* in 1947 they started a revolution that continues today. Tubes are power hogs, bulky and fragile. *Transistors*—also three-terminal devices that amplify—seem to have no lower limit of size and can run on picowatts (Figure 1.21).

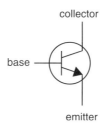

Figure 1.21: The schematic diagram of a bipolar NPN transistor with labeled terminals.

A transistor is made from a single crystal, normally of silicon, into which impurities are doped to change the nature of the material. The tube description showed how it's a voltage-controlled device; bipolar transistors are current-controlled.

Writers love to describe transistor operation by analogy to water flow or to the movement of holes and carriers within the silicon crystal. These are at best poor attempts to describe the quantum mechanics involved. Suffice to say that, in Figure 1.21, feeding current into the base allows current to flow between the collector and emitter.

And that's about all you need to know to get a sense of how a transistor amplifier works. The circuit shown in Figure 1.22 is a trivialized example of one. A microphone—which has a tiny output—drives current into the base of the transistor, which amplifies the signal, causing the lamp to fluctuate in rhythm with the speaker's voice.

A real amplifier might have many cascaded stages, each using a transistor to get a bit of amplification. A radio, for instance, might have to increase the antenna's signal by many millions before it gets to the speakers.

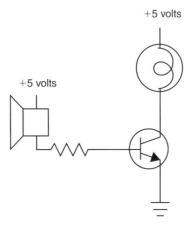

Figure 1.22: A very simple amplifier.

Transistors arc also switches, the basic element of digital circuits. The previous circuit is a simplified—but totally practical—NOR gate (Figure 1.23). When both inputs are zero, both transistors are off. No current flows from their collectors to emitters, so the output is 5 volts (as supplied by the resistor).

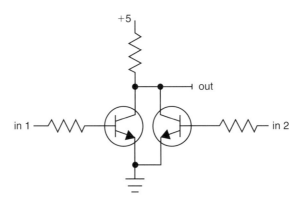

Figure 1.23: A NOR gate circuit.

If either input goes to a high level, the associated transistor turns on. This causes a conduction path through the transistor, pulling "out" low. In other words, any input going to a one gives an output of zero. Table 1.7 illustrates the circuit's behavior.

It's equally easy to implement any logic function.

The circuit we just analyzed would work; in the 1960s all "RTL" integrated circuits used exactly this design. But the gain of this approach is very low. If the input dawdles between a zero and a one, so will the output. Modern logic circuits use very high amplification factors,

Table 1.7: Truth table.

in1	in2	out
0	0	1
0	1	0
1	0	0
1	1	0

so the output is either a legal zero or one, not some in-between state, no matter what input is applied.

The silicon is a conductor, but a rather lousy one compared to a copper wire. The resistance of the device between the collector and the emitter changes as a function of the input voltage; for this reason active silicon components are called *semiconductors*.

Transistors come in many flavors; the one we just looked at is a bipolar part, characterized by high power consumption but (typically) high speeds. Modern ICs are constructed from MOSFET—Metal Oxide Semiconductor Field Effect Transistor—devices, or variants thereof (Figure 1.24). A mouthful? You bet. Most folks call these transistors FETs for short.

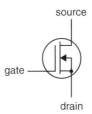

Figure 1.24: The schematic diagram of a MOSFET.

A FET is a strange and wonderful beast. The gate is insulated by a layer of oxide from a silicon channel running between the drain and source. No current flows from the gate to the silicon channel. Yet putting a bias voltage (like a tube, a FET is a voltage device) on the gate creates an electrostatic field that reduces current flow between the other two terminals. Again, *no current flows from the gate*. And when turned on, the source-drain resistance is much lower than in a bipolar transistor. This means the part dissipates little power, a critical concern when putting millions of these transistors on a single IC.

A *diode* is a two-terminal semiconductor that passes current in one direction only. In Figure 1.25, a positive voltage will flow from the left to the right, but not in the reverse

Figure 1.25: The schematic symbol for a diode.

direction. This seems a little thing, but it's incredibly useful. Figure 1.26 shows a circuit that implements an OR gate without a transistor.

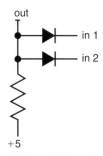

Figure 1.26: A diode OR circuit.

If both inputs are logic one, the output is a one (pulled up to +5 by the resistor). Any input going low will drag the output low as well. Yet the diodes ensure that a low-going input doesn't drag the other input down.

1.5 Putting It Together: A Power Supply

A power supply is a simple yet common circuit that uses many of the components we've discussed. The input is 110 volts AC (or 220 volts in Europe, 100 in Japan, 240 in the UK). Output might be 5 volts DC for logic circuits. How do we get from high voltage AC input to 5 volts DC?

The first step is to convert the AC mains to a lower voltage AC, as follows:

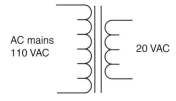

Now let's turn that lower voltage AC into DC. A diode does the trick nicely:

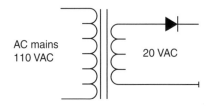

The AC mains are a sine wave, of course. Since the diode conducts in one direction only, its output looks like:

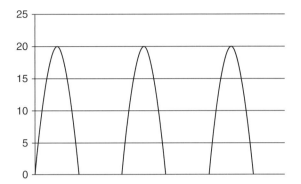

This isn't DC … but the diode has removed all the negative-going parts of the waveform.

But we've thrown away half the signal; it's wasted. A better circuit uses four diodes arranged in a *bridge* configuration as follows:

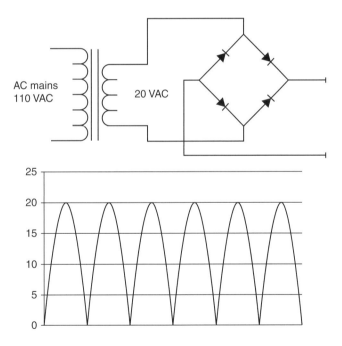

The bridge configuration ensures that two diodes conduct on each half of the AC input, as shown above. It's more efficient and has the added benefit of doubling the apparent frequency, which will be important when we're figuring out how to turn this moving signal into a DC level.

The average of this signal is clearly a positive voltage; if only we had a way to create an average value. Turns out that a capacitor does just that:

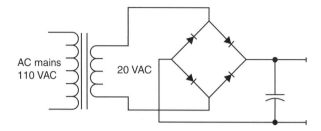

A huge-value capacitor filters best—typical values are in the thousands of microfarads.

The output is a pretty decent DC wave, but we're not done yet. The load—the device this circuit will power—will draw varying amounts of current. The diodes and transformer both have resistance. If the load increases, current flow goes up, so the drop across the parts will increase (Ohm's Law tells us $E = IR$, and as I goes up, so does E). Logic circuits are very sensitive to fluctuations in their power, so some form of *regulation* is needed.

A regulator takes varying DC in and produces a constant DC level out. For example:

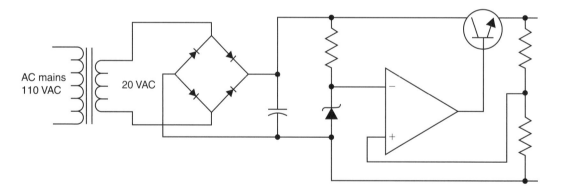

The odd-looking part in the middle is a *zener diode*. The voltage drop across the zener is always constant, so if, for example, this is a 3-volt part, the intersection of the diode and the resistor will *always* be 3 volts.

The regulator's operation is straightforward. The zener's output is a constant voltage. The triangle is a bit of magic—an error amplifier circuit—that compares the zener's constant voltage to the output of the power supply (at the node formed by the two resistors). If the output voltage goes up, the error amplifier applies less bias to the base of the transistor, making it conduct less … and lowering the supply's output. The transistor is key to the circuit; it's sort of like a variable resistor controlled by the error amp.

If, say, 20 volts of unregulated DC go into the transistor from the bridge and capacitor, and the supply delivers 5 volts to the logic, there's 15 volts dropped across the transistor. If the supply provides even just two amps of current, that's 30 watts (15 volts times two amps) dissipated by that semiconductor—a lot of heat! Careful heatsinking will keep the device from burning up.

1.5.1 The Scope

The oscilloscope (colloquially known as the "scope") is the most basic tool used for troubleshooting and understanding electronic circuits. Without some understanding of this most critical of all tools, you'll be like a blind person trying to understand color.

The scope has only one function: it displays a graph of the signal or signals you're probing (Figure 1.27). The horizontal axis is usually time; the vertical is amplitude, a fancy electronics term for voltage.

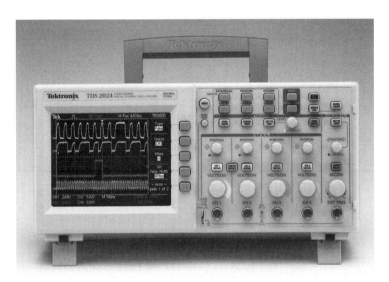

Figure 1.27: A sea of knobs. Don't be intimidated. There's a logical grouping to these. Master them and wow your friends and family. Photo courtesy of Tektronix, Inc.

1.5.2 Controls

In Figure 1.28, note first the two groups of controls labeled "vertical input 1" and "vertical input 2." This is a two-channel scope, by far the most common kind, which allows you to sample and display two different signals at the same time.

The vertical controls are simple. "Position" allows you to move the graphed signal up and down on the screen to the most convenient viewing position. When you're looking at two signals it allows you to separate them, so they don't overlap confusingly.

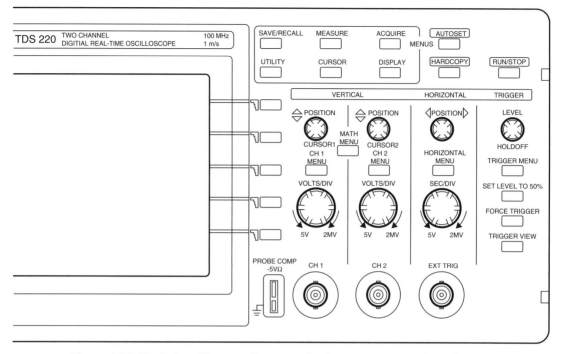

Figure 1.28: Typical oscilloscope front panel. Picture courtesy Tektronix, Inc.

"Volts/div" is short for volts-per-division. You'll note the screen is a matrix of 1 cm by 1 cm boxes; each is a "division." If the "volts/div" control is set to 2, then a two-volt signal extends over a single division. A five-volt signal will use 2.5 divisions. Set this control so the signal is easy to see. A reasonable setting for TTL (5-volt) logic is 2 volts/div.

The "coupling" control selects "DC"—which means what you see is what you get. That is, the signal goes unmolested into the scope. "AC" feeds the input through a capacitor; since caps cannot pass DC signals, this essentially subtracts DC bias (Figure 1.29).

The "mode" control lets us look at the signal on either channel, or both simultaneously.

Now check out the horizontal controls. These handle the scope's "time base," so called because the horizontal axis is always the time axis.

The "position" control moves the trace left and right, analogously to the vertical channel's knob of the same name.

"Time/div" sets the horizontal axis' scale. If set to 20 nsec/div, for example, each cm on the screen corresponds to 20 nsec of time. Figure 1.30 shows the same signal displayed using two different time base settings; it's more compressed in the left picture simply because at 2000 μsec/div more pulses occur in the 1 cm division mark.

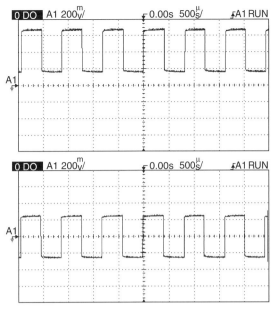

Figure 1.29: The signal is an AC waveform riding on top of a constant DC signal. On the left we're observing it with the scope set to DC coupling; note how the AC component is moved up by the amount of DC (in other words, the total signal is the DC component + the AC). On the right we've changed the coupling control to "AC"; the DC bias is removed and the AC component of the signal rides in the middle of the screen.

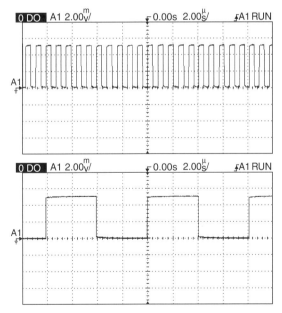

Figure 1.30: The left picture shows a signal with the time base set to 2000 μsec/division; the right is the same signal, but now we're sweeping at 200 μsec/division. Though the data is unchanged, the signal looks compressed. Also note that the 5-volt signal extends over 2.5 vertical boxes, since the gain is set to 2 volts/div. The first rule of scoping is to know the horizontal and vertical settings.

The last bank of knobs—those labeled "trigger"—are perhaps the most important of all. Though you see a line on the screen, it's formed by a dot swept across from left to right, repeatedly, at a very high speed. How fast? The dot moves at the speed you've set in the time/div knob. At 1 sec/div the dot takes 10 seconds to traverse the normal 10 cm-wide scope screen. More usual speeds for digital work are in the few microseconds to nanosecond range, so the dot moves faster than any eye can track.

Most of the signals we examine are more or less repetitive: it's pretty much the same old waveform over and over again. The trigger controls tell the scope when to start sweeping the dot across the screen. The alternative—if the dot started on the left side at a random time—would result in a very quickly scrolling screen, which no one could follow.

Twiddling the "trigger level" control sets the voltage at which the dot starts its inexorable left-to-right sweep. Set it to 6 volts and the normal 5-volt logic signal will never get high enough that the dot starts. The screen stays blank. Crank it to zero and the dot runs continuously, unsynchronized to the signal, creating a scrambled mess on the scope screen.

Set trigger level to 2 volts or so, and as the digital signal traverses from 0 to 5 volts the dot starts scanning, synchronizing to the signal.

It's most dramatic to learn how this control works when you're sampling a sine wave. As you twirl the knob clockwise (from a low trigger voltage to a higher one) the displayed sine wave shifts to the left. That is, the scan starts later and later since the triggering circuit waits for an ever-increasing signal voltage before starting.

"Trigger Menu" calls up a number of trigger selection criteria. Select "trigger on positive edge" and the scope starts sweeping when the signal goes from a low level through the trigger voltage set with the "Trigger Level" knob. "Trigger on negative edge" starts the sweep when the signal falls from a high level through the level.

Every scope today has more features than normal humans can possibly remember, let alone use. Various on-screen menus let you do math on the inputs (add them and so on), store signals that occur once, and much, much more. The instrument is just like a new PC application. Sure, it's nice to read the manual, but don't be afraid to punch a lot of buttons and see what happens. Most functions are pretty intuitive.

1.5.3 Probes

A "probe" connects the scope to your system. Experienced engineers' fingers are permanently bent a bit, warped from too many years holding the scope probe in hand while working on circuit boards. Though electrically the probe is just a wire, in fact there's a bit of electronics magic inside to propagate signals without distortion from your target system to the scope.

So too for any piece of test equipment. The tip of the scope probe is but one of the two connections required between the scope and your target system. A return path is needed, a ground (Figure 1.31). If there's no ground connection the screen will be nuts, a swirling mass of meaningless scrolling waveforms.

Figure 1.31: Always connect the probe's ground lead to the system.

Yet often we'll see engineers probing nonchalantly without an apparent ground connection. Oddly, the waves look fine on the scope. What gives? Where's the return path?

It's in the lab wall. Most electric cords, including the one to the scope and possibly to your target system, have three wires. One is ground. It's pretty common to find the target grounded to the scope via this third wire, going through the wall outlets. Of one thing be sure: even if this ground exists, it's ugly. It's a marginal connection at best, especially when dealing with high-speed logic signals or low level noise-sensitive analog inputs. Never, ever count on it even when all seems well. Every bit of gear in the lab, probably in the entire building, shares this ground. When the Xerox machine on the third floor kicks in, the big inductive spike from the motor starting up will distort the scope signal.

No scope will give decent readings on high-speed digital data unless it is *properly* grounded. I can't count the times technicians have pointed out a clock improperly biased 2 volts above ground, convinced they found the fault in a particular system, only to be bemused and embarrassed when a good scope ground showed the signal in its correct 0 to 5 volt glory. Ground the probe and thus the scope to your target using the little wire that emits from the end of the probe. As circuits get faster, shorten the wire. The very shortest ground lead results in the least signal distortion (see Figure 1.32.)

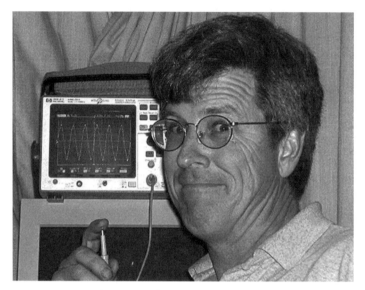

Figure 1.32: Here we probe a complex nonembedded circuit. Note the displayed waveform. A person is an antenna that picks up the 60 Hz hum radiated from the power lines in the walls around us. Some say engineers are particularly sensitive (though not their spouses).

Yet most scope probes come with crummy little lead alligator clips on the ground wire that are impossible to connect to an IC. The frustrated engineer might clip this to a clip lead that has a decent "grabber" end. Those extra 6–12 inches of ground may very well trash the display, showing a waveform that is not representative of reality. It's best to cut the alligator clip off the probe and solder a micrograbber on in its place.

One of the worst mistakes we make is neglecting probes. Crummy probes will turn that wonderful 1-GHz instrument into junk. After watching us hang expensive probes on the floor, mixed in with all sorts of other debris, few bosses are willing to fork over the $150 that Tektronix or Agilent demands. But the $50 alternatives are junk. Buy the best and take good care of them (see Figure 1.33.)

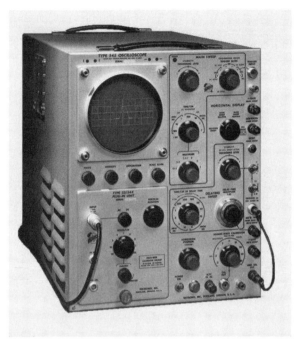

Figure 1.33: Tektronix introduced the 545 scope back in the dark ages; a half-century later, many are still going strong. Replace a tube from time to time and these might last forever. About the size of a two-drawer file cabinet and weighing almost 100 pounds, they're still favored by Luddites and analog designers.

Endnotes

[1.1] *Embedded Microcomputer Systems*, Valvano, p. 509.

[1.2] Net Silicon, "Net50BlockDiagram."

[1.3] *Beginner's Guide to Reading Schematics*, Traister and Lisk, p. 49.

[1.4] H. Malcolm, *Foundations of Computer Architecture.* Additional references include Stallings, W., *Computer Organization and Architecture*, Prentice Hall, fourth edition, 1995; Tanenbaum, A. S., *Structured Computer Organization*, Prentice Hall, third edition, 1990; Baron, R. J. and Higbie L., *Computer Architecture*, Addison-Wesley, 1992; Kane, G. and Heinrich, J., *MIPS RISC Architecture*, Prentice-Hall, 1992; and Patterson, D. A., and Hennessy, J. L., *Computer Organization and Design: The Hardware/Software Interface*, Morgan Kaufmann, third edition, 2005.

[1.5]   National Semiconductor, "Geode User Manual," Rev. 1, p. 13.

[1.6]   Net Silicon, "Net+ARM40 Hardware Reference Guide," pp. 1–5.

[1.7]   "EnCore M3 Embedded Processor Reference Manual," Revision A, p. 8.

[1.8]   "EnCore PP1 Embedded Processor Reference Manual," Revision A, p. 9.

CHAPTER 2
Logic Circuits

Jack Ganssle
Tammy Noergaard

2.1 Coding

The unhappy fact that most microprocessor books start with a chapter on coding and number systems reflects the general level of confusion on this, the most fundamental of all computer topics.

Numbers are existential nothings, mere representations of abstract quantitative ideas. We humans have chosen to measure the universe and itemize our bank accounts, so we have developed a number of arbitrary ways to count.

All number systems have a *base*, the number of unique identifiers combined to form numbers. The most familiar is decimal, base 10, which uses the 10 symbols 0 through 9. Binary is base 2 and can construct any integer using nothing more than the symbols 0 and 1. Any number system using any base is possible and in fact much work has been done in higher-order systems like base 64—which obviously must make use of a lot of odd symbols to get 64 unique identifiers. Computers mostly use binary, octal (base 8), and hexadecimal (base 16, usually referred to as "hex"; see Table 2.1).

Why binary? Simply because logic circuits are primitive constructs cheaply built in huge quantities. By restricting the electronics to two values only—on and off—we care little if the voltage drifts from 2 to 5. It's possible to build trinary logic, base 3, which uses 0, 1, and 2. The output of a device in certain ranges represents each of these quantities. But defining three bands means something like: 0 to 1 volt is a zero, 2 to 3 volts a 1, and 4 to 5 a 2. By contrast, binary logic says anything lower than (for TTL logic) 0.8 volts is a 0 and anything above 2 a 1. That's easy to make cheaply.

Why hex? Newcomers to hexadecimal find the use of letters baffling. Remember that "A" is as meaningless as "5"; both simply represent values. Unfortunately "A" viscerally means something that's not a number to those of us raised to read.

Hex combines four binary digits into a single number. It's compact. "8B" is much easier and less prone to error than "10001011."

Table 2.1: Various coding schemes. BCD is covered a bit later in the text.

Decimal	Binary	Octal	Hex	BCD
00	000000	00	00	0000 0000
01	000001	01	01	0000 0001
02	000010	02	02	0000 0010
03	000011	03	03	0000 0011
04	000100	04	04	0000 0100
05	000101	05	05	0000 0101
06	000110	06	06	0000 0110
07	000111	07	07	0000 0111
08	001000	10	08	0000 1000
09	001001	11	09	0000 1001
10	001010	12	0A	0001 0000
11	001011	13	0B	0001 0001
12	001100	14	0C	0001 0010
13	001101	15	0D	0001 0011
14	001110	16	0E	0001 0100
15	001111	17	0F	0001 0101
16	010000	20	10	0001 0110
17	010001	21	11	0001 0111
18	010010	22	12	0001 1000
19	010011	23	13	0001 1001
20	010100	24	14	0010 0000
21	010101	25	15	0010 0001
22	010110	26	16	0010 0010
23	010111	27	17	0010 0011
24	011000	30	18	0010 0100
25	011001	31	19	0010 0101
26	011010	32	1A	0010 0110
27	011011	33	1B	0010 0111
28	011100	34	1C	0010 1000
29	011101	35	1D	0010 1001
30	011110	36	1E	0011 0000
31	011111	37	1F	0011 0001
32	100000	40	20	0011 0010

Why octal? Base 8 is an aberration created by early programmers afraid of the implications of using letters to denote numbers. It's a grouping of three binary digits to represent the quantities 0 through 7. It's less compact than hex but was well suited to some early mainframe computers that used 36-bit words. Twelve octal digits exactly fills one 36-bit word (12 times 3 bits per digit). Hex doesn't quite divide into 36 bits evenly. Today, though, virtually all computers are 8, 16, 32, or 64 bits, all of which are cleanly divisible by 4, so the octal dinosaur is rarely used.

To convert from one base to another, just remember that the following rule constructs any integer in any number system:

$$Number = \cdots + C_4 \times b^4 + C_3 \times b^3 + C_2 \times b^2 + C_1 \times b^1 + C_0$$

Each of the C's are coefficients—the digit representing a value, and b is the base. So, the decimal number 123 really is three digits that represent the value:

$$123 = 1 \times 10^2 + 2 \times 10^1 + 3$$

D'oh, right? This pedantic bit of obviousness, though, tells us how to convert any number to base 10. For binary, the binary number 10110:

$$10110_2 = 1 \times 2^4 + 0 \times 2^3 + 1 \times 2^2 + 1 \times 2^1 + 0 \times 2^0$$
$$= 22_{10}$$

A1C in hex is:

$$A1C_{16} = A \times 16^2 + 1 \times 16^1 + C \times 16^0$$
$$= 10 \times 16^2 + 1 \times 16^1 + 12 \times 16^0$$
$$= 2588_{10}$$

Converting from decimal to another base is a bit more work. First, you need a cheat sheet, one that most developers quickly memorize for binary and hex, as shown in Table 2.2.

To convert 1234 decimal to hex, for instance, use the table to find the largest even power of 16 that goes into the number (in this case 16^2, or 256 base 10). Then see how many times you can subtract this number without the result going negative. In this case, we can take 256 from 1234 four times. The first digit of the result is thus 4.

$$\text{First digit} = 4. \text{ Remainder} = 1234 - 4 * 256 = 210$$

Now, how many times can we subtract 16^1 from the remainder (210) without going negative? The answer is 13, or D in hex.

$$\text{Second digit} = D. \text{ Remainder} = 210 - 13*16 = 2$$

Table 2.2: Binary and hex cheat sheet.

Decimal	Binary	Hex
1	2^0	16^0
2	2^1	
4	2^2	
8	2^3	
16	2^4	16^1
32	2^5	
64	2^6	
128	2^7	
256	2^8	16^2
512	2^9	
1024	2^{10}	
2048	2^{11}	
4096	2^{12}	16^3
8192	2^{13}	
16384	2^{14}	
32768	2^{15}	
65536	2^{16}	16^4

Following the same algorithm for the 160 placeholder, we see a final result of 4D2. For another example, convert 41007 decimal to hex:

16^3 goes into 41007 10 times before the remainder goes negative, so the first digit is 10 (A in hex).

The remainder is: $41007 - 10 * 16^3 = 47$

16^2 cannot go into 47 without going negative. The second digit is therefore 0.

16^1 goes into 47 twice. The next digit is 2.

Remainder $= 47 - 2 * 16^1 = 15$

The final digit is 15 (F in hex).

Final result: A02F

2.1.1 BCD

BCD stands for *binary coded decimal*. The BCD representation of a number is given in groups of four bits; each group expresses one decimal digit. The normal base 2 binary codes map to the 10 digits 0 through 9. Just as the decimal number 10 requires two digits, its equivalent in BCD uses two groups of four bits: 0001 0000. Each group maps to one of the decimal digits.

It's terribly inefficient, because the codes from 1010 to 1111 are never used. Yet BCD matches the base 10 way we count. It's often used in creating displays that show numerics.

2.2 Combinatorial Logic

Combinatorial logic is that whose state always reflects the inputs. There's no memory; past events have no impact on present outputs.

An adder is a typical combinatorial device: the output is always just the sum of the inputs—no more, no less.

The easiest way to understand how any combinatorial circuit works—be it a single component or a hundred interconnected ICs—is via a *truth table*, a matrix that defines every possible combination of inputs and outputs. We know, for example, that a wire's output is always the same as its input, as reflected in its table (see Table 2.3).

Table 2.3: The truth table for a wire's output and input.

In	Out
0	0
1	1

Gates are the basic building blocks of combinatorial circuits. Though there's no limit to the varieties available, most are derived from AND, OR, and NOT gates.

2.2.1 NOT Gate

The simplest of all gates inverts the input. It's the opposite of a wire, as shown by the truth table in Table 2.4.

Table 2.4: A gate's truth table.

In	Out
0	1
1	0

The Boolean expression is a bar over a signal: the NOT of an input A is $\overline{A}$. Any expression can have an inverse, $\overline{A + B}$ is the NOT of $A + B$.

The schematic symbol is:

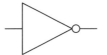

Note the circle on the device's output node. By convention a circle always means inversion. Without it, this symbol would be a buffer: a device that performs no logic function at all (rather like a piece of wire, though it does boost the signal's current). On a schematic, any circle appended to a gate means invert the signal.

2.2.2 AND and NAND Gates

An AND gate combines two or more inputs into a single output, producing a 1 if all the inputs are ones (see Table 2.5). If any input is zero, the output will be too.

Table 2.5: The truth table for an AND gate.

Input1	Input2	Output
0	0	0
0	1	0
1	0	0
1	1	1

The AND of inputs A and B is expressed as: output = AB

On schematics, a two-input AND looks like:

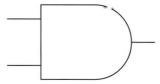

NAND is short for NOT-AND, meaning the output is zero when all inputs are one. It's an AND with an inverter on the output. So the NAND of inputs A and B is: output = $\overline{AB}$. Schematically a circle shows the inversion:

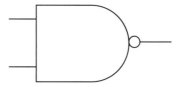

The NAND truth table is shown in Table 2.6.

Table 2.6: The truth table for a NAND gate.

Input1	Input2	Output
0	0	1
0	1	1
1	0	1
1	1	0

The AND and NAND gates we've looked at all have two inputs. Though these are very common, there's no reason not to use devices with three, four, or more inputs. Here's the symbol for a 13-input NAND gate ... its output is zero only when *all* inputs are one:

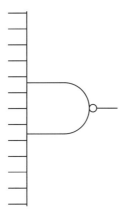

2.2.3 OR and NOR Gates

An OR gate's output is true if any input is a one (see Table 2.7). That is, it's zero only if every input is zero.

Table 2.7: An OR gate's truth table.

Input1	Input2	Output
0	0	0
0	1	1
1	0	1
1	1	1

The OR of inputs A and B is: output = $A + B$

Schematically:

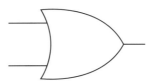

NOR means NOT-OR and produces outputs opposite that of OR gates (see Table 2.8).

Table 2.8: A NOR gate's truth table.

Input1	Input2	Output
0	0	1
0	1	0
1	0	0
1	1	0

The NOR equation is: output $= \overline{A + B}$ The gate looks like:

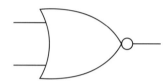

2.2.4 XOR

XOR is short for Exclusive-OR. Often used in error correction circuits, its output goes true if one of the inputs, but not both, is true (see Table 2.9). Another way of looking at it is that the XOR produces a true if the inputs are different.

Table 2.9: An XOR truth table.

Input1	Input2	Output
0	0	0
0	1	1
1	0	1
1	1	0

The exclusive OR of A and B is: output $= A \oplus B$

The XOR gate schematic symbol is:

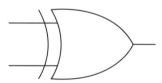

2.2.5 Circuits

Sometimes combinatorial circuits look frighteningly complex, yielding great job security for the designer. They're not. All can be reduced to a truth table that completely describes how each input affects the output(s). Though there are several analysis techniques, truth tables are usually the clearest and easiest to understand.

A proper truth table lists every possible input and output of the circuit. It's trivial to produce a circuit from the complete table. One approach is to ignore any row in the table for which the output is a zero. Instead, write an equation that describes each row with a true output, and then OR these.

Consider the XOR gate previously described. The truth table (Table 2.9) shows true outputs only when both inputs are different. The Boolean equivalent of this statement, assuming A and B are the inputs, is:

$$\text{XOR} = \bar{A}B + A\bar{B}$$

The circuit is just as simple:

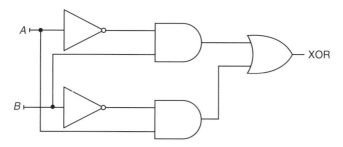

Note that an AND gate combines inputs A and B into AB; another combines the inversions of A and B. An OR gate combines the two product terms into the exclusive OR.

How about something that might seem harder? Let's build an adder, a device that computes the sum of two 16-bit binary numbers.

We could create a monster truth table of 32 inputs, but that's as crazy as the programmer who eschews subroutines in favor of a single, huge, monolithic `main()` function. Instead, realize that each of the 16 outputs (A_0 to A_{15}) is merely the sum of the two single-bit inputs, plus the carry from the previous stage. A 16-bit adder is really nothing more than 16 single-bit addition circuits. Each of those has a truth table as shown in Table 2.10.

Table 2.10: The single-bit addition circuit truth table.

A_n	B_n	CARRY$_{in}$	SUM$_n$	CARRY$_{out}$
0	0	0	0	0
0	1	0	1	0
1	0	0	1	0
1	1	0	0	1
0	0	1	1	0
0	1	1	0	1
1	0	1	0	1
1	1	1	1	1

The two outputs (the sum plus a carry bit) are the sum of the A and B inputs, plus the carry out from the previous stage. (The very first stage, for A_0 and B_0, has $CARRY_{in}$ connected to zero.)

The 1-bit adder has two outputs: sum and carry. Treat them independently; we'll have a circuit for each.

The trick to building combinatorial circuits is to minimize the amount of logic needed by not implementing terms that have no effect. In the truth table above we're only really interested in combinations of inputs that result in an output of 1, since any other combination results in 0, by default.

For each truth table row which has a one for the output, write a Boolean term, and then OR each of these as follows:

$$SUM_n = \overline{A_n} \overline{B_n} CARRY_{in} + \overline{A_n} B_n \overline{CARRY_{in}} + A_n \overline{B_n} \overline{CARRY_{in}} + A_n B_n CARRY_{in}$$
$$CARRY_{out} = \overline{A_n} B_n CARRY_{in} + A_n \overline{B_n} CARRY_{in} + A_n B_n \overline{CARRY_{in}} + A_n B_n CARRY_{in}$$

Each output is a different circuit, sharing only inputs. This implementation could be simplified—note that both outputs share an identical last product term—but in this example we'll pedantically leave it unminimized.

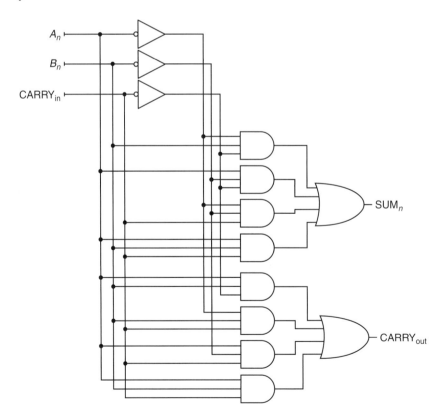

The drawing looks intricate, but does nothing more than the two simple equations above. Drawing the entire circuit for two 16-bit input numbers would cover a rather large sheet of paper, yet is merely a vast amount of duplicated simplicity.

And so, just as programmers manipulate the very smallest of things (ones and zeroes) in massive quantities to create applications, digital designers use gates and other brain-dead bits of minimalism to build entire computers.

This is the essence of all combinatorial design. Savvy engineers will work hard to reduce the complexity of a circuit and therefore reduce parts count, by noticing repetitive patterns, using truth tables, DeMorgan's theorem, and other tools. Sometimes it's hard to figure out how a circuit works, because we're viewing the result of lots of work done to minimize part count. It's analogous to trying to make sense of an object file, without the source. Possible, but tedious.

2.2.6 Tristate Devices

Though all practical digital circuits are binary, there are occasions when it's useful to have a state other than a zero or one. Consider busses: a dozen interconnected RAM chips, for example, all use the same data bus. Yet if more than one tries to drive that bus at a time, the result is babble, chaos. Bus circuits expect each connected component to behave itself, talking only when all other components are silent.

But what does the device do when it is supposed to be quiet? Driving a one or zero is a seriously Bad Thing because either will scramble any other device's attempt to talk. Yet ones and zeroes are the only legitimate binary codes.

Enter the tristate. This is a nonbinary state when a bus-connected device is physically turned off. It's driving neither a zero nor a one; rather, the output floats, electrically disconnected from the rest of the circuit.

Bus devices like memories have a control pin, usually named "Output Enable" (OE for short), that, when unasserted, puts the component's output pins into a tristate condition, floating them free of the circuit.

2.3 Sequential Logic

The output of sequential logic reflects both the inputs *and the previous state of the circuit*. That is, it remembers the past and incorporates history into the present. A counter whose output is currently 101, for instance, remembers that state to know the next value must be 110.

Sequential circuits are always managed by one or more clocks. A *clock* is a square wave (or at least one that's squarish) that repeats at a fixed frequency. Every sequential circuit is idle until the clock transitions; then, for a moment, everything changes. Counters count. Timers tick.

UARTs squirt a serial bit. The clock sequences a series of changes; the circuit goes idle after clock changes to allow signals to settle out.

The clock in a computer ensures that every operation has time to complete correctly. It takes time to access memory, for example. A 50 nsec RAM needs 50 nsec to retrieve data after being instructed as to which location to access. The system clock paces operations so the data has time to appear.

Just as gates are the basic units of combinatorial circuits, flip-flops form all sequential logic. A flip-flop (aka a "flop" or "bistable") changes its output based on one or more inputs, after the supplied clock transitions. Databooks show a veritable zoo of varieties. The simplest is the set-reset flop (SR for short), which looks like Figure 2.1.

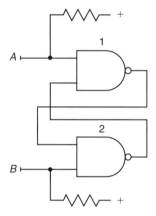

Figure 2.1: SR flip-flop.

To understand how this works, pretend input A is a zero. Leave B open. It's pulled to a one by the resistor. With A low, NAND gate 1 must go to a one, supplying a one to the input gate 2, which therefore, since B is high, must go low. Remove input A and gate 2 still drives a zero into gate 1, keeping output 1 high. The flop has remembered that A was low for a while. Now momentarily drive B low. Like a cat chasing its tail, the pattern reverses: output 1 goes high and 2 low.

What happens if we drive A and B low and release them at the same time? No one knows. Don't do that.

Flip flops are *latches*, devices that store information. A RAM is essentially an array of many latches.

Possibly the most common of all sequential components is the D flip-flop. As Figure 2.2 shows, it has two inputs and one output. The value at the D input is transferred to the Q output when clock transitions. Change the D input and nothing happens till clock comes again.

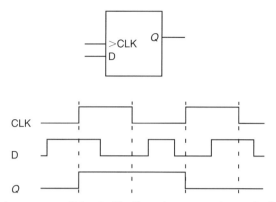

Figure 2.2: Note that the output of the D flip-flop changes only on the leading edge of clock.

(Some versions also have set and clear inputs that drive Q to a high or low regardless of the clock. It's not unusual to see a $\bar{Q}$ output as well, which is the inversion of Q.)

Clearly the D flop is a latch (also known as a *register*). String eight of these together, tied to one common clock, and you've created a byte-wide latch, a parallel output port.

Another common, though often misunderstood, synchronous device is the JK flip-flop, named for its inventor (John Kardash). Instead of a single data input (D on the D flip-flop), there are two, named J and K. Like the D, nothing happens unless the clock transitions.

But when the clock changes, if J is held low, then the Q output goes to a zero (it follows J). If J is one and K zero, Q also follows J, going to a one.

But if both J and K are one, then the output toggles—it alternates between zero and one every time a clock comes along.

The JK flip-flop can form all sorts of circuits, like the following counter:

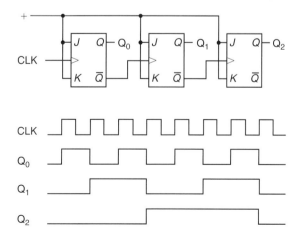

Every clock transition causes the three outputs to change state, counting through the binary numbers. Notice how the clock paces the circuit's operation; it keeps things happening at the rate the designer desires.

Counters are everywhere in computer systems; the program counter sequences fetches from memory. Timer peripherals and real-time clocks are counters.

The example above is a *ripple counter*, so called because a binary pattern ripples through each flip-flop. That's relatively slow. Worse, after the clock transitions it takes some time for the count to stabilize; there's a short time when the data hasn't settled. Synchronous counters are more complex but switch rapidly, with virtually no settle time (Figure 2.3).

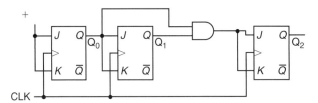

Figure 2.3: A 3-bit synchronous counter.

Cascading JK flip-flops in a different manner creates a shift register. The input bits march (shift) through each stage of the register as the clock operates (Figure 2.4).

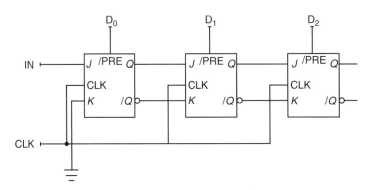

Figure 2.4: A 3-bit shift register.

Putting a lot of what we've covered together, let's build a simplified UART and driver for an RS-232 device (see Figure 2.5). This is the output part of the system only; considerably more logic is needed to receive serial data, and this drawing doesn't show the start and stop bits. But it shows the use of a counter, a shift register, and discrete parts.

RS-232 data is slow by any standard. Normal microprocessor clocks are far too fast for, say, 9600 baud operation. The leftmost chip is an eight-stage counter. It divides the input frequency by 256. So a 2.5 MHz clock, often found in slower embedded systems, divided as shown, provides 9600 Hz to the shift register.

The register is one that can be parallel-loaded; when the computer asserts the LOAD signal, the CPU's data bus is preset into the 8 stage shift register IC (middle of the drawing). The clock makes this 8-bit parallel word, loaded into the shift register, march out to the QH output, one bit at a time. A simple transistor amplifier translates the logic's 5 volt levels to 12 volts for the RS-232 device.

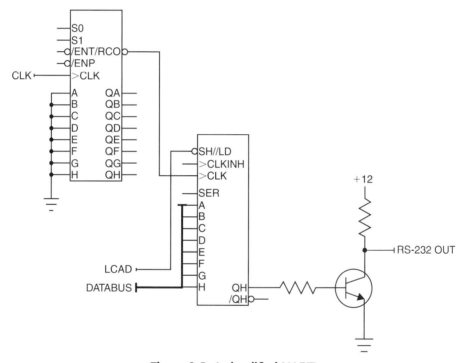

Figure 2.5: A simplified UART.

2.3.1 Logic Wrap-Up

Modern embedded systems do use all these sorts of components. However, most designers use integrated circuits that embody complex functions instead of designing with lots of gates and flops. A typical IC might be an 8-bit synchronous counter or a 4-bit arithmetic-logic unit (that does addition, subtraction, shifting, and more).

Yet you'll see gates and flops used as "glue" logic, parts needed to interface big complex ICs together.

2.4 Putting It All Together: The Integrated Circuit

Gates, along with the other electronic devices that can be located on a circuit, can be compacted to form a single device, called an *integrated circuit* (IC). ICs, also referred to as *chips*, are usually classified into groups according to the number of transistors and other electronic components they contain, as follows:

- *SSI* (small-scale integration) containing up to 100 electronic components per chip
- *MSI* (medium-scale integration) containing between 100–3,000 electronic components per chip.
- *LSI* (large-scale integration) containing 3,000–100,000 electronic components per chip
- *VLSI* (very large-scale integration) containing between 100,000–1,000,000 electronic components per chip
- *ULSI* (ultra large-scale integration) containing over 1,000,000 electronic components per chip

ICs are physically enclosed in a variety of packages that include SIP, DIP, flat pack, and others (see Figure 2.6). They basically appear as boxes with pins protruding from the body of the box. The pins connect the IC to the rest of the board.

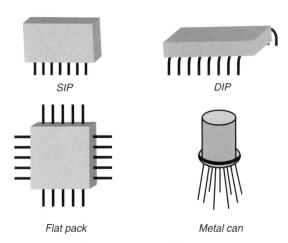

Figure 2.6: IC packages.

Physically packaging so many electronic components in an IC has its advantages as well as drawbacks. These include:

- *Size*. ICs are much more compact than their discrete counterparts, allowing for smaller and more advanced designs.

- *Speed.* The buses interconnecting the various IC components are much, much smaller (and thus faster) than on a circuit with the equivalent discrete parts.

- *Power.* ICs typically consume much less power than their discrete counterparts.

- *Reliability.* Packaging typically protects IC components from interference (dirt, heat, corrosion, etc.) far better than if these components were located discretely on a board.

- *Debugging.* It is usually simpler to replace one IC than try to track down one component that failed among 100,000 (for example) components.

- *Usability.* Not all components can be put into an IC, especially those components that generate a large amount of heat, such as higher value inductors or high-powered amplifiers.

In short, ICs are the master processors, slave processors, and memory chips located on embedded boards (see Figures 2.7a–e).

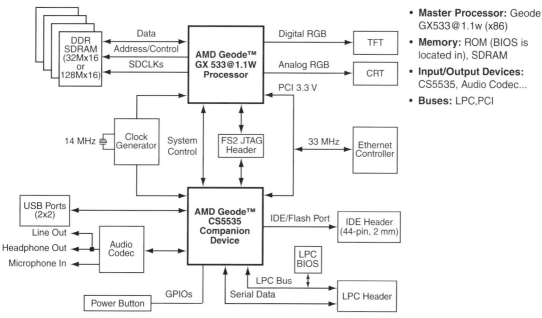

Figure 2.7a: AMD/National Semiconductor x86 reference board.[2.1]

© 2004 Advanced Micro Devices, Inc. Reprinted with permission.

60 Chapter 2

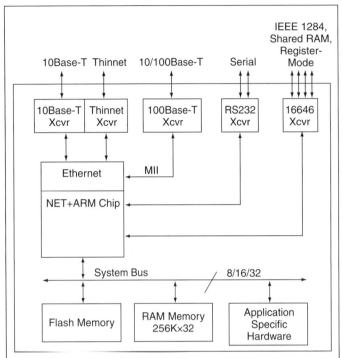

Figure 2.7b: Net Silicon ARM7 reference board.[2.2]

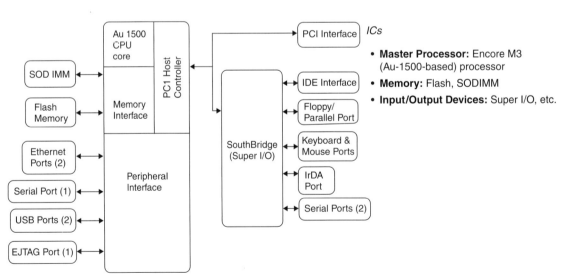

Figure 2.7c: Ampro MIPS reference board.[2.3]

Logic Circuits

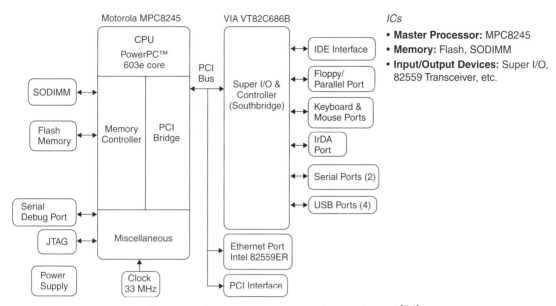

Figure 2.7d: Ampro PowerPC reference board.[2.4]

Copyright of Freescale Semiconductor, Inc., 2004. Used by permission.

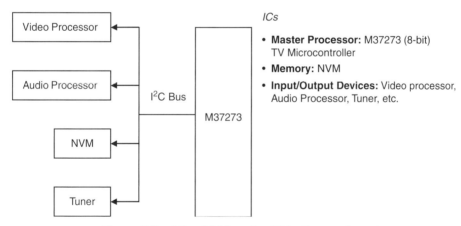

Figure 2.7e: Mitsubishi analog TV reference board.

Endnotes

[2.1] National Semiconductor, "Geode User Manual," Rev. 1, p. 13.

[2.2] Net Silicon, "Net+ARM40 Hardware Reference Guide," pp. 1–5.

[2.3] "EnCore M3 Embedded Processor Reference Manual," Revision A, p. 8.

[2.4] "EnCore PP1 Embedded Processor Reference Manual," Revision A, p. 9.

CHAPTER 3
Embedded Processors

Tammy Noergaard

3.1 Introduction

Processors are the main functional units of an embedded board and are primarily responsible for processing instructions and data. An electronic device contains at least one master processor, acting as the central controlling device, and can have additional slave processors that work with and are controlled by the master processor. These slave processors may either extend the instruction set of the master processor or act to manage memory, buses, and I/O (input/output) devices. In the block diagram of an x86 reference board, shown in Figure 3.1, the Atlas STPC is the master processor, and the super I/O and Ethernet controllers are slave processors.

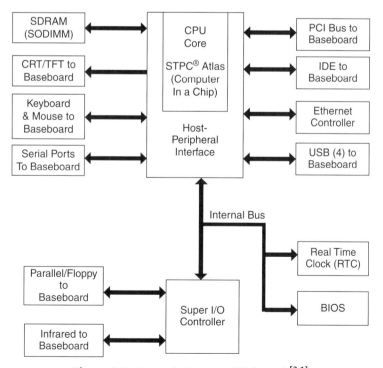

Figure 3.1: Ampro's Encore 400 board.[3.1]

As shown in Figure 3.1, embedded boards are designed around the master processor. The complexity of the master processor usually determines whether it is classified as a *microprocessor* or a *microcontroller*. Traditionally, microprocessors contain a minimal set of integrated memory and I/O components, whereas microcontrollers have most of the system memory and I/O components integrated on the chip. However, keep in mind that these traditional definitions may not strictly apply to recent processor designs. For example, microprocessors are increasingly becoming more integrated.

Why Use an Integrated Processor?

> Although *some* components, like I/O, may show a decrease in performance when integrated into a master processor as opposed to remaining a dedicated slave chip, many others show an increase in performance because they no longer have to deal with the latencies involved with transmitting data over buses between processors. An integrated processor also simplifies the entire board design since there are fewer board components, resulting in a board that is simpler to debug (fewer points of failure at the board level). The power requirements of components integrated into a chip are typically a lot less than those same components implemented at the board level. With fewer components and lower power requirements, an integrated processor may result in a smaller and cheaper board. On the flip side, there is less flexibility in adding, changing, or removing functionality, since components integrated into a processor cannot be changed as easily as if they had been implemented at the board level.

There are literally hundreds of embedded processors available, and not one of them currently dominates embedded system designs. Despite the sheer number of available designs, embedded processors can be separated into various "groups" called *architectures*. What differentiates one processor group's architecture from another is the set of machine code instructions that the processors within the architecture group can execute. Processors are considered to be of the same architecture when they can execute the same set of machine code instructions. Table 3.1 lists some examples of real-world processors and the architecture families they fall under.

Table 3.1: Real-world architectures and processors.

Architecture	Processor	Manufacturer
AMD	Au1xxx	Advanced Micro Devices, …
ARM	ARM7, ARM9, …	ARM, …
C16X	C167CS, C165H, C164CI, …	Infineon, …
ColdFire	5282, 5272, 5307, 5407, …	Motorola/Freescale, …

Table 3.1: Continued

Architecture	Processor	Manufacturer
1960	1960 Vmetro, …	
M32/R	32170, 32180, 32182, 32192, …	Renesas/Mitsubishi, …
M Core	MMC2113, MMC2114, …	Motorola/Freescale
MIPS32	R3K, R4K, 5K, 16, …	MTI4kx, IDT, MIPS Technologies, …
NEC	Vr55xx, Vr54xx, Vr41xx	NEC Corporation, …
PowerPC	82xx, 74xx,8xx,7xx,6xx,5xx,4xx	IBM, Motorola/Freescale, …
68k	680x0 (68K, 68030, 68040, 68060, …), 683xx	Motorola/Freescale, …
SuperH (SH)	SH3 (7702,7707, 7708,7709), SH4 (7750)	Hitachi, …
SHARC	SHARC	Analog Devices, Transtech DSP, Radstone, …
strongARM	strongARM	Intel, …
SPARC	UltraSPARC II	Sun Microsystems, …
TMS320C6xxx	TMS320C6xxx	Texas Instruments, …
x86	X86 [386,486,Pentium (II, III, IV)…]	Intel, Transmeta, National Semiconductor, Atlas, …
TriCore	TriCore1, TriCore2, …	Infineon, …

3.2 ISA Architecture Models

The *features* that are built into an architecture's instruction set are commonly referred to as the *Instruction Set Architecture,* or *ISA*. The ISA defines such features as the *operations* that can be used by programmers to create programs for that architecture, the *operands* (data) that are accepted and processed b+-y an architecture, *storage, addressing modes* used to gain access to and process operands, and the handling of *interrupts*. These features are described in more detail in this section because an ISA implementation is a determining factor in defining important characteristics of an embedded design, such as performance, design time, available functionality, and cost.

3.2.1 Operations

Operations are made up of one or more instructions that execute certain commands. (*Note that operations are commonly referred to simply as instructions*.) Different processors can execute the exact same operations using a different number and different types of instructions. An ISA typically defines the types and formats of operations.

3.2.1.1 Types of Operations

Operations are the functions that can be performed on the data, and they typically include computations (math operations), movement (moving data from one memory location/register to another), branches (conditional/unconditional moves to another area of code to process), input/output operations (data transmitted between I/O components and master processor), and context switching operations (where location register information is temporarily stored when switching to some routine to be executed and, after execution, by the recovery of the temporarily stored information, there is a switch back to executing the original instruction stream).

The instruction set on a popular lower-end processor, the 8051, includes just over 100 instructions for math, data transfer, bit variable manipulation, logical operations, branch flow and control, and so on. In comparison, a higher-end MPC823 (Motorola/Freescale PowerPC) has an instruction set a little larger than that of the 8051 but with many of the same types of operations contained in the 8051 set, along with an additional handful, including integer operations/floating-point (math) operations, load and store operations, branch and flow control operations, processor control operations, memory synchronization operations, PowerPC VEA operations, and so on. Figure 3.2a lists examples of common operations defined in an ISA.

Math and Logical	Shift/Rotate	Load/Store	Compare Instructions… Move Instructions… Branch Instructions …
Add Subtract Multiply Divide AND OR XOR …..	Logical Shift Right Logical Shift Left Rotate Right Rotate Left …..	Stack PUSH Stack POP Load Store …..	…..

Figure 3.2a: Sample ISA operations.

In short, different processors can have similar types of operations, but they usually have different overall instruction sets. As mentioned, what is also important to note is that different architectures can have operations with the same purpose (add, subtract, compare, etc.), but the operations may have different names or internally operate much differently, as seen in Figures 3.2b and c.

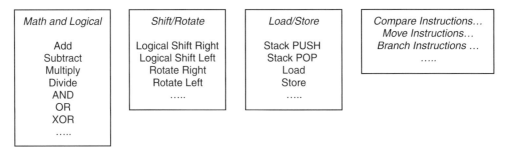

Figure 3.2b: MPC823 compare operation.[3.2]

Copyright of Freescale Semiconductor, Inc., 2004. Used by permission.

C.cond.S fs, ft
C.cond.D fs, ft ...

```
if SNaN(ValueFPR(fs, fmt)) or SNaN(ValueFPR(ft, fmt)) or
QNaN(ValueFPR(fs, fmt)) or QNaN(ValueFPR(ft, fmt)) then
less ← false
equal ← false
unordered ← true
if (SNaN(ValueFPR(fs,fmt)) or SNaN(ValueFPR(ft,fmt))) or
(cond3 and (QNaN(ValueFPR(fs,fmt)) or QNaN(ValueFPR(ft,fmt)))) then
SignalException(InvalidOperation)
endif
else
less ← ValueFPR(fs, fmt) <fmt ValueFPR(ft, fmt)
equal ← ValueFPR(fs, fmt) =fmt ValueFPR(ft, fmt)
unordered ← false
endif
condition ← (cond2 and less) or (cond1 and equal)
or (cond0 and unordered)
SetFPConditionCode(cc, condition)
```

The MIPS32/MIPS 1 compare operation is a floating point operation. The value in floating point register fs is compared to the value in floating point register ft. The MIPS I architecture defines a single floating point condition code, implemented as the coprocessor 1 condition signal (Cp1Cond) and the C bit in the FP Control/Status register.

Figure 3.2c: MIPS32/MIPS I – compare operation.[3.3]

3.2.1.2 Operation Formats

The format of an operation is the actual number and combination of bits (1's and 0's) that represent the operation and is commonly referred to as the operation code, or *opcode*. MPC823 opcodes, for instance, are structured the same and are all 6 bits long (0–63 decimal) (see Figure 3.3a). MIPS32/MIPS I opcodes are also 6 bits long, but the opcode can vary as to where it is located, as shown in Figure 3.3b. An architecture like the SA-1100, which is based on the ARM v4 Instruction Set, can have several instruction set formats, depending on the type of operation being performed (see Figure 3.3c).

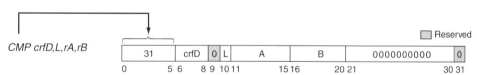

Figure 3.3a: MPC823 "CMP" operation size.[3.4]

Copyright of Freescale Semiconductor, Inc., 2004. Used by permission.

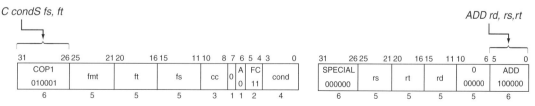

Figure 3.3b: MIPS32/MIPS I "CMP" and "ADD" operation sizes and locations.[3.5]

Instruction Type	31	2827				1615	87		0
Data Processing1/PSR Transfer	Cond	0 0	I	Opcode	S	Rn	Rd	Operand2	
Multiply	Cond	0 0 0 0 0 0	A	S		Rd	Rn	Rs	1 0 0 1	Rm
Long Multiply	Cond	0 0 0 0 1	U	A	S	RdHi	RdLo	Rs	1 0 0 1	Rm
Swap	Cond	0 0 0 1 0	B	0 0		Rn	Rd	0 0 0 0	1 0 0 1	Rm
Load & Store Byte/Word	Cond	0 1	I	P	U	B	W	L	Rn	Rd	Offset
Halfword Transfer : Immediate Offset	Cond	1 0 0	P	U	S	W	L	Rn	Register List
Halfword Transfer : Register Offset	Cond	0 0 0	P	U	1	W	L	Rn	Rd	Offset 1	1	S	H	1	Offset2
Branch	Cond	0 0 0	P	U	0	W	L	Rn	Rd	0 0 0 0	1	S	H	1	Rm
Branch Exchange	Cond	1 0 1	L	Offset
Coprocessor Data Transfer	Cond	0 0 0 1	0 0 1 0	1 1 1 1	1 1 1 1	1 1 1 1	0 0 0 1	Rn
Coprocessor Data Operation	Cond	1 1 0	P	U	N	W	L	Rn	CRd	CPNum	Offset
Coprocessor Register Transfer	Cond	1 1 1 0	Op1	CRn	CRd	CPNum	Op2	0	CRm
Software Interrupt	Cond	1 1 1 0	Op1	L	CRn	Rd	CPNum	Op2	1	CRm
...	Cond	1 1 1 1	SWI Number

1 - Data Processing OpCodes
0000 = AND – Rd: = Op1 AND Op2
0001 = EOR – Rd: = Op1 EOR Op2
0010 = SUR – Rd: = Op1 – Op2
0011 = RSB – Rd: = Op2 – Op1
0100 = ADD – Rd: = Op1 + Op2
0101 = ADC – Rd: = Op1 + Op2 + C
0110 = SEC – Rd: = Op2 – Op1 + C –1
0111 = RSC – Rd: = Op2 – Op1 + C – 1
1000 = TST – set condition codes on Op1 AND Op2
1001 = TEQ – set condition codes on Op1 EOR Op2
1010 = CMP – set condition codes on Op1 – Op2
1011 = CMN – set condition codes on Op1 + Op2
1100 = ORR – Rd: = Op1 OR Op2
1101 = MOV – Rd: = Op2
1110 = BIC – Rd: = Op1 AND NOT Op2
1111 = MVN – Rd: = NOT Op2

Figure 3.3c: SA-1100 instruction.[3.6]

3.2.2 Operands

Operands are the data that operations manipulate. An ISA defines the types and formats of operands for a particular architecture. For example, in the case of the MPC823 (Motorola/Freescale PowerPC), SA-1110 (Intel StrongARM), and many other architectures, the ISA defines simple operand types of bytes (8 bits), halfwords (16 bits), and words (32 bits). More complex data types such as integers, characters, or floating point are based on the simple types shown in Figure 3.4.

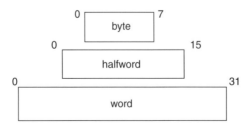

Figure 3.4: Simple operand types.

An ISA also defines the operand formats (the way the data looks) that a particular architecture can support, such as binary, decimal, and hexadecimal. Figure 3.5 shows an example illustrating the way an architecture can support various operand formats.

```
MOV        registerX, 10d             ; Move decimal value 10 into register X
MOV        registerX, $0Ah            ; Move hexadecimal value  A(decimal 10) to register X
MOV        registerX, 00001010b       ; Move binary value 00001010 (decimal 10) to register X
.....
```

Figure 3.5: Operand formats pseudocode example.

3.2.3 Storage

The ISA specifies the features of the programmable storage used to store the data being operated on, primarily:

A. *The organization of memory used to store operands.* *Memory* is simply an array of programmable storage, like that shown in Figure 3.6, that stores data, including operations, operands, and so on.

The indices of this array are locations referred to as *memory addresses*, where each location is a unit of memory that can be addressed separately. The actual physical or virtual range of addresses available to a processor is referred to as the *address space*.

An ISA defines specific characteristics of the address space, such as whether it is:

- *Linear*. A linear address space is one in which specific memory locations are represented incrementally, typically starting at 0 thru 2^{N-1}, where *N* is the address width in bits.
- *Segmented*. A segmented address space is a portion of memory that is divided into sections called *segments*. Specific memory locations can only be accessed by specifying a segment identifier, a segment number that can be explicitly defined or implicitly obtained from a register, and specifying the offset within a specific segment within the segmented address space.

  The offset within the segment contains a base address and a limit, which map to another portion of memory that is set up as a linear address space. If the offset is less than or equal to the limit, the offset is added to the base address, giving the unsegmented address within the linear address space.

- Containing any *special address regions*.
- *Limited* in any way.

An important note regarding ISAs and memory is that different ISAs not only define where data is stored but also *how* data is stored in memory—specifically in what order

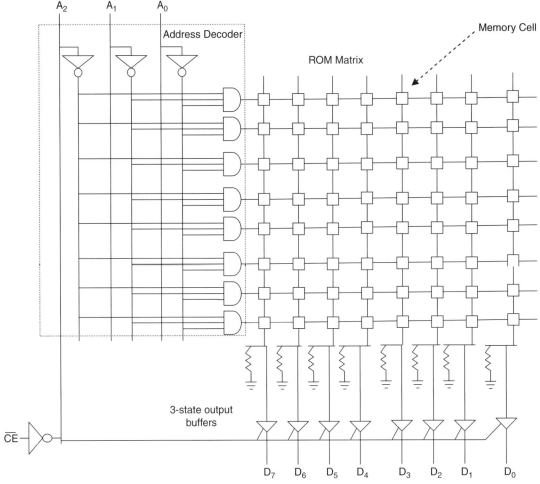

Figure 3.6: Block diagram of memory array.[3.7]

the bits (or bytes) that make up the data are stored, or *byte ordering*. The two byte-ordering approaches are big-endian, in which the most significant byte or bit is stored first, and little-endian, in which the least significant bit or byte is stored first.

For example:

- 68000 and SPARC are big-endian
- x86 is little-endian
- ARM, MIPS and PowerPC can be configured as either big-endian or little-endian using a bit in their machine state registers

B. *Register set.* A register is simply fast programmable memory normally used to store operands that are immediately or frequently used. A processor's set of registers is commonly referred to as the *register set* or the *register file*. Different processors have different register sets, and the number of registers in their sets varies between very few to several hundred (even over a thousand). For example, the SA-1110 register set has 37 32-bit registers, whereas the MPC823, on the other hand, has about a few hundred registers (general purpose, special purpose, floating-point, etc.).

C. *How registers are used.* An ISA defines which registers can be used for what transactions, such as special purpose or floating point, and which can be used by the programmer in a general fashion (general-purpose registers).

As a final note on registers, one of many ways processors can be referenced is according to the size (in bits) of *data* that can be processed and the *size* (in bits) of the *memory space* that can be addressed in a single instruction by that processor. This specifically relates back to the basic building block of registers, the flip-flop; we'll discuss this concept in more detail in Section 3.3.

Commonly used embedded processors support 4-bit, 8-bit, 16-bit, 32-bit, and/or 64-bit processing, as shown in Table 3.2. Some processors can process larger amounts of data and can access larger memory spaces in a single instruction, such as 128-bit architectures, but they are not commonly used in embedded designs.

Table 3.2: "X"-bit architecture examples.

"X"-Bit	Architecture
4	Intel 4004, …
8	Mitsubishi M37273, 8051, 68HC08, Intel 8008/8080/8086, …
16	ST ST10, TI MSP430, Intel 8086/286, …
32	68K, PowerPC, ARM, x86 (386+), MIPS32, …

3.2.4 Addressing Modes

Addressing modes define the way the processor can access operand storage. In fact, the use of registers is partly determined by the ISA's *Memory Addressing Modes*. The two most common types of addressing mode models are:

- *Load-store architecture*, which only allows operations to process data in registers, not anywhere else in memory. For example, the PowerPC architecture has only one addressing mode for load and store instructions: register plus displacement (supporting register indirect with immediate index, register indirect with index, etc.).

- *Register-memory architecture*, which allows operations to be processed within both registers and other types of memory. Intel's i960 Jx processor is an example of an addressing mode architecture that is based on the register-memory model (supporting absolute, register indirect, etc.).

3.2.5 Interrupts and Exception Handling

Interrupts (also referred to as *exceptions* or *traps,* depending on the type) are mechanisms that stop the standard flow of the program in order to execute another set of code in response to some event, such as problems with the hardware, resets, and so forth. The ISA defines what if any type of hardware support a processor has for interrupts.

> **Note:** Because of their complexity, interrupts are discussed in more detail in Section 3.3 later in this chapter.

Architectures are based on several different ISA models, each with its own definitions for the various features. The most commonly implemented ISA models are application-specific, general-purpose, instruction-level parallel, or some hybrid combination of these three ISAs.

3.2.6 Application-Specific ISA Models

Application-specific ISA models define processors that are intended for specific embedded applications, such as processors made only for TVs. There are several types of application-specific ISA models implemented in embedded processors, the most common models being the ones discussed in this section.

3.2.6.1 Controller Model

The Controller ISA is implemented in processors that are not required to perform complex data manipulation, such as video and audio processors that are used as slave processors on a TV board, for example (see Figure 3.7).

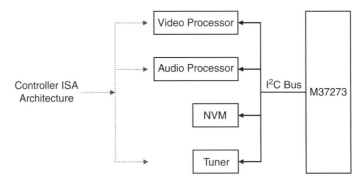

Figure 3.7: Analog TV board example with controller ISA implementations.

3.2.6.2 Datapath Model

The Datapath ISA is implemented in processors for which the purpose is to repeatedly perform fixed computations on different sets of data, a common example being digital signal processors (DSPs), shown in Figure 3.8.

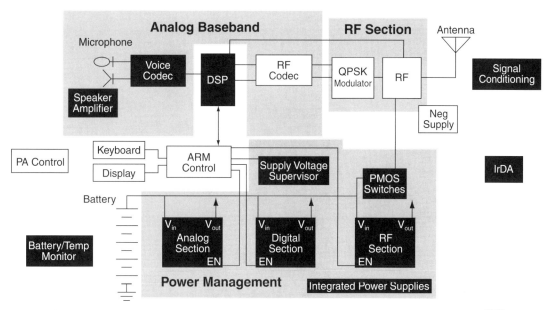

Figure 3.8: Board example with datapath ISA implementation: digital cellphone.[3.8]

3.2.6.3 Finite State Machine with Datapath (FSMD) Model

The FSMD ISA is an implementation based on a combination of the Datapath ISA and the Controller ISA for processors that are not required to perform complex data manipulation and must repeatedly perform fixed computations on different sets of data. Common examples of an FSMD implementation are application-specific integrated circuits (ASICs), shown in Figure 3.9; programmable logic devices (PLDs), and field-programmable gate arrays (FPGAs, which are essentially more complex PLDs).

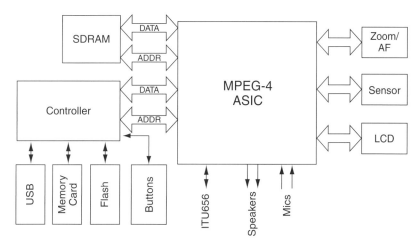

Figure 3.9: Board example with FSMD ISA implementation: solid-state digital camcorder.[3.9]

3.2.6.4 Java Virtual Machine (JVM) Model

The JVM ISA is based on one of the Java Virtual Machine standards. Real-world JVMs can be implemented in an embedded system via hardware, such as in aJile's aj-80 and aj-100 processors, for example (Figure 3.10).

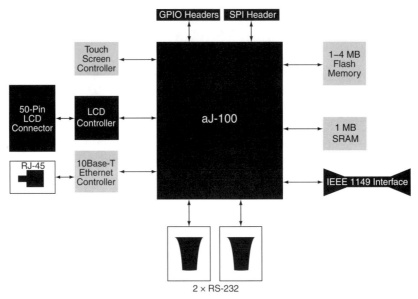

Figure 3.10: JVM ISA implementation example.[3.10]

3.2.7 General-Purpose ISA Models

General-purpose ISA models are typically implemented in processors targeted to be used in a wide variety of systems, rather than only in specific types of embedded systems. The most common types of general-purpose ISA architectures implemented in embedded processors are those discussed in this section.

3.2.7.1 Complex Instruction Set Computing (CISC) Model

The CISC ISA, as its name implies, defines complex operations made up of several instructions (see Figure 3.11). Common examples of architectures that implement a CISC ISA are Intel's ×86 and Motorola/Freescale's 68000 families of processors.

3.2.7.2 Reduced Instruction Set Computing (RISC) Model

In contrast to CISC, the RISC ISA (see Figure 3.12) usually defines:

- An architecture with simpler and/or fewer operations made up of fewer instructions
- An architecture that has a reduced number of cycles per available operation

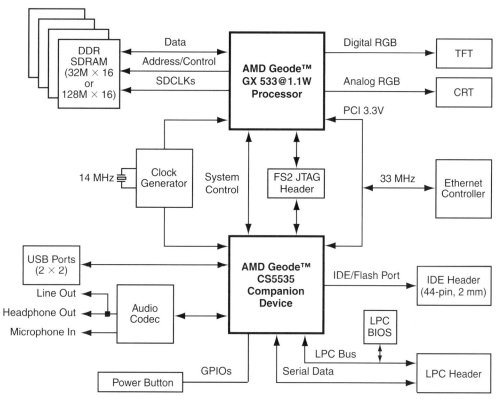

Figure 3.11: CISC ISA implementation example.[3.11]

© 2004 Advanced Micro Devices, Inc. Reprinted with permission.

Many RISC processors have only one-cycle operations, whereas CISCs typically have multiple-cycle operations. ARM, PowerPC, SPARC, and MIPS are just a few examples of RISC-based architectures.

A Final Note on CISC vs. RISC

> In the area of general-purpose computing, note that many current processor designs fall under the CISC or RISC category primarily because of their heritage. RISC processors have become more complex, while CISC processors have become more efficient to compete with their RISC counterparts, thus blurring the lines between the definition of a RISC versus a CISC architecture. Technically, these processors have both RISC and CISC attributes, regardless of their definitions.

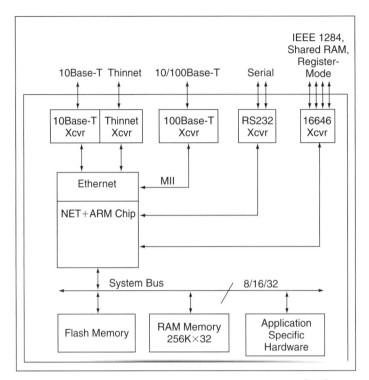

Figure 3.12: RISC ISA implementation example.[3.12]

3.2.8 Instruction-Level Parallelism ISA Models

Instruction-level parallelism ISA architectures are similar to general-purpose ISAs, except that they execute multiple instructions in parallel, as the name implies. In fact, instruction-level parallelism ISAs are considered higher evolutions of the RISC ISA, which typically has one-cycle operations, one of the main reasons that RISCs are the basis for parallelism. Examples of instruction-level parallelism ISAs include those discussed in this section.

3.2.8.1 Single Instruction, Multiple Data (SIMD) Model

The SIMD Machine ISA (see Figure 3.13) is designed to process an instruction simultaneously on multiple data components that require action to be performed on them.

3.2.8.2 Superscalar Machine Model

The superscalar ISA (see Figure 3.14) is able to process multiple instructions simultaneously within one clock cycle through the implementation of multiple functional components within the processor.

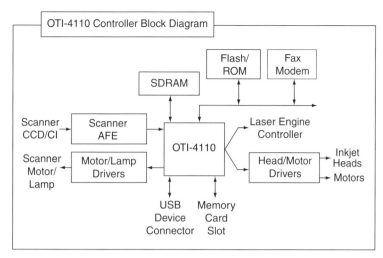

Figure 3.13: SIMD ISA implementation example.[3.13]

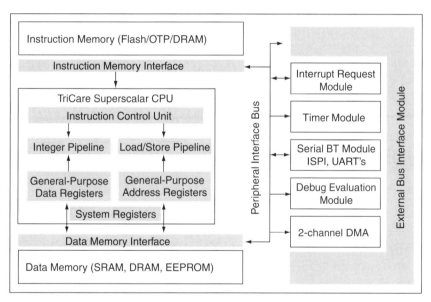

Figure 3.14: Superscalar ISA implementation example.[3.14]

3.2.8.3 Very Long Instruction Word Computing (VLIW) Model

The VLIW ISA (see Figure 3.15) defines an architecture in which a very long instruction word is made up of multiple operations. These operations are then broken down and processed in parallel by multiple execution units within the processor.

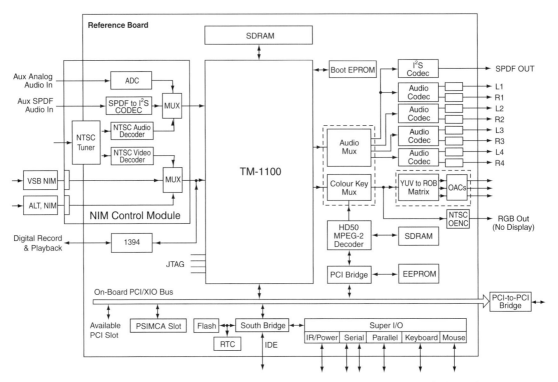

Figure 3.15: VLIW ISA implementation example: (VLIW) Trimedia-based DTV board.[3.15]

3.3 Internal Processor Design

The ISA defines *what* a processor can do, and it is the processor's internal interconnected hardware components that physically implement the ISA's features. Interestingly, the fundamental components that make up an embedded board are the same as those that implement an ISA's features in a processor: a CPU, memory, input components, output components, and buses. As mentioned in Figure 3.16, these components are the basis of the von Neumann model.

Of course, many current real-world processors are more complex in design than the von Neumann model has defined. However, most of these processors' hardware designs are still based on von Neumann components or a version of the von Neumann model called the *Harvard architecture model*. These two models differ in primarily one area, and that is memory. A von Neumann architecture defines a single memory space to store instructions and data. A Harvard architecture defines separate memory spaces for instructions and data; separate data and instruction buses allow simultaneous fetches

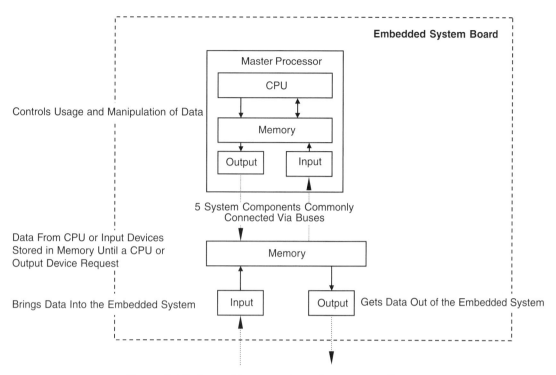

Figure 3.16: A von Neumann-based processor diagram.

and transfers to occur. The main reasoning behind using von Neumann versus a Harvard-based model for an architecture design is performance. Given certain types of ISAs, like Datapath model ISAs in DSPs, and their functions, such as continuously performing fixed computations on different sets of data, the separate data and instruction memory allows for an increase in the amount of data that can be processed per unit of time, given the lack of competing interests of space and bus accesses for transmissions of data and instructions.

As mentioned previously, most processors are based on some variation of the von Neumann model (in fact, the Harvard model itself is a variation of the von Neumann model; see Figure 3.17). Real-world examples of Harvard-based processor designs include ARM's ARM9/ARM10, MPC860, 8031, and DSPs (see Figure 3.18a), whereas ARM's ARM7 and x86 are von Neumann-based designs (see Figure 3.18b).

> **Note:** Although the MPC860 is a complex processor design, it is still based on the fundamental components of the Harvard model: the CPU, instruction memory, data memory, I/O, and buses.

Chapter 3

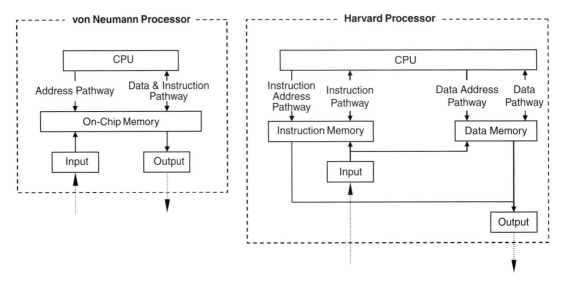

Figure 3.17: The von Neumann vs. Harvard architectures.

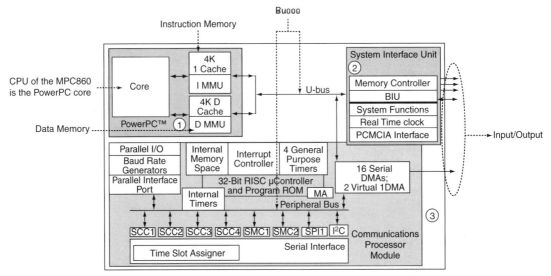

Figure 3.18a: Harvard architecture example: MPC860.[3.16]

Copyright of Freescale Semiconductor, Inc., 2004. Used by permission.

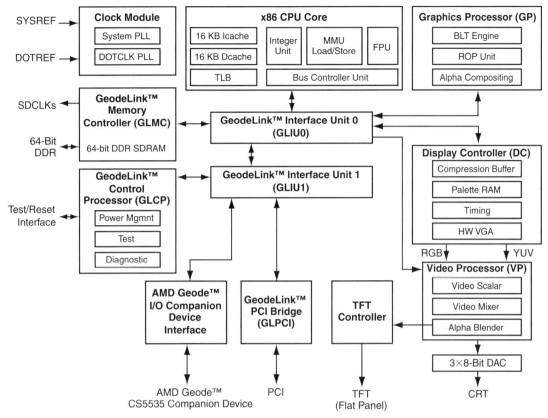

Figure 3.18b: A von Neumann architecture example: x86.[3.17]

Note: x86 is a complex processor design based on the von Neumann model in which, unlike the MPC860 processor, instructions and data share the same memory space.

Why Talk About the von Neumann Model?

The von Neumann model not only impacts the internals of a processor (what you don't see) but it shapes what you do see and what you can access within a processor. As discussed in Chapter 2, ICs—and a processor is an IC—have protruding pins that connect them to the board. Different processors vary widely in the number of pins and their associated signals, but the components of the von Neumann model, both at the board and at the internal processor level, also define the signals that all processors have. As shown in Figure 3.19, to accommodate board memory, processors typically have address and data signals to read and write data to and from memory. To communicate to memory or I/O, a processor usually has some type of READ and WRITE pins to indicate it wants to retrieve or transmit data.

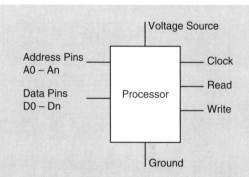

Figure 3.19: Von Neumann and processor pins.

Of course there are other pins not explicitly defined by von Neumann that are required for practical purposes, such as a synchronizing mechanism like a clock signal to drive a processor and some method of powering and grounding of the processor. However, regardless of the differences between processors, the von Neumann model essentially drives the external pins all processors have.

3.3.1 Central Processing Unit (CPU)

The semantics of this section can be a little confusing because processors themselves are commonly referred to as CPUs, but it is actually the processing *unit* within a processor that is the CPU. The CPU is responsible for executing the cycle of fetching, decoding, and executing instructions (see Figure 3.20). This three-step process is commonly referred to as a *three-stage pipeline,* and most recent CPUs are pipelined designs.

CPU designs can differ widely, but understanding the basic components of a CPU will make it easier to understand processor design and the cycle shown in Figure 3.20. As defined by the von Neumann model, this cycle is implemented through some combination of four major CPU components:

- *The arithmetic logic unit (ALU).* Implements the ISA's operations.
- *Registers.* A type of fast memory.
- *The control unit (CU).* Manages the entire fetching and execution cycle.
- *The internal CPU buses.* Interconnect the ALU, registers, and the CU.

Looking at a real-world processor, these four fundamental elements defined by the von Neumann model can be seen within the CPU of the MPC860 (see Figure 3.21).

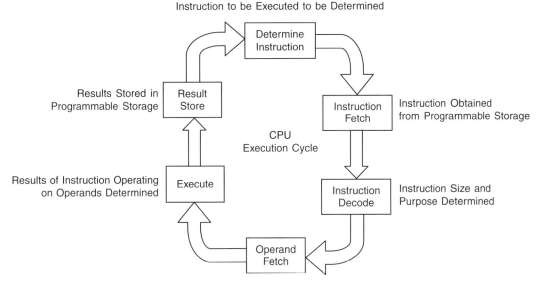

Figure 3.20: The fetch, decode, and execution cycle of CPU.

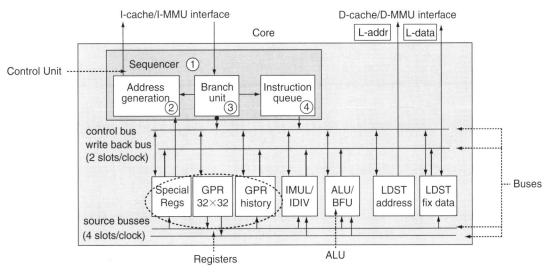

Figure 3.21: The MPC860 CPU: the PowerPC core.[3.18]

Copyright of Freescale Semiconductor, Inc., 2004. Used by permission.

Remember: Not all processors have these components as strictly defined by the von Neumann model, but they will have some combination of these components under various aliases somewhere on the processor. Remember that this model is a reference tool you can use to understand the major components of a CPU design.

3.3.1.1 Internal CPU Buses

The CPU buses are the mechanisms that interconnect the CPU's other components: the ALU, the CU, and registers (see Figure 3.22). Buses are simply wires that interconnect the various other components within the CPU. Each bus's wire is typically divided into logical functions, such as data (which carries data, bidirectionally, between registers and the ALU), address (which carries the locations of the registers that contain the data to be transferred), control (which carries control signal information, such as timing and control signals, between the registers, the ALU, and the CU), and so on.

> **Note:** To avoid redundancy, buses are discussed in more detail in Chapter 4.

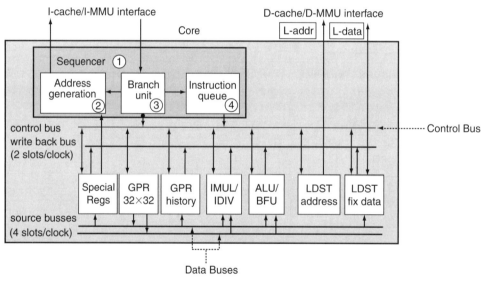

Figure 3.22: PowerPC core and buses.[3.19]

> **Note:** In the PowerPC Core, there is a control bus that carries the control signals between the ALU, CU, and registers. What the PowerPC calls "source buses" are the data buses that carry the data between registers and the ALU. There is an additional bus called the write-back that is dedicated to writing data received from a source bus directly back from the load/store unit to the fixed or floating-point registers.

3.3.1.2 Arithmetic Logic Unit (ALU)

The arithmetic logic unit (ALU) implements the comparison, mathematical, and logical operations defined by the ISA. The format and types of operations implemented in the CPU's ALU can vary depending on the ISA. Considered the core of any processor, the ALU is responsible for accepting multiple *n*-bit binary operands and performing any logical (AND, OR, NOT, etc.), mathematical (+, −, *, etc.), and comparison (=, <, >, etc.) operations on these operands.

The ALU is a combinational logic circuit that can have one or more inputs and only one output. An ALU's output is dependent only on inputs applied at that instant, as a function of time, and "no" past conditions (see Chapter 2 on gates). The basic building block of most ALUs (from the simplest to the multifunctional) is considered the *full adder*, a logic circuit that takes three 1-bit numbers as inputs and produces two 1-bit numbers. The way this actually works will be discussed in more detail later this section.

To understand how a full adder works, let us first examine the mechanics of adding binary numbers (0's and 1's) together:

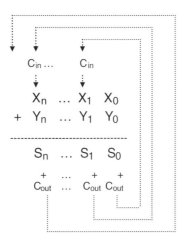

Starting with two 1-bit numbers, adding them will produce, at most, a 2-bit number:

X_0	Y_0	S_0	C_{out}
0	0	0	0
0	1	1	0
1	0	1	0
1	1	0	1

$\Rightarrow$ 0b + 0b = 0b

$\Rightarrow$ 0b + 1b = 0b

$\Rightarrow$ 1b + 0b = 1b

$\Rightarrow$ 1b + 1b = 10b (or 2d) In binary addition of 2 1-bit numbers, when the count exceeds 10 (the binary of 2 decimal), the 1 (C_{out}) is carried and added to the next row of numbers thus resulting in a 2-bit number.

This simple addition of two 1-bit numbers can be executed via a *half-adder* circuit, a logical circuit that takes two 1-bit numbers as inputs and produces a 2-bit output. Half-adder circuits, like all logical circuits, can be designed using several possible combinations of gates, such as the combinations shown in Figure 3.23.

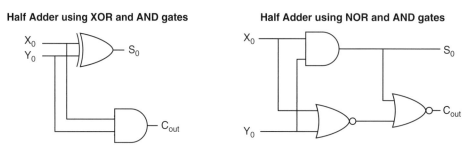

Figure 3.23a: Half-adder logic circuits.[3.20]

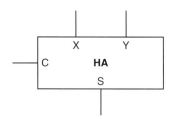

Figure 3.23b: Half-adder logic symbol.[3.20]

To add a larger number, the adder circuit must increase in complexity, and this is where the full adder comes into play. In trying to add two-digit numbers, for example, a full adder must be used in conjunction with a half adder. The half adder takes care of adding the first digits of the two numbers to be added (i.e., x_0, y_0, and so on); the full adder's three 1-bit inputs are the second digits of the two numbers to be added (i.e., x_1, y_1,...) along with the carry in (C_{in}) from the half adder's addition of the first digits. The half adder's output is the sum (S_0) along with the carry out (C_{out}) of the first digit's addition operation; the two 1-bit outputs of the full adder are the sum (S_1) along with the carry out (C_{out}) of the second digits' addition operation. Figure 3.24a shows the logic equations and truth table, Figure 3.24b shows the logic symbol, and Figure 3.24c shows an example of a full adder at the gate level, in this case, a combination XOR and NAND gate.

X	Y	C_{in}	S	C_{out}
0	0	0	0	0
0	0	1	1	0
0	1	0	1	0
0	1	1	0	1
1	0	0	1	0
1	0	1	0	1
1	1	0	0	1
1	1	1	1	1

Sum (S) = $XYC_{in} + XY'C_{in}' + X'YC_{in}' + X'Y'C_{in}'$
Carry Out (C_{out}) = $XY + X C_{in} = Y C_{in}$

Figure 3.24a: Full adder truth table and logic equations.[3.21]

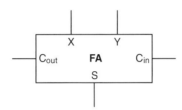

Figure 3.24b: Full adder logic symbol.[3.21]

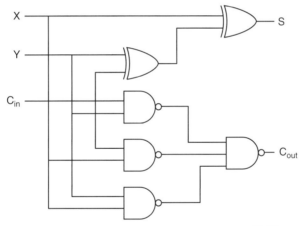

Figure 3.24c: Full adder gate-level circuit.[3.21]

To add larger numbers, additional full adders can then be integrated (cascaded) to the half-adder/full-adder hybrid circuit (see Figure 3.25). The example shown in this figure is the basis of the *ripple-carry adder* (one of many types of adders), in which *n* full adders are cascaded so that the carry produced by the lower stages propagates (ripples) through to the higher stages in order for the addition operation to complete successfully.

Multifunction ALUs that provide addition operations, along with other mathematical and logical operations, are designed around the adder circuitry, with additional circuitry incorporated

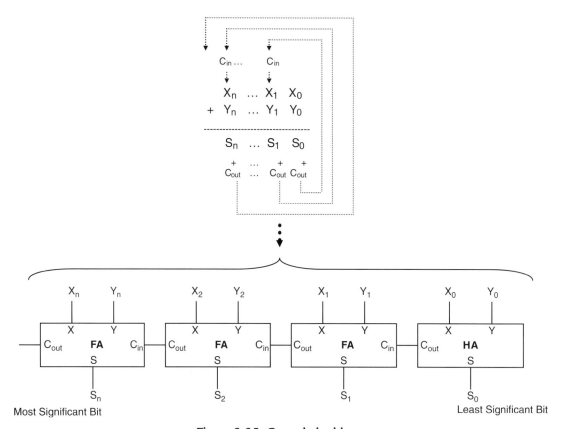

Figure 3.25: Cascaded adders.

for performing subtraction, logical AND, logical OR, and so on (see Figure 3.26a). The logic diagram shown in Figure 3.26b is an example of two stages of an n-bit multifunction ALU. The circuit in Figure 3.26 is based on the ripple-carry adder that was just described. In the logic circuit in Figure 3.26b, control inputs k_0, k_1, k_2, and c_{in} determine the function performed on the operand or operands. Operand inputs are $X = x_{n-1} \ldots x_1 x_0$ and $Y = y_{n-1} \ldots y_1 y_0$ and the output is sum (S) = $s_{n-1} \ldots s_1 s_0$ where the ALU saves the generated results varies with different architectures. With the PowerPC shown in Figure 3.27, results are saved in a register called an Accumulator. Results can also be saved in memory (on a stack or elsewhere) or in some hybrid combination of these locations.

> **Note:** In the PowerPC core, the ALU is part of the "Fixed Point Unit" that implements all fixed-point instructions other than load/store instructions. The ALU is responsible for fixed-point logic, add, and subtract instruction implementation. In the case of the PowerPC, generated results of the ALU are stored in an Accumulator. Also, note that the PowerPC has an IMUL/IDIV unit (essentially another ALU) specifically for performing multiplication and division operations.

Control Inputs				Result	Function
k_2	k_1	k_0	c_{in}		
0	0	0	0	S = X	Transfer X
0	0	0	1	S = X + 1	Increment X
0	0	1	0	S = X + Y	Addition
0	0	1	1	S = X + Y + 1	Add with Carry In
0	1	0	0	S = X − Y − 1	Subtract with Borrow
0	1	0	1	S = X − Y	Subtraction
0	1	1	0	S = X − 1	Decrement X
0	1	1	1	S = X	Transfer X
1	0	0	...	S = X OR Y	Logical OR
1	0	1	...	S = X XOR Y	Logical XOR
1	1	0	...	S = X AND Y	Logical AND
1	1	1	...	S = NOT X	Bit-wise Compliment

Figure 3.26a: Multifunction ALU truth table and logic equations.[3.22]

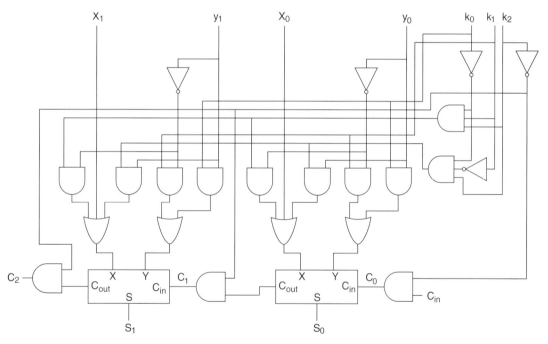

Figure 3.26b: Multifunction ALU gate-level circuitry.[3.22]

3.3.1.3 Registers

Registers are simply a combination of various flip-flops that can be used to temporarily store data or to delay signals. A *storage register* is a form of fast programmable internal processor memory usually used to temporarily store, copy, and modify operands that are immediately or

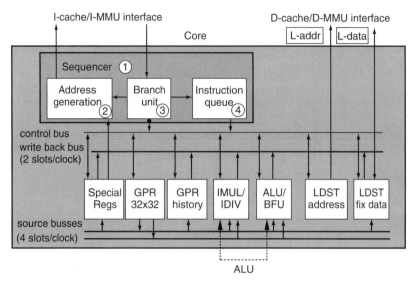

Figure 3.27: PowerPC core and the ALU.[3.23]

frequently used by the system. *Shift registers* delay signals by passing the signals between the various internal flip-flops with every clock pulse.

Registers are made up of a set of flip-flops that can be activated either individually or as a set. In fact, it is *the number of flip-flops in each register* that is actually used to describe a processor (for example, a 32-bit processor has working registers that are 32 bits wide containing 32 flip-flops, a 16-bit processor has working registers that are 16 bits wide containing 16 flip-flops, and so on). The number of flip-flops within these registers also determines the width of the data buses used in the system. Figure 3.28 shows an example of how eight flip-flops could comprise an 8-bit register and thus impact the size of the data bus. In short, registers are made up of one flip-flop for every bit being manipulated or stored by the register.

ISA designs do not all use registers in the same way to process data, but storage typically falls under one of two categories, either *general purpose* or *special purpose* (see Figure 3.29). *General-purpose registers* can be used to store and manipulate any type of data determined by the programmer, whereas *special-purpose registers* can *only* be used in a manner specified by the ISA, including holding results for specific types of computations, having predetermined *flags* (single bits within a register that can act and be controlled independently) acting as *counters* (registers that can be programmed to change states—that is, increment—asynchronously or synchronously after a specified length of time), and controlling *I/O ports* (registers managing the external I/O pins connected to the body of the processor and to board I/O). Shift registers are inherently special purpose because of their limited functionality.

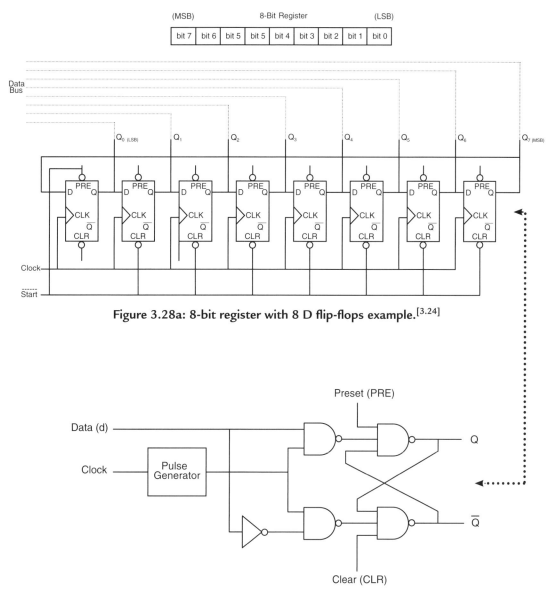

Figure 3.28a: 8-bit register with 8 D flip-flops example.[3.24]

Figure 3.28b: Example of a gate-level circuit of a flip-flop.[3.24]

Note: The PowerPC Core has a "Register Unit" that contains all registers visible to a user. PowerPC processors generally have two types of registers: general-purpose and special-purpose (control) registers.

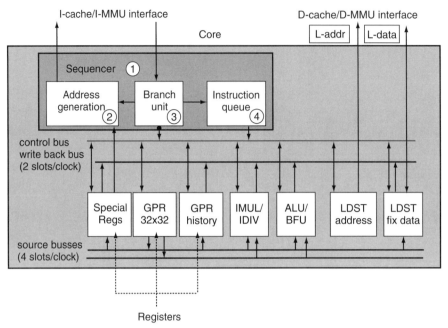

Figure 3.29: PowerPC core and register usage.[3.25]

The number of registers, the types of registers, and the size of the data that these registers can store (8-bit, 16-bit, 32-bit, and so forth) vary depending on the CPU, according to the ISA definitions. In the cycle of fetching and executing instructions, the CPU's registers have to be fast so as to quickly feed data to the ALU, for example, and to receive data from the CPU's internal data bus. Registers are also multiported so as to be able to both receive and transmit data to these CPU components. The next several pages of this section will give some real-world examples of how some common registers in architectures—specifically flags and counters—can be designed.

3.3.1.4 Example 1: Flags

Flags are typically used to indicate to other circuitry that an event or a state change has occurred. In some architectures, flags can be grouped together into specific flag registers, whereas in other architectures, flags comprise some part of several different types of registers.

To understand how a flag works, let's examine a logic circuit that can be used in designing a flag. Given a register, for instance, let's assume that bit 0 is a flag (see Figures 3.30a and b) and the flip-flop associated with this flag bit is a *set-reset (SR) flip-flop*, the simplest of data-storage asynchronous sequential digital logic. The (cross NAND) SR flip-flop is used in this example to asynchronously detect an event that has occurred in attached circuitry via the set

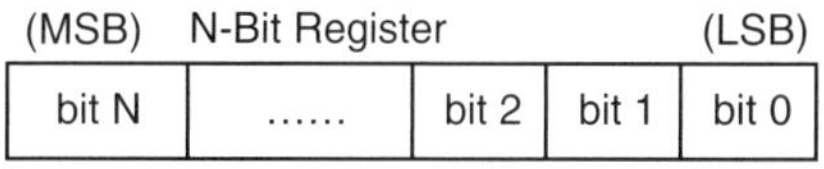

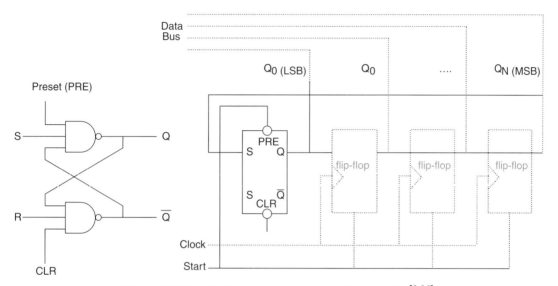

Figure 3.30a: *N*-bit register with flag and SR flip-flop example.[3.26]

Figure 3.30b: SR flip-flop gate-level circuit example.[3.26]

(S) or reset (R) input signal of the flip-flop. When the set/reset signal changes from 0 to 1 or 1 to 0, it immediately changes the state of the flip-flop, which results, depending on the input, in the flip-flop setting or resetting.

3.3.1.5 Example 2: Counters

As mentioned at the beginning of this section, registers can also be designed to be counters, programmed to increment or decrement either asynchronously or synchronously, such as with a processor's program counter (PC) or timers, which are essentially counters that count clock cycles. An *asynchronous counter* is a register whose flip-flops are not driven by the same central clock signal. Figure 3.31a shows an example of a 8-bit MOD-256 (modulus-256) asynchronous counter using JK flip-flops (which have 128 binary states—capable of counting between 0 and 255, 128 * 2 = 256). This counter is a binary counter, made up of 1's and 0's, with 8 digits, one flip-flop per digit. It loops counting between 00000000 and 11111111, recycling back to 00000000 when 11111111 is reached, ready to start over with the count. Increasing the size of the counter—the maximum number of digits the counter can count to—is only a matter of adding a flip-flop for every additional digit.

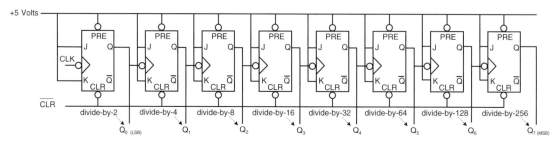

Figure 3.31a: An 8-bit MOD-256 asynchronous counter example.[3.27]

PRE	CLR	CLK	J	K	Q	$\overline{Q}$	Mode
0	1	x	x	x	1	0	Preset
1	0	x	x	x	0	1	Clear
0	0	x	x	x	1	1	Unused
1	1	—	0	0	Q_0	$\overline{Q}_0$	Hold
1	1	—	0	1	0	0	Reset
1	1	—	1	0	0	0	Set
1	1	—	1	1	$\overline{Q}_0$	Q_0	Toggle
1	1	−0.1	1	1	Q_0	$\overline{Q}_0$	Hold

Figure 3.31b: JK flip-flop truth table.[3.27]

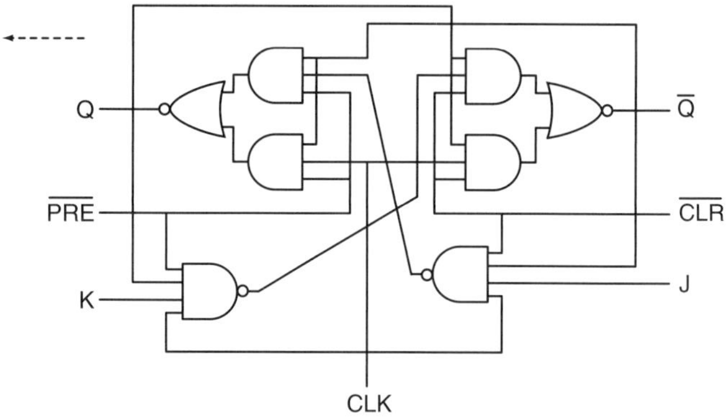

Figure 3.31c: JK flip-flop gate-level diagram.[3.27]

All the flip-flops of the counter are fixed in toggle mode; looking at the counter's truth table in Figure 3.31b under toggle mode, the flip-flop inputs (J and K) are both = 1 (HIGH). In toggle mode, the first flip-flop's output (Q_0) switches to the opposite of its current state at each active clock HIGH-to-LOW (falling) edge (see Figure 3.32).

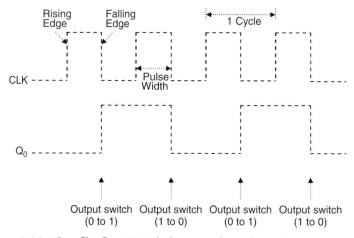

Figure 3.32: First flip-flop CLK timing waveform for MOD-256 counter.

As shown in Figure 3.32, the result of toggle mode is that Q_0, the output of the first flip-flop, has half the frequency of the CLK signal that was input into its flip-flop. Q_0 becomes the CLK signal for the next flip-flop in the counter. As shown in the timing diagram in Figure 3.33, Q_1, the output of the second flip-flop signal has half the frequency of the CLK signal that was input into it (one-quarter of the original CLK signal).

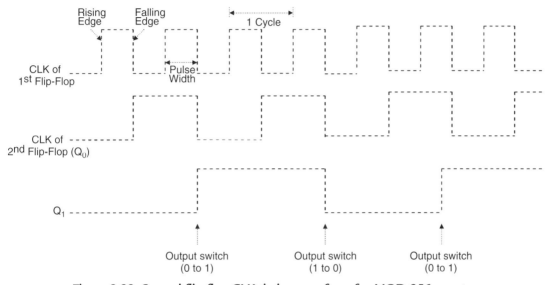

Figure 3.33: Second flip-flop CLK timing waveform for MOD-256 counter.

This cycle in which the output signals for the preceding flip-flops become the CLK signals for the next flip-flops continues until the last flip-flop is reached. The division of the CLK signal originally input into the first flip-flop can be seen in Figure 3.31a. The combination of output switching of all the flip-flops on the falling edges of the outputs of the previous flip-flop, which acts as their CLK signals, is how the counter is able to count from 00000000 to 11111111 (see Figure 3.34).

With synchronous counters, all flip-flops within the counter are driven by a *common* clock input signal. Again using JK flip-flops, Figure 3.35 demonstrates how a MOD-256 synchronous counter circuitry differs from a MOD-256 asynchronous counter (the previous example).

The five additional AND gates (two of which are not explicitly shown due to the scale of the diagram) in the synchronous counter example in Figure 3.35 serve to put the flip-flops either in HOLD mode if inputs J and K = 0 (LOW) or in TOGGLE mode if inputs J and K = 1 (HIGH). Refer to the JK flip-flop truth table in Figure 3.30b. The synchronous counter in this example works because the first flip-flop is always in TOGGLE mode at the start of the count 00000000,

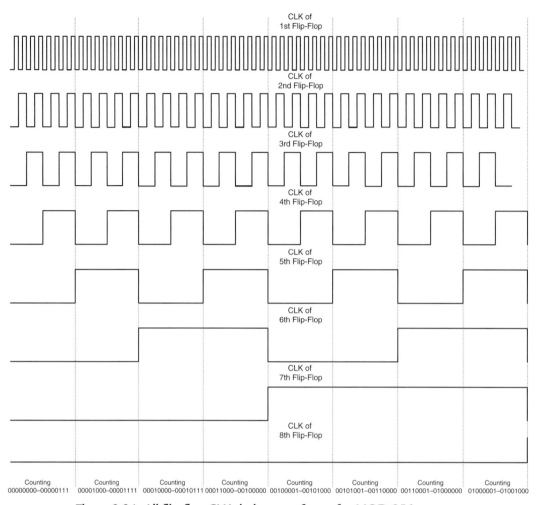

Figure 3.34: All flip-flop CLK timing waveforms for MOD-256 counter.

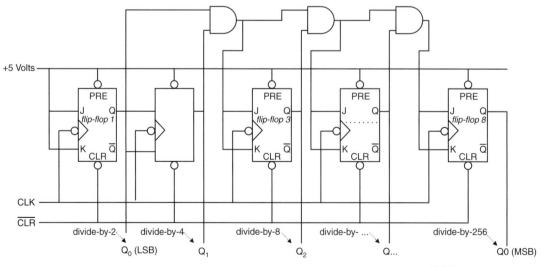

Figure 3.35: An 8-bit MOD-256 synchronous counter example.[3.28]

whereas the rest are in HOLD mode. When counting (0 to 1 for the first flip-flop), the next flip-flop is then TOGGLED, leaving the remaining flip-flops on HOLD. This cycle continues (2–4 for the second flip-flop, 4–8 for the third flip-flop, 8–15 for the fourth flip-flop, 15–31 for the fifth flip-flop, and so on) until all counting is completed to 11111111 (255). At that point, all the flip-flops have been toggled and held accordingly.

3.3.1.6 Control Unit (CU)

The control unit (CU) is primarily responsible for generating timing signals as well as controlling and coordinating fetching, decoding, and executing instructions in the CPU. After the instruction has been fetched from memory and decoded, the control unit then determines what operation will be performed by the ALU and selects and writes signals appropriate to each functional unit within or outside the CPU (i.e., memory, registers, ALU, etc.). To better understand how a processor's control unit functions, let's examine more closely the control unit of a PowerPC processor.

As shown in Figure 3.36, the PowerPC core's CU is called a "sequencer unit" and is the heart of the PowerPC core. The sequencer unit is responsible for managing the continuous cycle of

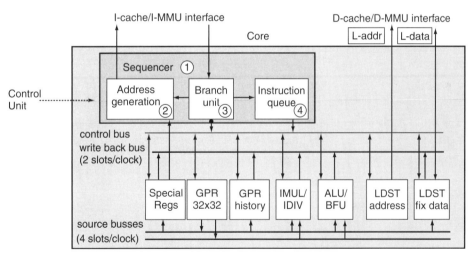

Figure 3.36: PowerPC core and the CU.[3.29]

fetching, decoding, and executing instructions while the PowerPC has power, including such tasks as:

- Providing the central control of the data and instruction flow among the other major units within the PowerPC core (CPU), such as registers, ALU, and buses

- Implementing the basic instruction pipeline
- Fetching instructions from memory to issue these instructions to available execution units
- Maintaining a state history for handling exceptions

Like many CUs, the PowerPC's sequencer unit isn't one physically separate, explicitly defined unit; rather, it is made up of several circuits distributed within the CPU that all work together to provide the managing capabilities. Within the sequencer unit these components are mainly an *address generation unit* (provides address of next instruction to be processed), a *branch prediction unit* (processes branch instructions), a *sequencer* (provides information and centralized control of instruction flow to the other control subunits), and an *instruction queue* (stores the next instructions to be processed and dispatches the next instructions in the queue to the appropriate execution unit).

3.3.1.7 The CPU and the System (Master) Clock

A processor's execution is ultimately synchronized by an external *system* or *master clock*, located on the board. The master clock is an oscillator along with a few other components, such as a crystal. It produces a fixed frequency sequence of regular on/off pulse signals (square waves), as shown in Figure 3.37. The CU, along with several other components on an embedded board, depends on this master clock to function. Components are driven by either the actual level of the signal (a "0" or a "1"), the rising edge of a signal (the transition from "0" to "1"), and/or the falling edge of the signal (the transition from "1" to "0"). Different master clocks, depending on the circuitry, can run at a variety of frequencies but typically must run so that the slowest component on the board has its timing requirements met. In some cases, the master clock signal is divided by the components on the board to create other clock signals for their own use.

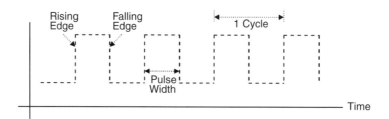

Figure 3.37: Clock signal.

In the case of the CU, for instance, the signal produced by the master clock is usually divided or multiplied within the CPU's CU to generate at least one internal clock signal. The CU then uses internal clock signals to control and coordinate the fetching, decoding, and executing of instructions.

3.3.2 On-Chip Memory

The CPU goes to memory to get what it needs to process, because it is in memory that all the data and instructions to be executed by the system are stored. Embedded platforms have a *memory hierarchy*, a collection of different types of memory, each with unique speeds, sizes, and usages (see Figure 3.38). Some of this memory can be physically integrated on the processor, such as registers, read-only memory (ROM), certain types of random access memory (RAM), and level-1 cache.

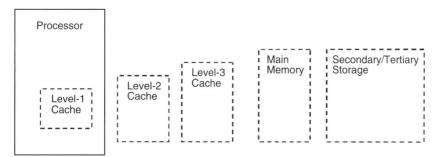

Figure 3.38: Memory hierarchy.

3.3.2.1 Read-Only Memory (ROM)

On-chip ROM is memory integrated into a processor that contains data or instructions that remain even when there is no power in the system, due to a small, longer-life battery, and therefore is considered to be *nonvolatile memory* (NVM). The content of on-chip ROM usually can only be read by the system it is used in.

To get a clearer understanding of how ROM works, let's examine a sample logic circuit of 8×8 ROM, shown in Figure 3.39. This ROM includes three address lines ($\log_2 8$) for all eight words, meaning that the 3-bit addresses ranging from 000 to 111 will each represent one of the 8 bytes.

> Note that different ROM designs can include a wide variety of addressing configurations for the exact same matrix size, and this addressing scheme is just an example of one such scheme.

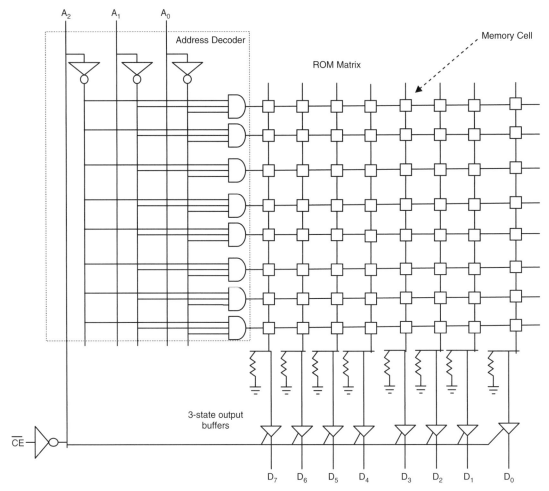

Figure 3.39: 8 × 8 ROM logic circuit.[3.30]

D_0 through D_7 are the output lines from which data is read, one output line for each bit. Adding rows to the ROM matrix increases its size in terms of the number of address spaces, whereas adding columns increases a ROM's data size (the number of bits per address) it can store. ROM size specifications are represented in the real world identically to what is used in this example, where the matrix reference (8 × 8, 16k × 32, and so on) reflects the actual size of ROM where the first number, preceding the "×", is the number of addresses and the second number, after the "×", reflects the size of the data (number of bits) at each address location—8 = byte, 16 = halfword, 32 = word, and so on. Also, note that in some design documentation, the ROM matrix size may be summarized. For example, 16 kB (kBytes) of ROM is 16 K × 8 ROM, 32 MB of ROM is 32 M × 8 ROM, and so on.

In this example, the 8 × 8 ROM is an 8 × 8 matrix, meaning that it can store eight different 8-bit bytes, or 64 bits of information. Every intersection of a row and column in this matrix is a memory location, called a *memory cell*. Each memory cell can contain either a bipolar or MOSFET transistor (depending on the type of ROM) and a fusible link (see Figure 3.40).

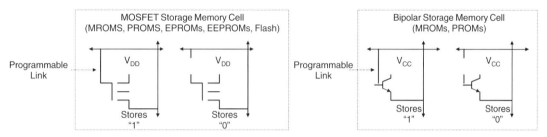

Figure 3.40: 8 × 8 MOSFET and bipolar memory cells.[3.31]

When a programmable link is in place, the transistor is biased ON, resulting in a 1 being stored. All ROM memory cells are typically manufactured in this configuration. When writing to ROM, a "0" is stored by breaking the programmable link. The way links are broken depends on the type of ROM. The way to read from a ROM depends on the ROM, but in this example, the chip enable (CE) line is toggled (HIGH to LOW) to allow the data stored to be output via D_0–D_7 after having received the 3-bit address requesting the row of data bits (see Figure 3.41).

Finally, the most common types of on-chip ROM include:

- *MROM* (mask ROM), which is ROM (with data content) that is permanently etched into the microchip during the manufacturing of the processor and cannot be modified later.

- *PROMs* (programmable ROM) or OTPs (one-time programmables), which is a type of ROM that can be integrated on-chip and that is one-time programmable by a PROM programmer (in other words, it can be programmed outside the manufacturing factory).

- *EPROM* (erasable programmable ROM), which is ROM that can be integrated on a processor, in which content can be erased and reprogrammed more than once. The number of times erasure and reuse can occur depends on the processor. The content of EPROM is written to the device using special separate devices and erased, either selectively or in its entirety, using other devices that output intense ultraviolet light into the processor's built-in window.

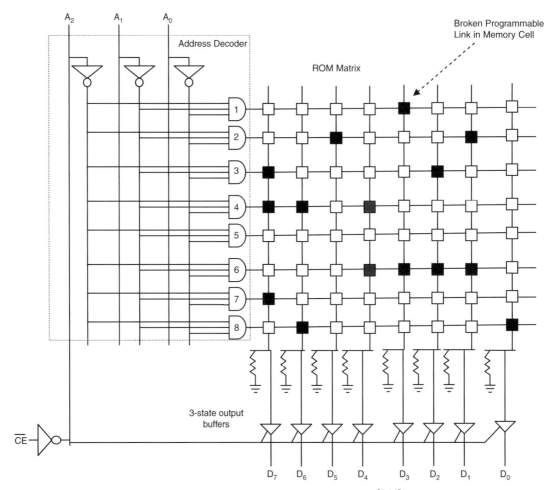

Figure 3.41: 8 × 8 reading ROM circuit.[3.32]

- *EEPROM* (electrically erasable programmable ROM), which, like EPROM, can be erased and reprogrammed more than once. The number of times erasure and reuse can occur depends on the processor. Unlike EPROMs, the content of EEPROM can be written and erased without using any special devices while the embedded system is functioning. With EEPROMs, erasing must be done in its entirety, unlike EPROMs, which can be erased selectively.

A cheaper and faster variation of the EEPROM is Flash memory. Where EEPROMs are written and erased at the byte level, Flash can be written and erased in blocks or sectors (a group of bytes). Like EEPROM, Flash can be erased while still in the embedded device.

3.3.2.2 Random Access Memory (RAM)

RAM (random access memory), commonly referred to as *main memory*, is memory in which any location within it can be accessed directly (randomly, rather than sequentially from some starting point) and whose content can be changed more than once (the number depending on the hardware). Unlike ROM, contents of RAM are erased if RAM loses power, meaning that RAM is *volatile*. The two main types of RAM are *static RAM (SRAM)* and *dynamic RAM (DRAM)*.

As shown in Figure 3.42a, SRAM memory cells are made up of transistor-based flip-flop circuitry that typically holds its data due to a moving current being switched bidirectionally on a pair of inverting gates in the circuit until power is cut off or the data is overwritten.

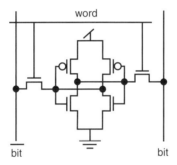

Figure 3.42a: A six-transistor SRAM cell.[3.33]

To get a clearer understanding of how SRAM works, let's examine a sample logic circuit of 4K × 8 SRAM shown in Figure 3.42b.

In this example, the 4K × 8 SRAM is a 4K × 8 matrix, meaning that it can store 4096 (4 × 1024) different 8-bit bytes, or 32768 bits of information. As shown in the diagram, 12 address lines (A_0–A_{11}) are needed to address all 4096 (000000000000b–111111111111b) possible addresses, one address line for every address digit of the address. In this example, the 4K × 8 SRAM is set up as a 64 × 64 array of rows and columns where addresses A_0–A_5 identify the row and A_6–A_{11} identify the column. As with ROM, every intersection of a row and column in the SRAM matrix is a memory cell, and in the case of SRAM memory cells, they can contain flip-flop circuitry mainly based on semiconductor devices such as polysilicon load resistors, bipolar transistors, and/or CMOS transistors. There are eight output lines (D_0–D_7), a byte for every byte stored at an address.

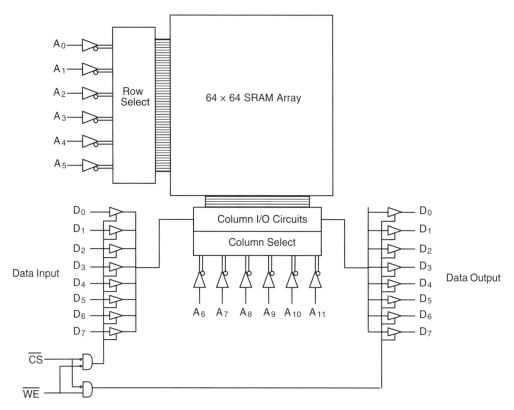

Figure 3.42b: 4K × 8 SRAM logic circuit.[3.34]

In this SRAM example, when the chip select (CS) is HIGH, then memory is in standby mode (no read or writes are occurring). When CS is toggled to LOW and write-enable input (WE) is LOW, then a byte of data is written through the data input lines (D_0–D_7) at the address indicated by the address lines. Given the same CS value (LOW) and WE is HIGH, then a byte of data is being read from the data output lines (D_0–D_7) at the address indicated by the address lines (A_0–A_7).

As shown in Figure 3.43, DRAM memory cells are circuits with *capacitors* that hold a charge in place (the charges or lack thereof reflecting data). DRAM capacitors need to be refreshed frequently with power in order to maintain their respective charges and to recharge capacitors after DRAM is read. (Reading DRAM discharges the capacitor.) The cycle of discharging and recharging of memory cells is why this type of RAM is called *dynamic*.

Given a sample logic DRAM circuit of 16K × 8, this RAM configuration is a two-dimensional array of 128 rows and 128 columns, meaning that it can store 16384 (16 × 1024) different 8-bit bytes, or 131072 bits of information. With this address configuration, larger DRAMs can either be designed with 14 address lines (A_0–A_{13}) needed to address all 16384

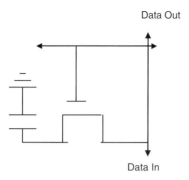

Figure 3.43: DRAM (capacitor-based) memory cell.[3.35]

(00000000000b–11111111111b) possible addresses—one address line for every address digit of the address—or these address lines can be *multiplexed*, or combined into fewer lines to share, with some type of data selection circuit managing the shared lines. Figure 3.44 demonstrates how a multiplexing of address lines could occur in this example.

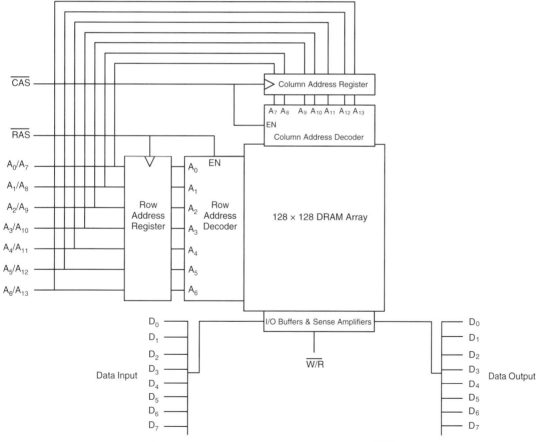

Figure 3.44: 16K × 8 SRAM logic circuit.[3.36]

The 16 K × 8 DRAM is set up with addresses A_0–A_6 identifying the row and A_7–A_{13} identifying the column. In this example, the ROW address strobe (RAS) line is toggled (from HIGH to LOW) for A_0–A_6 to be transmitted, and then the Column Address Strobe (CAS) line is toggled (from HIGH to LOW) for A_7–A_7 to be transmitted. After this point the memory cell is latched and ready to be written to or read from.

There are eight output lines (D_0–D_7), a byte for every byte stored at an address. When the write-enable (WE) input line is HIGH, data can be read from output lines D_0–D_7, and when WE is LOW, data can be written to input lines D_0–D_7.

One of the major differences between SRAM and DRAM lies in the makeup of the DRAM memory array itself. The capacitors in the memory array of DRAM are not able to hold a charge (data). The charge gradually dissipates over time, thus requiring some additional mechanism to *refresh* DRAM in order to maintain the integrity of the data. This mechanism *reads* the data in DRAM before it is lost, via a sense amplification circuit that senses a charge stored within the memory cell, and *writes* it back onto the DRAM circuitry. Ironically, the process of reading the cell also discharges the capacitor, even though reading the cell is part of the process of correcting the problem of the capacitor gradually discharging in the first place. A *memory controller* (see Section 5.4, "Memory Management," for more information) in the embedded system typically manages a DRAM's recharging and discharging cycle by initiating refreshes and keeping track of the refresh sequence of events. It is this refresh cycling mechanism that discharges and recharges memory cells that gives this type of RAM its name—"dynamic" RAM (DRAM)—and the fact that the charge in SRAM stays put is the basis for its name, "static" RAM (SRAM). It is this same additional recharge circuitry that makes DRAM slower in comparison to SRAM. (Note that SRAM is usually slower than registers because the transistors within the flip-flop are usually smaller and thus do not carry as much current as those typically used within registers.)

SRAMs also usually consume less power than DRAMs, since no extra energy is needed for a refresh. On the flip side, DRAM is typically cheaper than SRAM because of its capacitance-based design, in comparison to its SRAM flip-flop counterpart (more than one transistor). DRAM also can hold more data than SRAM, since DRAM circuitry is much smaller than SRAM circuitry and more DRAM circuitry can be integrated into an IC.

DRAM is usually the "main" memory in larger quantities and is also used for video RAM and cache. DRAMs used for display memory are also commonly referred to as *frame buffers*. SRAM, because it is more expensive, is typically used in smaller quantities, but because it is also the fastest type of RAM, it is used in external cache (see Section 5.2) and video memory (when processing certain types of graphics, and given a more generous budget, a system can implement a better-performing RAM).

Table 3.3 summarizes some examples of different types of integrated RAM and ROM used for various purposes in ICs.

Table 3.3: On-chip memory.[3.37]

	Main Memory	Video Memory	Cache
SRAM	NA	Random Access Memory Digital-to-Analog Converter (RAMDAC) processors are used in video cards for display systems without true color, to convert digital image data into analog display data for analog displays such as cathode ray tubes (CRTs). The built-in SRAM contains the color palette table that provides the red/green/blue (RGB) on version values used by the digital-to-analog converters (DACs), also built into the RAMDAC, to change the digital image data into analog signals for the display units.	SRAM has been used for both level-1 and level-2 caches. A type of SRAM, called Burst/SynchBurst Static Random-Access Memory (BSRAM), which is synchronized with either the system clock or a cache bus clock, has been primarily used for level-2 cache memory. (See Section 3.3.)
DRAM	Synchronous Dynamic Random Access Memory (SDRAM) is DRAM that is synchronized with the microprocessor's clock speed (in MHz). Several types of SDRAMs are used in various systems, such as the JDEC SDRAM (JEDEC Synchronous Dynamic Random Access Memory), PC100 SDRAM (PC100 Synchronous Dynamic Random Access Memory), and DDR SDRAM (Double Data Rate Synchronous Dynamic Random Access Memory). Enhanced Synchronous Dynamic Random Access Memory (ESDRAM) is SDRAM	On-Chip Rambus Dynamic Random Access Memory (RDRAM) and On-Chip Multibank Dynamic Random Access Memory (MDRAM) are DRAMs commonly used as display memory that store arrays of bit values (pixels of the image on the display). The resolution of the image is determined by the number of bits that have been defined per each pixel.	Enhanced Dynamic Random Access Memory (EDRAM) integrates SRAM within the DRAM and is usually used as level-2 cache (see Section 3.3). The faster SRAM portion of EDRAM is searched first for the data, and if it's not found there, the DRAM portion of EDRAM is searched.

Table 3.3: Continued

	Main Memory	Video Memory	Cache
	that integrates SRAM within the SDRAM, allowing for faster SDRAM. (Basically, the faster SRAM portion of the ESDRAM is checked first for data, then if not found, the remaining SDRAM portion is searched.)		
	Direct Rambus Dynamic Random Access Memory (DRDRAM) and SyncLink Dynamic Random Access Memory (SLDRAM) are DRAMs whose bus signals can be integrated and accessed on one line, thus decreasing the access time (since synchronizing operations on multiple lines is not necessary).	Fast Page Mode Dynamic Random Access Memory (FPM DRAM), Data Output Random Access/Dynamic Random Access Memory (EDORAM/EDO DRAM), and Data Burst Extended Data Output Dynamic Random-Access Memory (BEDO DRAM) ...	

3.3.2.3 Cache (Level-1 Cache)

Cache is the level of memory between the CPU and main memory in the memory hierarchy (see Figure 3.45). Cache can be integrated into a processor or can be off-chip. Cache existing on-chip is commonly referred to as *level-1 cache*, and SRAM memory is usually used as level-1 cache. Because (SRAM) cache memory is typically more expensive due to its speed, processors usually have a small amount of cache, whether on-chip or off-chip.

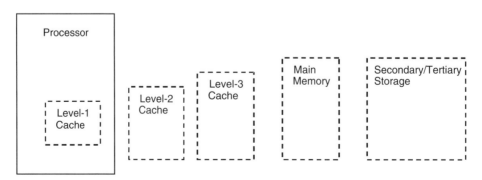

Figure 3.45: Level-1 cache in the memory hierarchy.

Using cache has become popular in response to systems that display a good *locality of reference*, meaning that these systems in a given time period access most of their data from a limited section of memory. Cache is used to store subsets of main memory that are used or accessed often. Some processors have one cache for both instructions and data; others have separate on-chip caches for each.

A variety of strategies are used in writing to and reading data from level-1 cache and main memory (Figure 3.46). These strategies include transferring data between memory and cache in either one-word or multiword *blocks*. These blocks are made up of data from main memory as well as the location of that data in main memory (called *tags*).

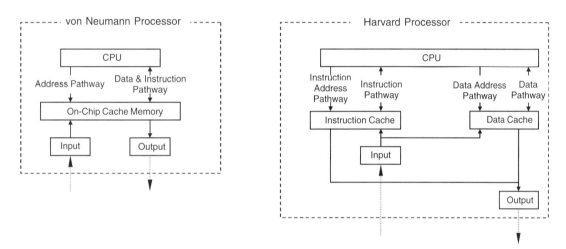

Figure 3.46: Level-1 cache in the von Neumann and Harvard models.

In the case of writing to memory, given some memory address from the CPU, this address is translated to determine its equivalent location in level-1 cache, since cache is a snapshot of a subset of memory. Writes must be done in both cache and main memory to ensure that cache and main memory are *consistent* (have the same value). The two most common write strategies to guarantee consistency are *write-through*, in which data is written to both cache and main memory every time, and *write-back*, in which data is initially only written into cache, and only when it is to be bumped and replaced from cache is it written into main memory.

When the CPU wants to read data from memory, level-1 cache is checked first. If the data is in cache, it is called a *cache hit*. The data is returned to the CPU and the memory access process is complete. If the data is not located in level-1 cache, it is called *cache miss*. Off-chip caches (if any) are then checked for the data desired. If this is a miss, then main memory is accessed to retrieve and return the data to the CPU.

Data is usually stored in cache in one of three schemes:

- *Direct mapped*, where data in cache is located by its associated block address in memory (using the "tag" portion of the block)

- *Set associative*, where cache is divided into *sets* into which multiple blocks can be placed; blocks are located according to an index field that maps into a cache's particular set

- *Full associative*, where blocks are placed anywhere in cache and must be located by searching the entire cache every time

In systems with memory management units (MMU) to perform the translation of addresses (see Section 3.3), cache can be integrated between the CPU and the MMU or between the MMU and main memory. There are advantages and disadvantages to both methods of cache integration with an MMU, mostly surrounding the handling of DMA (direct memory access), which is the direct access of off-chip main memory by slave processors on the board without going through the main processor. When cache is integrated between the CPU and MMU, only the CPU accesses to memory affect cache; therefore DMA writes to memory can make cache inconsistent with main memory unless CPU access to memory is restricted while DMA data is being transferred or cache is being kept updated by other units within the system besides the CPU. When cache is integrated between the MMU and main memory, more address translations need to be done, since cache is affected by both the CPU and DMA devices.

3.3.2.4 On-Chip Memory Management

Many different types of memory can be integrated into a system, and there are also differences in the way software running on the CPU views memory addresses (*logical/virtual addresses*) and the actual *physical memory addresses* (the two-dimensional array or row and column). *Memory managers* are ICs designed to manage these issues and in some cases are integrated onto the master processor.

The two most common types of memory managers that are integrated into the master processor are memory controllers (MEMC) and memory management units (MMUs). A *memory controller* (MEMC) is used to implement and provide glueless interfaces to the different types of memory in the system, such as cache, SRAM, and DRAM, synchronizing access to memory and verifying the integrity of the data being transferred. Memory controllers access memory directly with the memory's own physical (two-dimensional) addresses. The controller manages the request from the master processor and accesses the appropriate banks, awaiting feedback and returning that feedback to the master processor. In some cases, where the memory controller is mainly managing one type of memory, it may be referred to by that memory's name (such as DRAM controller, cache controller, and so forth).

Memory management units (MMUs) are used to translate logical addresses into physical addresses (*memory mapping*) as well as handle memory security, control cache, handle bus arbitration between the CPU and memory, and generate appropriate exceptions. Figure 3.47 shows the MPC860, which has both an integrated MMU (in the core) and an integrated memory controller (in the system interface unit).

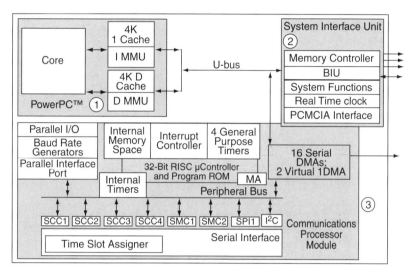

Figure 3.47: Memory management and the MPC860.[3.38]
Copyright of Freescale Semiconductor, Inc., 2004. Used by permission.

In the case of translated addresses, the MMU can use level-1 cache on the processor, or portions of cache allocated as buffers for caching address translations, commonly referred to as the *translation lookaside buffer*, or TLB, to store the mappings of logical addresses to physical addresses. MMUs also must support the various schemes in translating addresses, mainly segmentation, paging, or some combination of both schemes. In general, *segmentation* is the division of logical memory into large variable-size sections, whereas *paging* is the dividing of logical memory into smaller fixed-size units.

The memory protection schemes then provide shared, read/write, or read-only accessibility to the various pages and/or segments. If a memory access is not defined or allowed, an interrupt is typically triggered. An interrupt is also triggered if a page or segment isn't accessible during address translation (i.e., in the case of a paging scheme, a page fault, etc.). At that point the interrupt would need to be handled; the page or segment would need to be retrieved from secondary memory, for example.

The scheme supporting segmentation and/or paging of the MMU typically depends on the software—that is, the operating system.

3.3.2.5 Memory Organization

Memory organization includes not only the makeup of the memory hierarchy of the particular platform but also the internal organization of memory, specifically what different portions of memory may or may not be used for as well as how all the different types of memory are organized and accessed by the rest of the system. For example, some architectures may split memory so that a portion stores only instructions and another only stores data. The SHARC DSP contains integrated memory that is divided into separate *memory spaces* (sections of memory) for data and programs (instructions). In the case of the ARM architectures, some are based on the von Neumann model (for example, ARM7), which means that it has one memory space for instructions and data, whereas other ARM architectures (namely ARM9) are based on the Harvard model, meaning memory is divided into a section for data and a separate section for instructions.

The master processor, along with the software, treats memory as one large one-dimensional array, called a *memory map* (see Figure 3.48). This map serves to clearly define what address or set of addresses are occupied by what components.

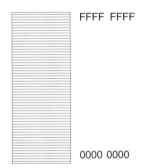

Figure 3.48a: Memory map.

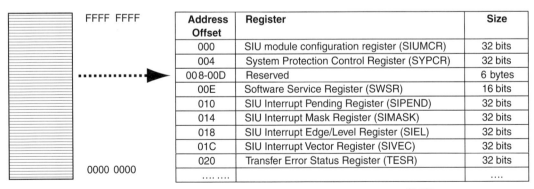

Address Offset	Register	Size
000	SIU module configuration register (SIUMCR)	32 bits
004	System Protection Control Register (SYPCR)	32 bits
008-00D	Reserved	6 bytes
00E	Software Service Register (SWSR)	16 bits
010	SIU Interrupt Pending Register (SIPEND)	32 bits
014	SIU Interrupt Mask Register (SIMASK)	32 bits
018	SIU Interrupt Edge/Level Register (SIEL)	32 bits
01C	SIU Interrupt Vector Register (SIVEC)	32 bits
020	Transfer Error Status Register (TESR)	32 bits
………		….

Figure 3.48b: MPC860 registers within a memory map.[3.39]

Copyright of Freescale Semiconductor, Inc., 2004. Used by permission.

Within this memory map, an architecture may define multiple address spaces accessible to only certain types of information. For example, some processors may require at a specific location—or given a random location—a set of offsets to be reserved as space for its own internal registers (see Figure 3.48b). The processor may also allow specific address spaces accessible to only internal I/O functionality, instructions (programs), and/or data.

3.3.3 Processor Input/Output (I/O)

Input/output components of a processor are responsible for moving information to and from the processor's other components to any memory and I/O outside the processor, on the board (see Figure 3.49). Processor I/O can be input components that only bring information into the master processor, output components that bring information out of the master processor, or components that do both (refer back to Figure 3.48).

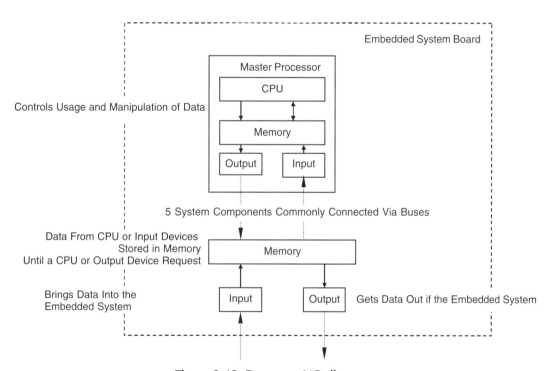

Figure 3.49: Processor I/O diagram.

Virtually any electromechanical system, embedded and nonembedded, conventional (keyboard, mouse, etc.) as well as unconventional (power plants, human limbs, etc.), can be connected to an embedded board and act as an I/O device. I/O is a high-level group that can be subdivided into smaller groups of either output devices, input devices, or devices that are both input and output devices. Output devices can receive data from board I/O components and display that data in some manner, such as printing it to paper, to a disk, or to a screen or a blinking LED light for a person to see. An input device transmits data to board I/O components, such as a mouse, keyboard, or remote control. I/O devices can do both, such as a networking device that can transmit data to and from the Internet, for instance. An I/O device can be connected to an embedded board via a wired or wireless data transmission medium, such as a keyboard or remote control, or can be located on the embedded board itself, such as an LED.

Because I/O devices can be such a wide variety of electromechanical systems, ranging from simple circuits to another embedded system entirely, processor I/O components can be organized into categories based on the functions they support, the most common including:

- Networking and communications I/O (the physical layer of the OSI model)
- Input (keyboard, mouse, remote control, voice, etc.)
- Graphics and output I/O (touch screen, CRT, printers, LEDs, etc.)
- Storage I/O (optical disk controllers, magnetic disk controllers, magnetic tape controllers, etc.)
- Debugging I/O (BDM, JTAG, serial port, parallel port, etc.)
- Real-time and miscellaneous I/O (timers/counters, analog-to-digital converters and digital-to-analog converters, key switches, and so on)

In short, an I/O subsystem can be as simple as a basic electronic circuit that connects the master processor directly to an I/O device (such as a master processor's I/O port to a clock or LED located on the board) to more complex I/O subsystem circuitry that includes several units, as shown in Figure 3.50. I/O hardware is typically made up of all or some combination of six main logical units:

- The *transmission medium*, wireless or wired medium connecting the I/O device to the embedded board for data communication and exchanges.

- A *communication port*, which is what the transmission medium connects to on the board, or if a wireless system, what receives the wireless signal.

- A *communication interface*, which manages data communication between master CPU and I/O device or I/O controller; also responsible for encoding data and decoding data to and from the logical level of an IC and the logical level of the I/O port. This interface can be integrated into the master processor, or can be a separate IC.

- An *I/O controller*, a slave processor that manages the I/O device.

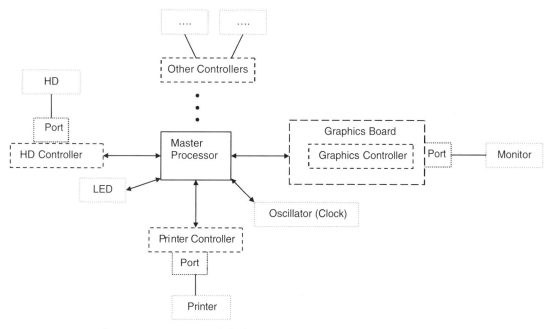

Figure 3.50: Ports and device controllers on an embedded board.

- *I/O buses*, the connection between the board I/O and master processor.
- The *master processor integrated I/O*.

This means that the I/O on the board can range from a complex combination of components, as shown in Figure 3.51a, to a few integrated I/O board components, as shown in Figure 3.51b.

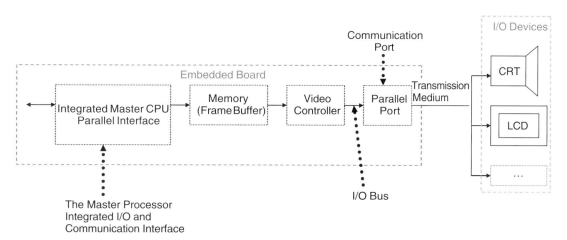

Figure 3.51a: Complex I/O subsystem.

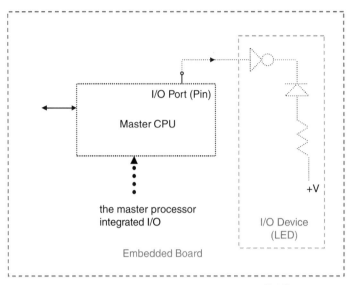

Figure 3.51b: Simple I/O subsystem.[3.40]

I/O controllers are a type of processor (see Section 3.2, "ISA Architecture Models"). An I/O device can be connected directly to the master processor via *I/O ports* (processor pins) if the I/O devices are located on the board, or it can be connected indirectly via a *communication interface* integrated into the master processor or a separate IC on the board.

As shown in the sample circuit in Figure 3.52, an I/O pin is typically connected to some type of current source and switching device. In this example it's a MOSFET transistor. This sample circuit allows for the pin to be used for both input and output. When the transistor is turned OFF (open switch), the pin acts as an input pin, and when the switch is ON it operates as an output port.

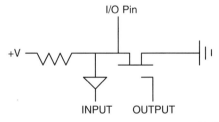

Figure 3.52: I/O port sample circuit.[3.41]

A pin or sets of pins on the processor can be programmed to support particular I/O functions (for example, Ethernet port receiver, serial port transmitter, bus signals, etc.), through a master processor's control registers (see Figure 3.53).

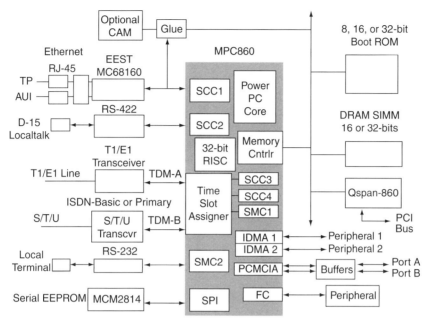

Figure 3.53: MPC860 reference platform and I/O.[3.42]
Copyright of Freescale Semiconductor, Inc., 2004. Used by permission.

> **Note:** In the case of the MPC860, the I/O such as Ethernet and RS-232 is implemented by the SCC registers, RS-232 by SMC2, and so on.

Within the various I/O categories (networking, debugging, storage, and so forth), processor I/O is typically subgrouped according to the way data is managed. Note that the actual subgroups may be entirely different, depending on the architecture viewpoint, as related to the embedded systems model. Here "viewpoint" means that hardware and software can view (and hence subgroup) I/O differently. Within software, the subgroups can even differ depending on the level of software (i.e., system software versus application software, operating system versus device drivers, and so on). For example, in many operating systems, I/O is considered to be either block or character I/O. Block I/O stores and transmits data in fixed block sizes and is addressable only in blocks. Character I/O, on the other hand, manages data in streams of characters, the size of the character depending on the architecture—such as one byte, for example.

From a hardware viewpoint, I/O manages (transmits and/or stores) data in *serial*, in *parallel*, or *both*.

3.3.3.1 Managing I/O Data: Serial vs. Parallel I/O

Processor I/O that can transmit and receive serial data is made up of components in which data is stored, transferred, and/or received *one bit at a time*. Serial I/O hardware is typically made up of some combination of the six main logical units outlined at the start of the chapter; serial communication then includes within its I/O subsystem a *serial port* and a *serial interface*.

Serial interfaces manage the serial data transmission and reception between the master CPU and either the I/O device or its controller. They include reception and transmission buffers to store and encode or decide the data they are responsible for transmitting to either the master CPU or an I/O device. In terms of serial data transmission and reception schemes, they generally differ as to the *direction* in which data can be transmitted and received, as well as in the actual *process* of how the data bits are transmitted (and thus received) within the data stream.

Data can be transmitted between two devices in one of three directions: one way, in both directions but at separate times because they share the same transmission line, and in both directions simultaneously. A *simplex* scheme for serial I/O data communication is one in which a data stream can only be transmitted—and thus received—in one direction (see Figure 3.54a). A *half-duplex* scheme is one in which a data stream can be transmitted and received in either direction, but in only one direction at any one time (see Figure 3.54b). A *full-duplex* scheme is

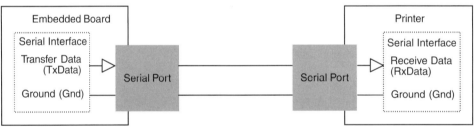

Figure 3.54a: Simplex transmission scheme example.[3.43]

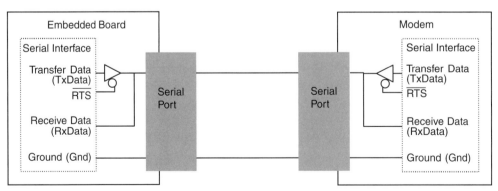

Figure 3.54b: Half-duplex transmission scheme example.[3.43]

one in which a data stream can be transmitted and received in either direction, simultaneously (see Figure 3.54c).

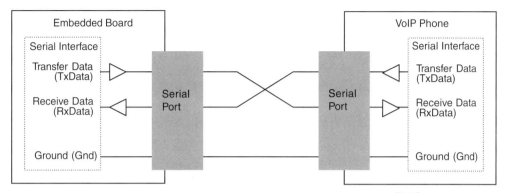

Figure 3.54c: Full-duplex transmission scheme example.[3.43]

Within the actual data stream, serial I/O transfers can occur either as a steady (continuous) stream at regular intervals regulated by the CPU's clock, referred to as a *synchronous* transfer, or intermittently at irregular (random) intervals, referred to as an *asynchronous* transfer.

In an asynchronous transfer (shown in Figure 3.55), the data being transmitted can be stored and modified within a serial interface's transmission buffer or registers. The serial interface at the transmitter divides the data stream into *packets* that typically range from either 4–8 or 5–9 bits, the number of bits per character. Each of these packets is then encapsulated in frames to be transmitted separately. The frames are packets that are modified before transmission by the serial interface to include a START bit at the start of the stream, and a STOP bit or bits (i.e., can be 1, 1.5, or 2 bits in length to ensure a transition from "1" to "0" for the START bit of the next frame) at the end of the data stream being transmitted. Within the frame, after the

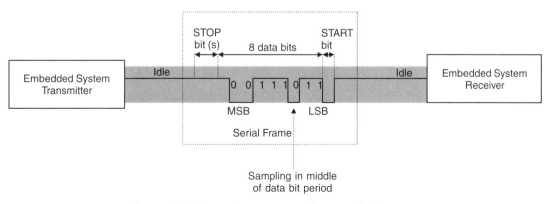

Figure 3.55: Asynchronous transfer sample diagram.

data bits and before the STOP bit, a *parity* bit may also be appended. A START bit indicates the start of a frame, the STOP bit(s) indicates the end of a frame, and the parity is an optional bit used for very basic error checking. Basically, parity for a serial transmission can be NONE (for no parity bit and thus no error checking), EVEN (where the total number of bits set to "1" in the transmitted stream, excluding the START and STOP bits, needs to be an even number for the transmission to be a success), or ODD (where the total number of bits set to "1" in the transmitted stream, excluding the START and STOP bits, needs to be an odd number for the transmission to be a success).

Between the transmission of frames, the communication channel is kept in an idle state, meaning a logical level "1" or non-return to zero (NRZ) state is maintained.

The serial interface of the receiver then receives frames by synchronizing to the START bit of a frame, delays for a brief period, and then shifts in bits, one at a time, into its receive buffer until reaching the STOP bit (s). For asynchronous transmission to work, the *bit rate* (bandwidth) has to be synchronized in all serial interfaces involved in the communication. Bit rate is defined as:

(number of actual data bits per frame / total number of bits per frame) * baud rate

The baud rate is the total number of bits, regardless of type, per unit of time (kbits/sec, Mbits/sec, etc.) that can be transmitted.

Both the transmitter's serial interface and the receiver's serial interface synchronize with separate bit-rate clocks to sample data bits appropriately. At the transmitter, the clock starts when transmission of a new frame starts, and it continues until the end of the frame so that the data stream is sent at intervals the receiver can process. At the receiving end, the clock starts with the reception of a new frame, delaying when appropriate, in accordance with the bit rate, sampling the middle of each data bit period of time, and then stopping when the frame's STOP bit(s) are received.

In a synchronous transmission (as shown in Figure 3.56), there are no START or STOP bits appended to the data stream, and there is no idle period. As with asynchronous transmissions,

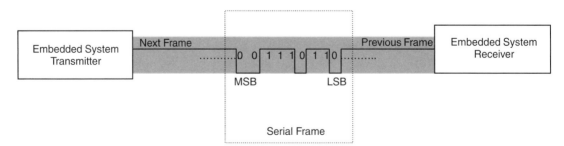

Figure 3.56: Synchronous transfer sample diagram.

the data rates on receiving and transmitting have to be in sync. However, unlike the separate clocks used in an asynchronous transfer, the devices involved in a synchronous transmission are synchronizing off one common clock that does not start and stop with each new frame (and on some boards there may be an entirely separate clock line for the serial interface to coordinate the transfer of bits). In some synchronous serial interfaces, if there is no separate clock line, the clock signal may even be transmitted along with the data bits.

The universal asynchronous receiver-transmitter (UART) is an example of a serial interface that does asynchronous serial transmission, whereas serial peripheral interface (SPI) is an example of a synchronous serial interface.

> **Note:** Different architectures that integrate a UART or other types of serial interfaces can have varying names for the same type of interface, such as the MPC860, which has serial management controller (SMC) UARTs, for example. Review the relevant documentation to understand the specifics.

Serial interfaces can either be separate slave ICs on the board or integrated onto the master processor. The serial interface transmits data to and from an I/O device via a *serial port* (see Chapter 4). Serial ports are serial communication (COM) interfaces that are typically used to interconnect off-board serial I/O devices to on-board serial board I/O. The serial interface is then responsible for converting data that is coming to and from the serial port at the logic level of the serial port into data that the logic circuitry of the master CPU can process.

3.3.3.2 Processor Serial I/O Example 1:

An Integrated Universal Asynchronous Receiver-Transmitter

The UART is an example of a full-duplex serial interface that can be integrated into the master processor and that does asynchronous serial transmission. As mentioned earlier, the UART can exist in many variations and under many names; however, they are all based on the same design: the original 8251 UART controller implemented in older PCs. A UART (or something like it) must exist on both sides of the communication channel, in the I/O device as well as on the embedded board, in order for this communication scheme to work.

In this example, we look at the MPC860 internal UART scheme, since it has more than one way to implement a UART. The MPC860 allows for two methods to configure a UART, either using a serial communication controller (SCC) or a serial management controller (SMC). Both of these controllers reside in the PowerPC's Communication Processor Module (shown in Figure 3.57) and can be configured to support a variety of communication schemes, such as Ethernet, HDLC, and the like for the SCC and transparent, GCI, and so on

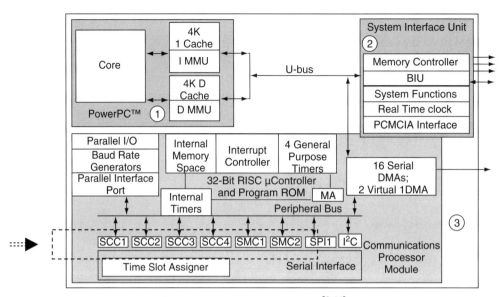

Figure 3.57: MPC860 UARTs.[3.44]

Copyright of Freescale Semiconductor, Inc., 2004. Used by permission.

for SMCs. In this example, however, we are only examining both being configured and functioning as a UART.

MPC860 SCC in UART Mode

As introduced at the start of this section, in an asynchronous transfer, the data being transmitted can be stored and modified within a serial interface's transmission buffer. With the SCCs on the MPC860, there are two UART first-in/first-out (FIFO) buffers, one for receiving data for the processor and one for transmitting data to external I/O (see Figures 3.58a and b). Both buffers are typically allocated space in main memory.

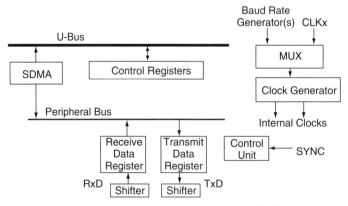

Figure 3.58a: SCC in receive mode.[3.45]

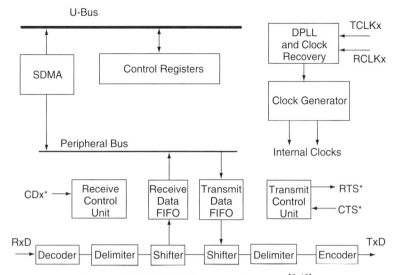

Figure 3.58b: SCC in transmit mode.[3.45]

As shown in Figures 3.58a and b, along with the reception and transmission buffers there are control registers to define the baud rate, the number of bits per character, the parity, and the length of the stop bit, among other things. As shown in Figures 3.58a and b as well as 3.59, there are five pins, extending out from the PowerPC chip, that the SCC is connected to for data transmission and reception: transmit (TxD), receive (RxD), carrier detect (CDx), collision on the transceiver (CTSx), and request-to-send (RTS). The way these pins work together is described in the next few paragraphs.

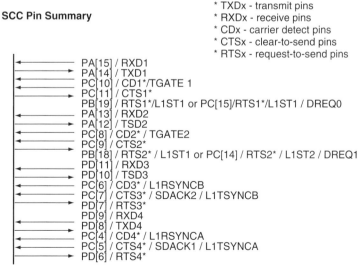

Figure 3.59: SCC pinouts.[3.46]

In either receive or transmit modes, the internal SCC clock is activated. In asynchronous transfers, every UART has its own internal clock that, though unsynchronized with the clock in the UART of the external I/O device, is set at the same baud rate as that of the UART it is in communication with. The carrier detect (CDx) is then asserted to allow the SCC to receive data, or the collision on the transceiver (CTSx) is asserted to allow the SCC to transmit data.

As mentioned, data is encapsulated into frames in asynchronous serial transmissions. When transmitting data, the SDMA transfers the data to the transmit FIFO and the request-to-send pin asserts (because it is a transmit control pin and asserts when data is loaded into the transmit FIFO). The data is then transferred (in parallel) to the shifter. The shifter shifts the data (in serial) into the delimiter, which appends the framing bits (i.e., start bits, stop bits, and so on). The frame is then sent to the encoder for encoding before transmission. In the case of an SCC receiving data, the framed data is then decoded by the decoder and sent to the delimiter to strip the received frame of nondata bits, such as start bit, stop bit(s), and so on. The data is then shifted serially into the shifter, which transfers (in parallel) the received data into the receive data FIFO. Finally, the SDMA transfers the received data to another buffer for continued processing by the processor.

MPC860 SMC in UART Mode

As shown in Figure 3.60a, the internal design of the SMC differs greatly from the internal design of the SCC (shown in Figures 3.58a and b), and in fact has fewer capabilities than an SCC. An SMC has no encoder, decoder, delimiter, or receive/transmit FIFO buffers. It uses registers instead. As shown in Figure 3.60b, there are only three pins that an SMC is connected to: a transmit pin (SMTXDx), a receive pin (SMRXDx), and sync signal pin (SMSYN). The sync pin is used in transparent transmissions to control receive and transmit operations.

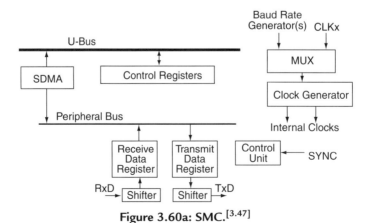

Figure 3.60a: SMC.[3.47]

*SMTXDx - transmit pins
*SMRXDx - receive pins
*SMSYNx - synch signal pins for transparent

```
← PB[24]/SMRXD1
→ PB[25]/SMTXD1
← PB[23]/SMSYN1/SDACK1
← PB[20]/SMRXD2/L1CLKOA
→ PB[21]/SMTXD2/L1CLKOB
← PB[22]/SMSYN2/SDACK2
```

Figure 3.60b: SMC pins.[3.47]

Data is received via the receive pin into the receive shifter, and the SDMA then transfers the received data from the receive register. Data to be transmitted is stored in the transmit register and then moved into the shifter for transmission over the transmit pin. Note that the SMC does not provide the framing and stripping of control bits (i.e., start bit, stop bit[s], and so on) that the SCC provides.

Processor Serial I/O Example: An Integrated Serial Peripheral Interface (SPI)
The SPI is an example of a full-duplex serial interface that can be integrated into the master processor and that does synchronous serial transmission. Like the UART, an SPI needs to exist on both sides of the communication channel (in the I/O device as well as on the embedded board) in order for this communication scheme to work. In this example, we examine the MPC860 internal SPI, which resides in the PowerPC's Communication Processor Module (shown in Figure 3.61).

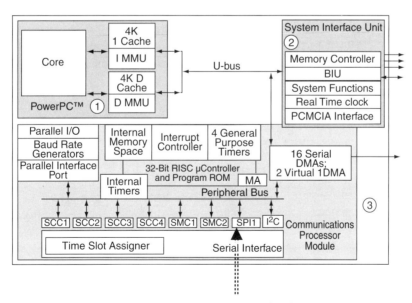

Figure 3.61: MPC860 SPI.[3.48]

Copyright of Freescale Semiconductor, Inc., 2004. Used by permission.

In a synchronous serial communication scheme, both devices are synchronized by the same clock signal generated by one of the communicating devices. In such a case, a master-slave relationship develops in which the master generates the clock signal which it and the slave device, adheres to. It is this relationship that is the basis of the four pins that the MPC860 SPI is connected to (as shown in Figure 3.62b): the master out/slave in or transmit (SPIMOSI), master in/slave out or receive (SPIMISO), clock (SPICLK), and slave select (SPISEL).

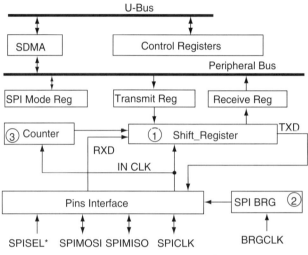

Figure 3.62a: SPI.[3.49]

* SPIMOSI - master out, slave in pin
* SPIMOSI - master in, slave out pin
* SPICLK - SPI clock pin
* SPISEL - SPI slave select pin, used when 860 SPI is in slave mode

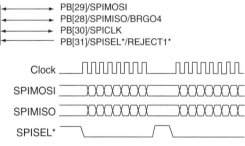

Figure 3.62b: SPI pins.[3.49]

When the SPI operates in a master mode, it generates the clock signals, while in slave mode, it receives clock signals as input. SPIMOSI in master mode is an output pin, SPMISO in master mode is an input pin, SPICLK supplies an output clock signal in master mode that synchronizes the shifting of received data over the SPIMISO pin or shifts out transmitted data over SPIMOSI. In slave mode, SPIMOSI is an input pin, SPIMISO is an output pin, and SPICLK receives a clock signal from the master synchronizing the shifting of data over the transmit and receive pins. The SPISEL is also relevant in slave mode because it enables input into the slave.

The way these pins work together, along with the internal components of the SPI, is shown in Figure 3.62a. Essentially, data is received or transmitted via one shift register. If data is received, it is then moved into a receive register. The SDMA then transfers the data into a receive buffer that usually resides in main memory. In the case of a data transmission, the SDMA moves the data to be transmitted from the transfer buffer in main memory to the transmit register. SPI transmission and reception occurs simultaneously; as data is received into the shift register, it shifts out data that needs to be transmitted.

3.3.3.2 Parallel I/O

I/O components that transmit data in parallel allow data to be transferred in multiple bits simultaneously. Just as with serial I/O, parallel I/O hardware is also typically made up of some combination of six main logical units, as introduced at the start of this chapter, except that the port is a *parallel port* and the communication interface is a *parallel interface*.

Parallel interfaces manage the parallel data transmission and reception between the master CPU and either the I/O device or its controller. They are responsible for decoding data bits received over the pins of the parallel port, transmitted from the I/O device, and receiving data being transmitted from the master CPU, and then encoding these data bits onto the parallel port pins.

They include reception and transmission buffers to store and manipulate the data they are responsible for transmitting either to the master CPU or an I/O device. Parallel data transmission and reception schemes, like serial I/O transmission, generally differ in terms of the *direction* in which data can be transmitted and received as well as the actual *process* of how the data bits are transmitted (and thus received) within the data stream. In the case of direction of transmission, as with serial I/O, parallel I/O uses simplex, half-duplex, or full-duplex modes. Again, like serial I/O, parallel I/O can be transmitted asynchronously or synchronously. Unlike serial I/O, parallel I/O does have a greater capacity to transmit data, because multiple bits can be transmitted or received simultaneously. Examples of I/O devices that transfer and receive data in parallel include IEEE1284 controllers (for printer/display I/O devices), CRT ports, and SCSI (for storage I/O devices).

3.3.3.3 Interfacing the Master Processor with an I/O Controller

When the communication interface is integrated into the master processor, as is the case with the MPC860, it is a matter of connecting the identical pins for transmitting data and receiving data from the master processor to an I/O controller. The remaining control pins are then connected according to their function. In Figure 3.63a, for instance, the request to send (RTS) on the PowerPC is connected to transmit enable (TENA) on the Ethernet controller, since RTS is automatically asserted if data is loaded into the transmit FIFO, indicating to the controller that data is on its way. The collision on the transceiver (CTS) on the PowerPC is connected to the clear to send (CLSN) on the Ethernet controller, and the carrier detect (CD) is connected to the receive enable (RENA) pin, since when either CD or CTS is asserted, a transmission or data reception can take place. If the controller does not clear to send or receive enable to indicate data is on its way to the PowerPC, no transmission or reception can take place. Figure 3.63b shows a MPC860 SMC interfaced to an RS-232 IC, which takes the SMC signals (transmit pin [SMTXDx] and receive pin [SMRXDx]) and maps them to an RS-232 port in this example.

Finally, Figure 3.63c shows an example of a PowerPC SPI in master mode interfaced with some slave IC, in which the SPIMISO (master in/slave out) is mapped to SPISO (SPI slave

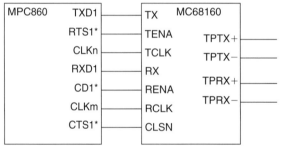

Figure 3.63a: MPC860 SCC UART interfaced to Ethernet controller.[3.50]

Copyright of Freescale Semiconductor, Inc., 2004. Used by permission.

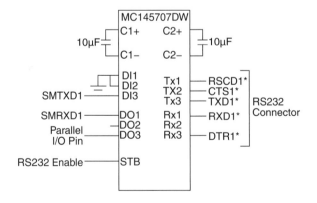

Figure 3.63b: MPC860 SMC interfaced to RS-232.[3.50]

Copyright of Freescale Semiconductor, Inc., 2004. Used by permission.

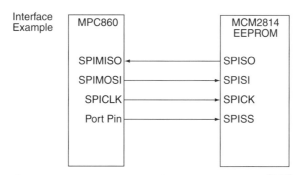

Figure 3.63c: MPC860 SPI interfaced to ROM.[3.50]
Copyright of Freescale Semiconductor, Inc., 2004. Used by permission.

out). Since in master mode SPIMISO is an input port, SPIMOSI (master out/slave in) is mapped to SPISI (slave in). Since SPIMOSI in master mode is an output port, SPICLK is mapped to SPICK (clock) because both ICs are synchronized according to the same clock, and SPISEL is mapped to SPISS (Slave Select input), which is only relevant if the PowerPC is in slave mode. If it were the other way around (that is, PowerPC in slave mode and slave IC in master mode), the interface would map identically.

Finally, for a subsystem that contains an I/O controller to manage the I/O device, the interface between an I/O controller and master CPU (via a communications interface) is based on four requirements:

- *An ability for the master CPU to initialize and monitor the I/O controller*. I/O controllers can typically be configured via *control registers* and monitored via *status registers*. These registers are all located on the I/O controller itself. Control registers can be modified by the master processor to configure the I/O controller. Status registers are read-only registers in which the master processor can get information as to the state of the I/O controller. The master CPU uses these status and control registers to communicate and/or control attached I/O devices via the I/O controller.

- *A way for the master processor to request I/O*. The most common mechanisms used by the master processor to request I/O via the I/O controller are special I/O instructions (I/O mapped) in the ISA and *memory-mapped I/O*, in which the I/O controller registers have reserved spaces in main memory.

- *A way for the I/O device to contact the master CPU*. I/O controllers that have the ability to contact the master processor via an interrupt are referred to as *interrupt-driven I/O*. Generally, an I/O device initiates an asynchronous interrupt requesting signaling to indicate (for example) that control and status registers can be read from or written to. The master CPU then uses its interrupt scheme to determine when an interrupt will be discovered.

Chapter 3

- *Some mechanism for both to exchange data.* This refers to the process by which data is actually exchanged between the I/O controller and the master processor. In a *programmed transfer*, the master processor receives data from the I/O controller into its registers, and the CPU then transmits this data to memory. For memory-mapped I/O schemes, DMA (direct memory access) circuitry can be used to bypass the master CPU entirely. DMA has the ability to manage data transmissions or receptions directly to and from main memory and an I/O device. On some systems, DMA is integrated into the master processor, and on others there is a separate DMA controller.

3.3.4 Processor Buses

Like the CPU buses, the processor's buses interconnect the processor's major internal components (in this case the CPU, memory, and I/O, as shown in Figure 3.64), carrying signals between the various components.

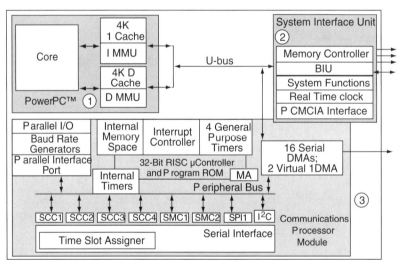

Figure 3.64: MPC860 processor buses.[3.51]

Copyright of Freescale Semiconductor, Inc., 2004. Used by permission.

Note: In the case of the MPC860, the processor buses include the U-bus interconnecting the system interface unit (SIU), the communications processor module (CPM), and the PowerPC core. Within the CPM there is a peripheral bus as well. Of course, this includes the buses within the CPU.

A key feature of processor buses is their *width*, which is the number of bits that can be transmitted at any one time. This can vary depending on both the buses implemented within the processor—for example: x86 contains bus widths of 16/32/64, 68K has 8/16/32/ 64 bit buses, MIPS 32 has 32 bit buses, and so forth—as well as the ISA register size definitions. Each bus also has a bus *speed* (in MHz) that impacts the performance of the processor. Buses implemented in real-world processor designs include the U, peripheral, and CPM buses in the MPC8xx family of processors and the C and X buses in the x86 Geode.

To avoid redundancy, buses are covered in more detail in Chapter 4, and more examples are provided there.

3.4 Processor Performance

There are several measures of processor performance, but are all based on the processor's behavior over a given length of time. One of the most common definitions of processor performance is a processor's *throughput*, the amount of work the CPU completes in a given period of time.

A processor's execution is ultimately synchronized by an external *system* or *master clock*, located on the board. The master clock is simply an oscillator producing a fixed frequency sequence of regular on/off pulse signals that is usually divided or multiplied within the CPU's CU (control unit) to generate at least one internal clock signal running at a constant number of clock cycles per second, or *clock rate*, to control and coordinate the fetching, decoding, and execution of instructions. The CPU's clock rate is expressed in MHz (megahertz).

Using the clock rate, the CPU's *execution time*, which is the total time the processor takes to process some program in seconds per program (total number of bytes), can be calculated. From the clock rate, the length of time a CPU takes to complete a clock cycle is the inverse of the clock rate (1/clock rate), called the *clock period* or *cycle time* and expressed in seconds per cycle. The processor's clock rate or clock period is usually located in the processor's specification documentation.

Looking at the instruction set, the *CPI* (average number of clock cycles per instruction) can be determined in several ways. One way is to obtain the CPI for each instruction (from the processor's instruction set manual) and multiply that by the frequency of that instruction, then add up the numbers for the total CPI.

```
CPI = Σ (CPI per instruction * instruction frequency)
```

At this point the total CPU's execution time can be determined by:

```
CPU execution time in seconds per program = (total number of
instructions per program or instruction count) * (CPI in number
```

of cycle cycles / instruction) * (clock period in seconds per
cycle) = ((instruction count) * (CPI in number of cycle cycles /
instruction)) / (clock rate in MHz)
```

The processor's average execution rate, also referred to as *throughput* or *bandwidth*, reflects the amount of work the CPU does in a period of time and is the inverse of the CPU's execution time:

```
CPU throughput (in bytes/sec or MB/sec) = 1 / CPU execution
time = CPU performance
```

Knowing the performance of two architectures (Geode and SA-1100, for example), the *speedup* of one architecture over another can then be calculated as follows:

```
Performance(Geode) / Performance (SA-1100) = Execution Time
(SA-1100) / Execution Time (Geode) = X
```

Therefore, Geode is X times faster than SA-1100.

Other definitions of performance besides throughput include:

- A processor's responsiveness, or *latency*, which is the length of elapsed time a processor takes to respond to some event

- A processor's *availability*, which is the amount of time the processor runs normally without failure; *reliability*, the average time between failures or MTBF (mean time between failures); and *recoverability*, the average time the CPU takes to recover from failure or mean time to recover (MTTR)

On a final note, a processor's internal design determines a processor's clock rate and the CPI; thus a processor's performance depends on which ISA is implemented and how the ISA is implemented. For example, architectures that implement Instruction-level Parallelism ISA models have better performance over the application-specific and general-purpose based processors due to the parallelism that occurs within these architectures. Performance can be improved because of the actual physical implementations of the ISA within the processor, such as implementing pipelining in the ALU.

> **Note:** There are variations on the full adder that provide additional performance improvements, such as the carry lookahead adder (CLA), carry completion adder, conditional sum adder, carry select adder, and so on. In fact, some algorithms that can improve the performance of a processor do so by designing the ALU to be able to process logical and mathematical instructions at a higher throughput—a technique called pipelining.
```

The increasing gap between the performance of processors and memory can be improved by cache algorithms that implement instruction and data prefetching (especially algorithms that use branch prediction to reduce stall time) and lockup-free caching. Basically, any design feature that allows for either an increase in the clock rate or a decrease in the CPI will increase the overall performance of a processor.

3.4.1 Benchmarks

One of the most common performance measures used for processors in the embedded market is *millions of instructions per seconds,* or *MIPS.*

```
MIPS = Instruction Count / (CPU execution time * 10⁶) = Clock
Rate / (CPI * 10⁶)
```

The MIPS performance measure gives the impression that faster processors have higher MIPS values, since part of the MIPS formula is inversely proportional to the CPU's execution time. However, MIPS can be misleading in terms of this assumption for a number of reasons, including:

- Instruction complexity and functionality aren't taken into consideration in the MIPS formula, so MIPS cannot compare the capabilities of processors with different ISAs.

- MIPS can vary on the same processor running different programs (with varying instruction count and different types of instructions).

Software programs called *benchmarks* can be run on a processor to measure its performance.

Endnotes

[3.1] "EnCore 400 Embedded Processor Reference Manual," Revision A, p. 9.

[3.2] "MPC8xx Instruction Set Manual," Motorola, p. 28.

[3.3] *MIPS32™ Architecture for Programmers Volume II: The MIPS32™ Instruction Set*, Rev 0.95, MIPS Technologies, p. 91.

[3.4] *MPC8xx Instruction Set Manual*, Motorola, p. 28.

[3.5] *MIPS32™ Architecture for Programmers Volume II: The MIPS32™ Instruction Set*, Rev 0.95, MIPS Technologies, pp. 39 and 90.

[3.6] *ARM Architecture*, Pietikainen, Ville, pp. 12 and 15.

[3.7] *Practical Electronics*, Scherz, Paul, p. 538.

[3.8] Texas Instruments website: http://focus.ti.com/docs/apps/catalog/resources/blockdiagram.jhtml?appId=178&bdId=112.

[3.9] "A Highly Integrated MPEG-4 ASIC for SDCAM Application," Chung-Ta Lee, Jun Zhu, Yi Liu, and Kou-Hu Tzou, p. 4.

[3.10] aJile Systems website: www.ajile.com.

[3.11] National Semiconductor, "Geode User's Manual," Rev. 1.

[3.12] Net Silicon "Net+ARM40 Hardware Reference Guide."

[3.13] Zoran website: www.zoran.com.

[3.14] Infineon Technologies website: www.infineon.com.

[3.15] Philips Semiconductor website: www.semiconductors.philips.com.

[3.16] Freescale, "MPC860 PowerQUICC User's Manual."

[3.17] National Semiconductor, "Geode User's Manual," Rev. 1.

[3.18] Freescale, "MPC860 PowerQUICC User's Manual."

[3.19] Freescale, "MPC860 PowerQUICC User's Manual."

[3.20] *Practical Electronics,* Scherz, Paul.

[3.21] *The Electrical Engineering Handbook*, Dorf, p. 1742.

[3.22] *The Electrical Engineering Handbook*, Dorf, p. 1742.

[3.23] Freescale, "MPC860 PowerQUICC User's Manual."

[3.24] *Practical Electronics*, Scherz, Paul.

[3.25] Freescale, "MPC860 PowerQUICC User's Manual."

[3.26] *Practical Electronics*, Scherz, Paul.

[3.27] *Practical Electronics*, Scherz, Paul.

[3.28] *Practical Electronics*, Scherz, Paul.

[3.29] Freescale, "MPC860 PowerQUICC User's Manual."

[3.30] *Practical Electronics*, Scherz, Paul, p. 538.

[3.31] *Practical Electronics*, Scherz, Paul.

[3.32] *Practical Electronics*, Scherz, Paul.

[3.33] *Computer Organization and Programming*, Ramm, Dietolf, p. 14.

[3.34] *Practical Electronics*, Scherz, Paul.

[3.35] *Practical Electronics*, Scherz, Paul.

[3.36] *Practical Electronics*, Scherz, Paul.

[3.37] "This RAM, That RAM, Which Is Which?" Robbins, Justin.

[3.38] Freescale, "MPC860 PowerQUICC User's Manual."

[3.39] Freescale, "MPC860 PowerQUICC User's Manual."

[3.40] *Computers as Components*, Wolf, Wayne, p. 206.

[3.41] *Embedded Controller Hardware Design,* Arnold, Ken, Newnes Press.

[3.42] Freescale, "MPC860 Training Manual."

[3.43] *Embedded Microcomputer Systems*, Valvano.

[3.44] Freescale, "MPC860 Training Manual."

[3.45] Freescale, "MPC860 PowerQUICC User's Manual."

[3.46] Freescale, "MPC860 Training Manual."

[3.47] Freescale, "MPC860 Training Manual."

[3.48] Freescale, "MPC860 PowerQUICC User's Manual."

[3.49] Freescale, "MPC860 Training Manual."

[3.50] Freescale, "MPC860 Training Manual."

[3.51] Freescale, "MPC860 PowerQUICC User's Manual."

CHAPTER 4
Embedded Board Buses and I/O

Tammy Noergaard

4.1 Board I/O

Input/output (I/O) components on a board are responsible for moving information into and out of the board to I/O devices connected to an embedded system. Board I/O can consist of input components, which only bring information from an input device to the master processor; output components, which take information out of the master processor to an output device; or components that do both (see Figure 4.1).

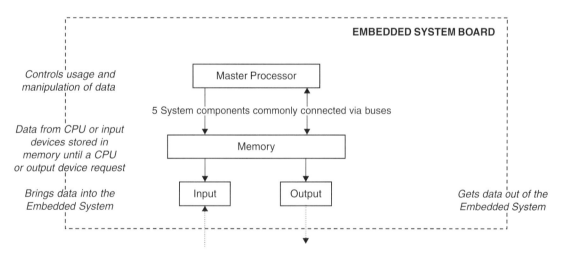

Figure 4.1: Von Neumann-based I/O block diagram.

Any electromechanical system, both embedded and nonembedded and whether conventional or unconventional, can be connected to an embedded board and act as an I/O device. I/O is a high-level group that can be subdivided into smaller groups of output devices, input devices, and devices that are both input and output devices. Output devices receive data from board I/O components and display that data in some manner, such as printing it to paper, to a disk, or to a screen or a blinking LED light for a person to see. An input device such as a mouse, keyboard, or remote control transmits data to board I/O components. Some I/O devices can do both, such as a networking device that can transmit data to and from the Internet, for instance. An I/O device can be connected to an embedded board via a wired or wireless data transmission medium such as a keyboard or remote control or can be located on the embedded board itself, such as an LED.

Because I/O devices are so varied, ranging from simple circuits to other complete embedded systems, board I/O components can fall under one or more of several different categories, the most common including:

- Networking and communications I/O (the physical layer of the OSI model)
- Input (keyboard, mouse, remote control, vocal, etc.)
- Graphics and output I/O (touch screen, CRT, printers, LEDs, etc.)
- Storage I/O (optical disk controllers, magnetic disk controllers, magnetic tape controllers, etc.)
- Debugging I/O (BDM, JTAG, serial port, parallel port, etc.)
- Real-time and miscellaneous I/O (timers/counters, analog-to-digital converters and digital-to-analog converters, key switches, and so on)

In short, board I/O can be as simple as a basic electronic circuit that connects the master processor directly to an I/O device, such as a master processor's I/O port to a clock or LED located on the board, to more complex I/O subsystem circuitry that includes several units, as shown in Figure 4.2. I/O hardware is typically made up of all or some combination of six main logical units:

- The *transmission medium*, a wireless or wired medium connecting the I/O device to the embedded board for data communication and exchanges
- A *communication port*, to which the transmission medium connects on the board or, if a wireless system, which receives the wireless signal
- A *communication interface*, which manages data communication between master CPU and I/O device or I/O controller and is responsible for encoding data and decoding data to and from the logical level of an IC and the logical level of the I/O port; this interface can be integrated into the master processor or can be a separate IC
- An *I/O controller*, a slave processor that manages the I/O device
- *I/O buses*, the connection between the board I/O and master processor
- The *master processor integrated I/O*

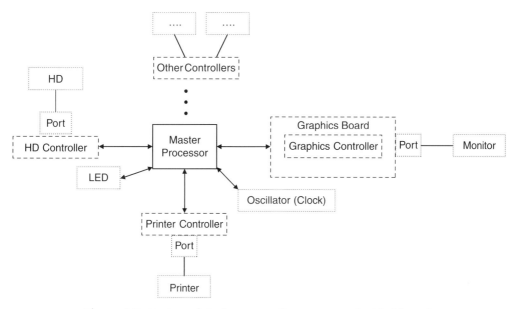

Figure 4.2: Ports and device controllers on an embedded board.

The I/O on a board can thus range from complex combination of components, as shown in Figure 4.3a, to a few integrated I/O board components, as shown in Figure 4.3b.

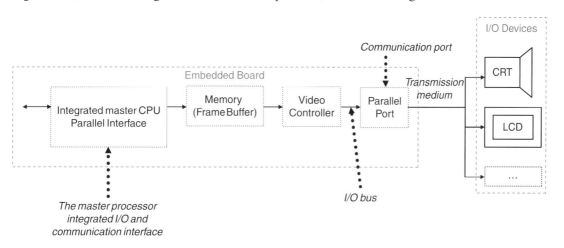

Figure 4.3a: Complex I/O subsystem.

The actual make-up of an I/O system implemented on an embedded board, whether using connectors and ports or using an I/O device controller, is dependent on the type of I/O device connected to, or located on, the embedded board. This means that, although other factors such as reliability and expandability are important in designing an I/O subsystem, what mainly dictates the details behind an I/O design are the *features* of the I/O device—its purpose within the system—and the *performance* of the I/O subsystem, discussed in Section 4.4.

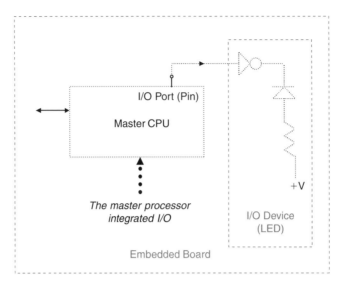

Figure 4.3b: Simple I/O subsystem.

Within the various I/O categories—networking, debugging, storage, and so forth—board I/O is typically subgrouped according to the way data is managed (transmitted). Note that the actual subgroups may be entirely different depending on the architecture viewpoint, as related to the embedded systems model. "Viewpoint" means that hardware and software can view, and hence subgroup, board I/O differently. Within software, the subgroups can even differ depending on the level of software—system software versus application software, operating system versus device drivers, and so on. For example, in many operating systems board I/O is considered either as block or character I/O. In short, block I/O manages in fixed block sizes and is addressable only in blocks. Character I/O, on the other hand, manages data in streams of characters, the size of the character depending on the architecture—such as one byte, for example.

From the hardware viewpoint, I/O manages (transmits and/or stores) data in *serial*, in *parallel*, or *both*.

4.2 Managing Data: Serial vs. Parallel I/O

Board I/O that can transmit and receive data in *serial* is made up of components in which data (characters) are stored, transferred, and received *one bit at a time*. Serial I/O hardware is typically made up of some combination of the six main logical units outlined at the start of the chapter. Serial communication includes within its I/O subsystem a *serial port* and a *serial interface*.

Serial interfaces manage the serial data transmission and reception between the master CPU and either the I/O device or its controller. They include reception and transmission buffers to store and encode or decode the data they are responsible for transmitting to either the master CPU or an I/O device. Serial data transmission and reception schemes generally differ in terms of the direction in which data can be transmitted and received as well as the actual transmission/reception process—in other words, the way the data bits are transmitted and received within the data stream.

Data can be transmitted between two devices in one of three directions: in a one-way direction, in both directions but at separate times because they share the same transmission line, and in both directions simultaneously. Serial I/O data communication that uses a *simplex* scheme is one in which a data stream can only be transmitted—and thus received—in one direction (see Figure 4.4a). A *half-duplex* scheme is one in which a data stream can be transmitted and received in either direction but in only one direction at any one time (see Figure 4.4b). A *full-duplex* scheme is one in which a data stream can be transmitted and received in either direction simultaneously (see Figure 4.4c).

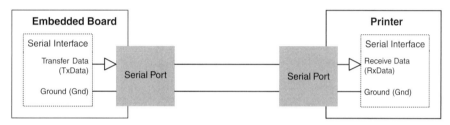

Figure 4.4a: Simplex transmission scheme example.

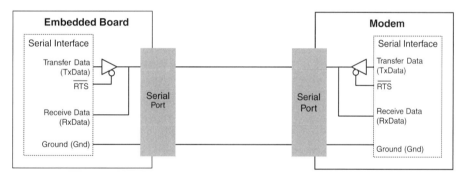

Figure 4.4b: Half-duplex transmission scheme example.

Within the actual data stream, serial I/O transfers can occur either as a steady (continuous) stream at regular intervals regulated by the CPU's clock, referred to as a *synchronous transfer*, or intermittently at irregular (random) intervals, referred to as an *asynchronous transfer*.

In an asynchronous transfer (shown in Figure 4.5), the data being transmitted is typically stored and modified within a serial interface's transmission buffer. The serial interface at the transmitter divides the data stream into groups, called *packets*, that typically range from either 4 to 8 bits per character or 5 to 9 bits per character. Each of these packets is then encapsulated in frames to be transmitted separately. The frames are packets modified (before transmission)

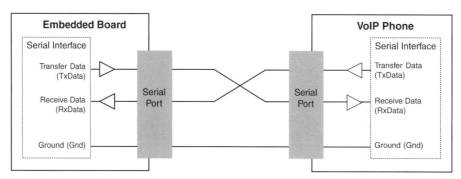

Figure 4.4c: Full-duplex transmission scheme example.

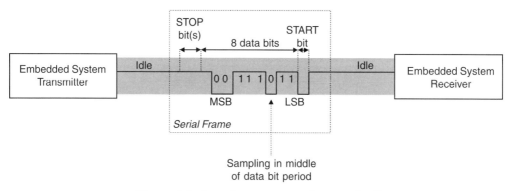

Figure 4.5: Asynchronous transfer sample diagram.

by the serial interface to include a START bit at the start of the stream and a STOP bit or bits (this can be 1, 1.5, or 2 bits in length to ensure a transition from "1" to "0" for the START bit of the next frame) at the end of the data stream being transmitted. Within the frame, after the data bits and before the STOP bit, a *parity* bit may also be appended. A START bit indicates the start of a frame, the STOP bit(s) indicate the end of a frame, and the parity is an optional bit used for very basic error checking. Basically, parity for a serial transmission can be NONE, for no parity bit and thus no error checking; EVEN, where the total number of bits set to "1" in the transmitted stream, excluding the START and STOP bits, must be an even number in order for the transmission to be a success; and ODD, where the total number of bits set to "1" in the transmitted stream, excluding the START and STOP bits, must be an odd number in order for the transmission to be a success. Between the transmission of frames, the communication channel is kept in an idle state, meaning that a logical level "1" or nonreturn to zero (NRZ) state is maintained.

The serial interface of the receiver then receives frames by synchronizing to the START bit of a frame, delays for a brief period, and then shifts in bits, one at a time, into its receive buffer until reaching the STOP bit (s). For asynchronous transmission to work, the *bit rate* (bandwidth) has to be synchronized in all serial interfaces involved in the communication. The bit rate is defined as:

(number of actual data bits per frame / total number of bits per frame) * baud rate

The baud rate is the total number of bits (regardless of type) per some unit of time (kbits/sec, Mbits/sec, etc.) that can be transmitted.

Both the transmitter's serial interface and the receiver's serial interface synchronize with separate bit-rate clocks to sample data bits appropriately. At the transmitter, the clock starts when transmission of a new frame starts and continues until the end of the frame so that the data stream is sent at intervals the receiver can process. At the receiving end, the clock starts with the reception of a new frame, delaying when appropriate (in accordance with the bit rate) and then sampling the middle of each data bit period of time and then stopping when receiving the frame's STOP bit(s).

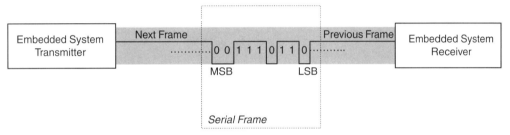

Figure 4.6: Synchronous transfer sample diagram.

In a synchronous transmission (as shown in Figure 4.6), there are no START or STOP bits appended to the data stream, and there is no idle period. As with asynchronous transmissions, the data rates for receiving and transmitting must be in sync. However, unlike the separate clocks used in an asynchronous transfer, the devices involved in a synchronous transmission are synchronizing off one common clock, which does not start and stop with each new frame. On some boards, there may be an entirely separate clock line for the serial interface to coordinate the transfer of bits. In some synchronous serial interfaces, if there is no separate clock line, the clock signal may even be transmitted along with the data bits. The universal asynchronous receiver-transmitter (UART) is an example of a serial interface that does asynchronous serial transmission, whereas serial peripheral interface (SPI) is an example of a synchronous serial interface.

> **Note:** Various architectures that integrate a UART or other types of serial interfaces may have different names and types for the same type of interface, such as the MPC860, which has serial management controller (SMC) UARTs, for example. Review the relevant documentation to understand the specifics.

Serial interfaces can either be separate slave ICs on the board or integrated onto the master processor. The serial interface transmits data to and from an I/O device via a *serial port* (shown in Figures 4.4a, b, and c). Serial ports are serial communication (COM) interfaces that are typically used to interconnect off-board serial I/O devices to on-board serial board I/O. The serial interface is then responsible for converting data that is coming to and from the serial port at the logic level of the serial port into data that the logic circuitry of the master CPU can process.

One of the most common serial communication protocols defining how the serial port is designed and what signals are associated with the different bus lines is RS-232.

4.2.1 Serial I/O Example 1: Networking and Communications: RS-232

One of the most widely implemented serial I/O protocols for either synchronous or asynchronous transmission is the *RS-232* or EIA-232 (Electronic Industries Association-232), which is primarily based on the EIA family of standards. These standards define
the major components of any RS-232 based system, which is implemented almost
entirely in hardware.

The hardware components can all be mapped to the physical layer of the OSI model (see Figure 4.7). The firmware (software) required to enable RS-232 functionality maps to the lower portion of the data link but will not be discussed in this section.

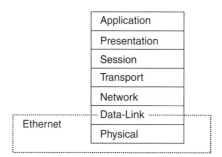

Figure 4.7: OSI model.

According to the EIA-232 standards, RS-232 compatible devices (shown in Figure 4.8) are called either Data Terminal Equipment (DTE) or Data Circuit-terminating Equipment (DCE). DTE devices are the initiators of a serial communication, such as a PC or embedded board. DCE is the device that the DTE wants to communicate with, such as an I/O device connected to the embedded board.

Figure 4.8: Serial network diagram.

The core of the RS-232 specification is called the *RS-232 interface* (see Figure 4.9). The RS-232 interface defines the details of the serial port and the signals, along with some additional circuitry that maps signals from a synchronous serial interface (such as SPI) or an asynchronous serial interface (such as UART) to the serial port and by extension to the I/O device itself. By defining the details of the serial port, RS-232 also defines the transmission medium, which is the serial cable. The same RS-232 interface must exist on both sides of a serial communication transmission (DTE and DCE or embedded board and I/O device), connected by an RS-232 serial cable, in order for this scheme to work.

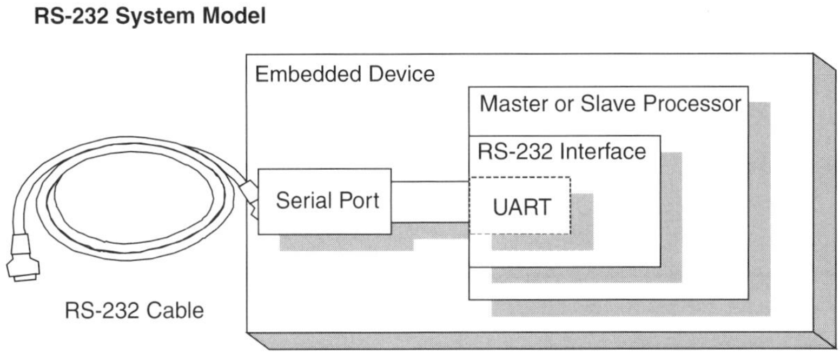

Figure 4.9: Serial components block diagram.

The actual physics behind the serial port—the number of signals and their definitions—differs among the different EIA232 standards. The parent RS-232 standard defines a total of 25 signals, along with a connector, called a DB25 connector, on either end of a wired transmission medium, shown in Figure 4.10a. The EIA RS-232 Standard EIA574 defines only nine signals (a subset of the original 25) that are compatible with a DB9 connector (shown in Figure 4.10b), whereas the EIA561 standard defines eight signals (again a subset of the original RS-232 25 signals) compatible with an RJ45 connector (see Figure 4.10c).

Two DTE devices can interconnect to each other using an internal wiring variation on serial cables called *null modem* serial cables. Since DTE devices transmit and receive data on the same pins, these null modem pins are swapped so that the transmit and receive connections on each DTE device are coordinated.

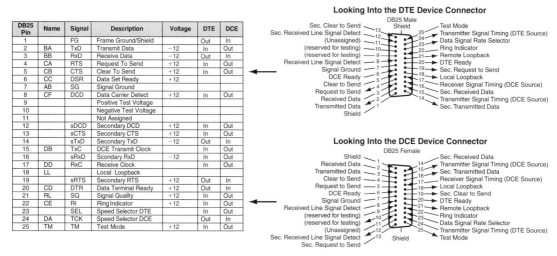

Figure 4.10a: RS-232 signals and DB25 connector.

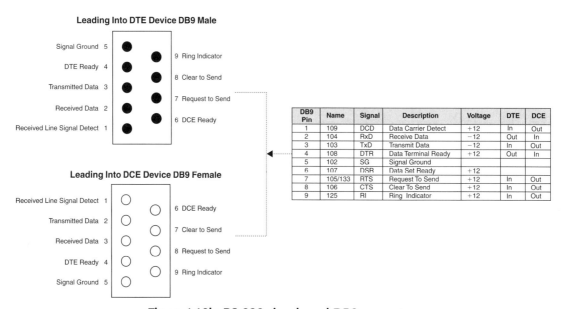

Figure 4.10b: RS-232 signals and DB9 connector.

4.2.2 Example: Motorola/Freescale MPC823 FADS Board RS-232 System Model

The serial interface on the Motorola/Freescale FADS board (a platform for hardware and software development around the MPC8xx family of processors) is integrated in the master processor, in this case the MPC823. To understand the serial port, the other major serial component located on the board, one only has to read the board's hardware manual.

Same Leading Into DTE Device and DCE Device

	Request to Send	8
	Clear to Send	7
	Transmit Data	6
	Receive Data	5
	Signal Ground	4
	Data Terminal Ready	3
	Data Carrier Detect	2
	Ring Indicator	1

DB9 Pin	Name	Signal	Description	Voltage	DTE	DCE
1	125	RI	Ring Indicator	+12	In	Out
2	109	DCD	Data Carrier Detect	+12	In	Out
3	108	DTR	Data Terminal Ready	+12	Out	In
4	102	SG	Signal Ground			
5	104	RxD	Receive Data	−12	Out	In
6	103	TxD	Transmit Data	−12	In	Out
7	106	CTS	Clear To Send	+12	In	Out
8	105/133	RTS	Request To Send	+12	In	Out

Figure 4.10c: RS-232 signals and RJ45 connector.

Section 4.9.3 of *The Motorola/Freescale 8xxFADS User's Manual* (*Rev. 1*) details the RS-232 system on the Motorola/Freescale FADS board as follows:

> To assist user's applications and to provide convenient communication channels with both a terminal and a host computer, two identical RS232 ports are provided on the FADS. ......
>
> Use is done with 9 pins, female D-type stack connector, configured to be directly (via a flat cable) connected to a standard IBM-PC-like RS232 connector.

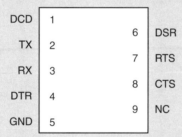

Figure 4.11: RS-232 serial port connector.

4.9.3.1 RS-232 Signal Description

In the following list:

DCD (O) – Data Carrier Detect

TX (O) – Transmit Data

...

From this manual, we can see that the FADS RS-232 port definition is based on the EIA574 DB9 DCE female device connector definition.

4.2.3 Serial I/O Example 2: Networking and Communications: IEEE 802.11 Wireless LAN

The IEEE 802.11 family of networking standards are serial wireless LAN standards and are summarized in Table 4.1. These standards define the major components of a wireless LAN system.

Table 4.1: 802.11 standards.

IEEE 802.11 Standard	Description
802.11-1999 Root Standard for Information Technology—Telecommunications and Information Exchange between Systems—Local and Metropolitan Area Network—Specific Requirements—Part 11: Wireless LAN Medium Access Control (MAC) and Physical Layer (PHY) Specifications	The 802.11 standard was the first attempt to define the way wireless data from a network should be sent. The standard defines operations and interfaces at the MAC (Media Access Control) and PHY (physical interface) levels in a TCP/IP network. There are three PHY layer interfaces defined (one IR and two radio: Frequency-Hopping Spread Spectrum [FHSS] and Direct Sequence Spread Spectrum [DSSS]), and the three do not interoperate. Use CSMA/CA (carrier sense multiple access with collision avoidance) as the basic medium access scheme for link sharing, phase-shift keying (PSK) for modulation.
802.11a-1999 "WiFi5" Amendment 1: High-speed Physical Layer in the 5 GHz band	Operates at radio frequencies between 5 GHz and 6 GHz to prevent interference with many consumer devices. Uses CSMA/CA as the basic medium access scheme for link sharing. As opposed to PSK, it uses a modulation scheme known as orthogonal frequency-division multiplexing (OFDM) that provides data rates as high as 54 Mbps maximum.
802.11b-1999 "WiFi" Supplement to **802.11a-1999**, Wireless LAN MAC and PHY Specifications: Higher-speed Physical Layer (PHY) extension in the 2.4 GHz band	Backward compatible with 802.11. 11Mbps speed, one single PHY layer (DSSS), uses CSMA/CA as the basic medium access scheme for link sharing and complementarycode keying (CCK), which allows higher data rates and is less susceptible to multipath-propagation interference.
802.11b-1999/Cor1-2001 Amendment 2: Higher-speed Physical Layer (PHY) extension in the 2.4 GHz band—Corrigendum 1	To correct deficiencies in the MIB definition of 802.11b.
802.11c IEEE Standard for Information Technology—Telecommunications and information exchange between systems—Local area networks—Media access control (MAC) bridges—Supplement for support by IEEE 802.11	Designated in 1998 to add a subclass under 2.5 Support of the Internal Sublayer Service by specific MAC Procedures to cover bridge operation with IEEE 802.11 MACs. Allows the use of 802.11 access points to bridge across networks within relatively short distances from each other (i.e., where there was a solid wall dividing a wired network).

Table 4.1: (continued)

IEEE 802.11 Standard	Description
802.11d-2001 Amendment to IEEE 802.11-1999 (ISO/IEC 8802-11), Specification for Operation in Additional Regulatory Domains	Internationalization—defines the physical layer requirements (channelization, hopping patterns, new values for current MIB attributes, and other requirements) to extend the operation of 802.11 WLANs to new regulatory domains (countries).
802.11e Amendment to STANDARD [for] Information Technology-Telecommunications and information exchange between systems-Local and metropolitan area networks-Specific requirements-Part 11: Wireless LAN Medium Access Control (MAC) and Physical Layer (PHY) specifications: Medium Access Method (MAC) Quality of Service Enhancements	Enhance the 802.11 Medium Access Control (MAC) to improve and manage quality of service (QoS), provide classes of service and efficiency enhancements in the areas of the Distributed Coordination Function (DCF) and Point Coordination Function (PCF). Defining a series of extensions to 802.11 networking to allow for QoS operation (i.e., to allow for adaptation for streaming audio or video via a preallocated dependable portion of the bandwidth.)
802.11f-2003 IEEE Recommended Practice for Multi-Vendor Access Point Interoperability via an Inter-Access Point Protocol Across Distribution Systems Supporting IEEE 802.11 Operation	Standard to enable handoffs (constant operation while the mobile terminal is actually moving) to be done in such a way as to work across access points from a number of vendors. Includes recommended practices for an Inter-Access Point Protocol (IAPP), which provides the necessary capabilities to achieve multivendor Access Point interoperability across a distribution system supporting IEEE P802.11 Wireless LAN Links. This IAPP will be developed for the following environment(s): (1) a distribution system consisting of IEEE 802 LAN components supporting an IETF IP environment; (2) others as deemed appropriate.
802.11g-2003 Amendment 4: Further Higher-Speed Physical Layer Extension in the 2.4 GHz Baud	A higher-speed(s) PHY extension to 802.11b—offering wireless transmission over relatively short distances at up to 54 Mbps compared to the maximum 11 Mbps of the 802.11 standard and operating in the 2.4 GHz range. Uses CSMA/CA as the basic medium access scheme for link sharing.
802.11h-2001 Spectrum and Transmit Power Management Extensions in the 5 GHz band in Europe	Enhancing the 802.11 MAC standard and 802.11a High Speed PHY in the 5 GHz Band supplement to the standard; to add indoor and outdoor channel selection for 5 GHz license exempt bands in Europe; and to enhance channel energy measurement and reporting mechanisms to improve spectrum and transmit power management (per CEPT and subsequent EU committee or body ruling incorporating CEPT Recommendation ERC 99/23). Looking into the tradeoffs involved in creating reduced-power transmission modes for networking in the 5 GHz space—essentially allowing 802.11a to be used by handheld computers and other devices with limited battery power available to them. Also, examining the possibility of allowing access points to reduce power to shape the geometry of a wireless network and reduce interference outside the desired influence of such a network.

Table 4.1: (continued)

IEEE 802.11 Standard	Description
802.11i Amendment to STANDARD [for] Information Technology-Telecommunications and information exchange between systems-Local and metropolitan area networks-Specific requirements-Part 11: Wireless LAN Medium Access Control (MAC) and Physical Layer (PHY) specifications: Medium Access Method (MAC) Security Enhancements	Enhances the 802.11 MAC to enhance security and authentication mechanisms and improve the PHY-level security that is used on these networks.
802.11j Amendment to STANDARD [for] Information Technology-Telecommunications and information exchange between systems-Local and Metropolitan networks-Specific requirements—Part 11: Wireless LAN Medium Access Control (MAC) and Physical Layer (PHY) specifications: 4.9–5 GHz Operation in Japan	The scope of the project is to enhance the 802.11 standard and amendments, to add channel selection for 4.9 GHz and 5 GHz in Japan to additionally conform to the Japanese rules for radio operation, to obtain Japanese regulatory approval by enhancing the current 802.11 MAC and 802.11a PHY to additionally operate in newly available Japanese 4.9 GHz and 5 GHz bands.
802.11k Amendment to STANDARD [for] Information Technology-Telecommunications and information exchange between systems-Local and Metropolitan networks- Specific requirements-Part 11: Wireless LAN Medium Access Control (MAC) and Physical Layer (PHY) specifications: Radio Resource Measurement of Wireless LANs	This project will define Radio Resource Measurement enhancements to provide interfaces to higher layers for radio and network measurements.
802.11ma Standard for Information Technology–Telecommunications and information exchange between systems–Local and Metropolitan networks–Specific requirements–Part 11: Wireless LAN Medium Access Control (MAC) and Physical Layer (PHY) specifications– Amendment x: Technical corrections and clarifications	Incorporates accumulated maintenance changes (editorial and technical corrections) into 802.11-1999, 2003 edition (incorporating 802.11a-1999, 802.11b-1999, 802.11b-1999 corrigendum 1-2001, and 802.11d-2001).
802.11n Amendment to STANDARD [for] Information Technology-Telecommunications and information exchange between systems-Local and Metropolitan networks- Specific requirements-Part 11: Wireless LAN Medium Access Control (MAC) and Physical Layer (PHY) specifications: Enhancements for Higher Throughput	The scope of this project is to define an amendment that shall define standard modifications to both the 802.11 physical layers (PHY) and the 802.11 Medium Access Control Layer (MAC) so that modes of operation can be enabled that are capable of much higher throughputs, with a maximum throughput of at least 100 Mbps, as measured at the MAC data service access point (SAP).

The first step is to understand the main components of an 802.11 system, regardless of whether these components are implemented in hardware or software. This is important because different embedded architectures and boards implement 802.11 components differently. On most platforms today, 802.11 standards are made up of root components that are implemented almost entirely in hardware. The hardware components can all be mapped to the physical layer of the OSI model, as shown in Figure 4.12. Any software required to enable 802.11 functionality maps to the lower section of the OSI data-link layer but will not be discussed in this section.

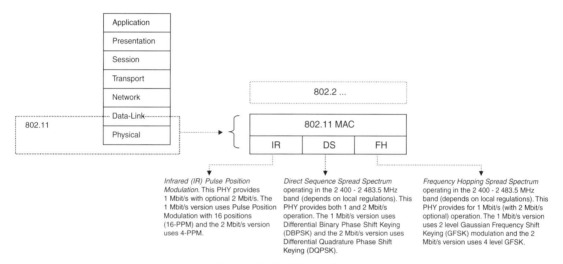

Figure 4.12: OSI model.

Off-the-shelf wireless hardware modules supporting one or some combination of the 802.11 standards (i.e., 802.11a, 802.11b, 802.11g, etc.) have in many ways complicated the efforts to commit to one wireless LAN standard. These modules also come in a wide variety of forms, including embedded processor sets, PCMCIA, Compact Flash, and PCI formats. In general, as shown in Figures 4.13a and b, embedded boards need to either integrate 802.11 functionality as a slave controller or into the master chip or the board needs to support one of the standard connectors for the other forms (PCI, PCMCIA, Compact Flash, etc.). This means that either (1) 802.11 chipset vendors can produce or port their PC Card firmware for an 802.11 embedded solution, which can be used for lower volume/more expensive devices or during product development, or (2) the same vendor's chipset on a standard PC card could be placed on the embedded board, which can be used for devices that will be manufactured in larger volumes.

On top of the 802.11 chipset integration, an embedded board design needs to take into consideration wireless LAN antenna placement and signal transmission requirements. The designer

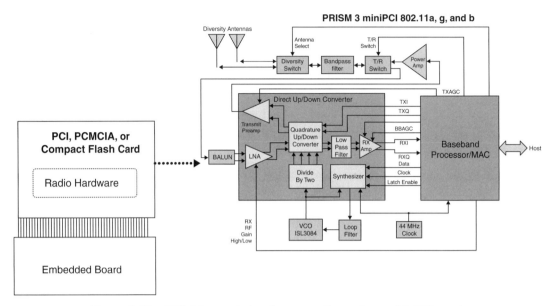

Figure 4.13a: 802.11 sample hardware configurations with PCI card.

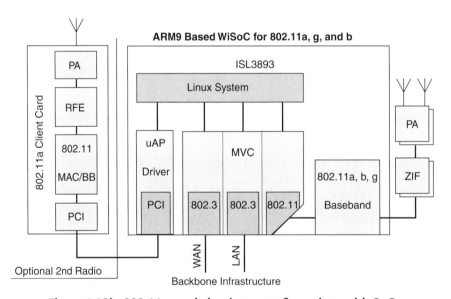

Figure 4.13b: 802.11 sample hardware configurations with SoC.

must ensure that there are no obstructions to prevent receiving and transmitting data. When 802.11 is not integrated into the master CPU, such as with the System-on-Chip (SoC) shown in Figure 4.13b, the interface between the master CPU and the 802.11 board hardware also needs to be designed.

4.2.4 Parallel I/O

Components that transmit data in parallel are devices that can transfer data in multiple bits simultaneously. Just as with serial I/O, parallel I/O hardware is also typically made up of some combination of six main logical units, as introduced at the start of this chapter, except that the port is a parallel port and the communication interface is a parallel interface.

Parallel interfaces manage the parallel data transmission and reception between the master CPU and either the I/O device or its controller. They are responsible for decoding data bits received over the pins of the parallel port (transmitted from the I/O device) and receiving data being transmitted from the master CPU, then encoding these data bits onto the parallel port pins.

They include reception and transmission buffers to store and manipulate the data being transferred. In terms of parallel data transmission and reception schemes, like serial I/O transmission, they generally differ in terms of the *direction* in which data can be transmitted and received as well as the actual *process* of transmitting/receiving data bits within the data stream. In the case of direction of transmission, as with serial I/O, parallel I/O uses simplex, half-duplex, or full-duplex modes. Also, as with serial I/O, parallel I/O devices can transmit data asynchronously or synchronously. However, parallel I/O does have a greater capacity to transmit data than serial I/O, because multiple bits can be transmitted or received simultaneously. Examples of board I/O that transfer and receive data in parallel include IEEE 1284 controllers (for printer/display I/O devices—see Example 3), CRT ports, and SCSI (for storage I/O devices). A protocol that can potentially support both parallel and serial I/O is Ethernet, presented in Example 4.

4.2.5 Parallel I/O Example 3: "Parallel" Output and Graphics I/O

Technically, the models and images that are created, stored, and manipulated in an embedded system are the graphics. There are typically three logical components (engines) of I/O graphics on an embedded board, as shown in Figure 4.14:

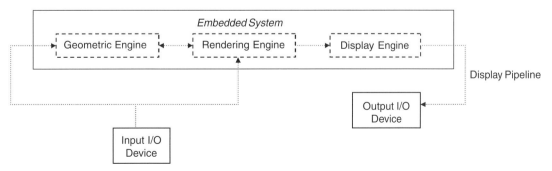

Figure 4.14: Graphical design engines.

- The *geometric engine*, which is responsible for defining what an object is. This includes implementing color models, an object's physical geometry, material and lighting properties, and so on.

- The *rendering engine*, which is responsible for capturing the description of objects. This includes providing functionality in support of geometric transformations, projections, drawing, mapping, shading, illumination, and so on.

- The *raster and display engine*, which is responsible for physically displaying the object. It is in this engine that the output I/O hardware comes into play.

An embedded system can output graphics via softcopy (video) or hardcopy (on paper) means. The contents of the display pipeline differ according to whether the output I/O device outputs hard or soft graphics, so the display engine differs accordingly, as shown in Figures 4.15a and b.

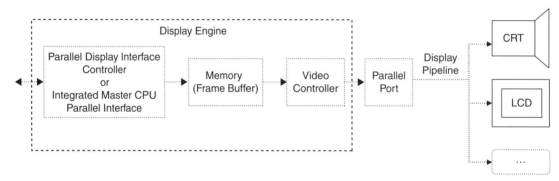

Figure 4.15a: Display engine of softcopy (video) graphics example.

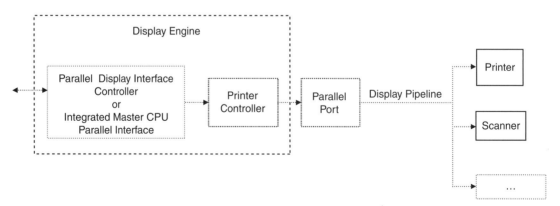

Figure 4.15b: Display engine of hardcopy graphics example.

The actual parallel port configuration differs from standard to standard in terms of the number of signals and the required cable. For example, on Net Silicon's NET+ARM50 embedded board (see Figure 4.16), the master processor (an ARM7-based architecture) has an integrated IEEE 1284 interface, a configurable MIC controller integrated in the master processor, to transmit parallel I/O over four on-board parallel ports.

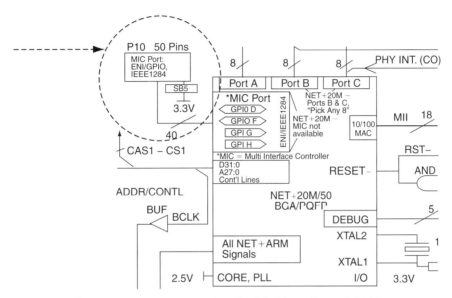

Figure 4.16: NET+ARM50 embedded board parallel I/O.

The IEEE 1284 specification defines a 40-signal port, but on the Net+ARM50 board, data and control signals are multiplexed to minimize the master processor's pin count. Aside from eight data signals DATA[8:1] ($D_0 - D_7$), IEEE 1284 control signals include:

- *PDIR*, which is used for bidirectional modes and defines the direction of the external data transceiver. Its state is directly controlled by the BIDIR bit in the IEEE 1284 Control register (0 state, data is driven from the external transceiver toward 1285, the cable, and in the 1 state, data is received from the cable).

- *PIO*, which is controlled by firmware. Its state is directly controlled by the PIO bit in the IEEE 1284 Control register.

- *LOOPBACK*, which configures the port in external loopback mode and can be used to control the mux line in the external FCT646 devices (set to 1, the FCT646 transceivers drive inbound data from the input latch and not the real-time cable interface). Its state is directly controlled by the LOOP bit in the IEEE 1284 Control register. The LOOP strobe signal is responsible for writing outbound data into the inbound latch (completing the loop back path). The LOOP strobe signal is an inverted copy of the STROBE* signal.

- *STROBE** (nSTROBE), *AUTOFD** (nAUTOFEED), *INIT** (nINIT), *HSELECT** (nSELECTIN), **ACK* (nACK), *BUSY*, *PE*, *PSELECT* (SELECT), **FAULT* (nER-ROR), …

4.2.6 Parallel and Serial I/O Example 4: Networking and Communications—Ethernet

One of the most widely implemented LAN protocols is Ethernet, which is primarily based on the IEEE 802.3 family of standards. These standards define the major components of any Ethernet system. Thus, to fully understand an Ethernet system design, you first need to understand the IEEE specifications. (Remember, this is not a book about Ethernet, and there is a lot more involved than is covered here. This example is about understanding a networking protocol and then being able to understand the design of a system based on a networking protocol such as Ethernet.)

The first step is understanding the main components of an Ethernet system, regardless of whether these components are implemented in hardware or software. This is important since different embedded architectures and boards implement Ethernet components differently. On most platforms, however, Ethernet is implemented almost entirely in hardware.

The hardware components can all be mapped to the physical layer of the OSI model. The firmware (software) required to enable Ethernet functionality maps to the lower section of the OSI data-link layer but will not be discussed in this section.

Several Ethernet system models are described in the IEEE 802.3 specification, so let's look at a few to get a clear understanding of some of the most common Ethernet hardware components.

Ethernet devices are connected to a network via *Ethernet cables:* thick coax (coaxial), thin coax, twisted-pair, or fiber optic cables. These cables are commonly referred to by their IEEE names. These names are made up of three components: the data transmission rate, the type of signaling used, and either the cable type or cable length.

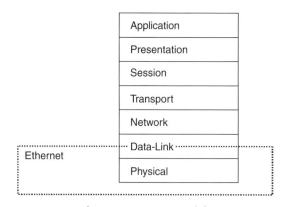

Figure 4.17: OSI model.

For example, a 10Base-T cable is an Ethernet cable that handles a data transmission rate of 10 Mbps (million bits per second), will only carry Ethernet signals (baseband signaling), and is a twisted-pair cable. A 100Base-F cable is an Ethernet cable that handles a data transmission rate of 100 Mbps, supports baseband signaling, and is a fiber optic cable. Thick or thin coax cables transmit at speeds of 10 Mbps and support baseband signaling but differ in the length of maximum segments cut for these cables (500 meters for thick coax, 200 meters for thin coax). Thus, these thick coax cables are called 10Base-5 (short for 500), and thin coax cables are called 10Base-2 (short for 200).

The Ethernet cable must then be connected to the embedded device. The type of cable, along with the board I/O (communication interface, communication port, etc.), determines whether the Ethernet I/O transmission is *serial* or *parallel*. The *Medium Dependent Interface (MDI)* is the network port on the board into which the Ethernet cable plugs. Different MDIs exist for the different types of Ethernet cables. For example, a 10Base-T cable has a RJ-45 jack as the MDI. In the system model of Figure 4.18, the MDI is an integrated part of the transceiver.

1 Mbps and 10 Mbps Ethernet System Model

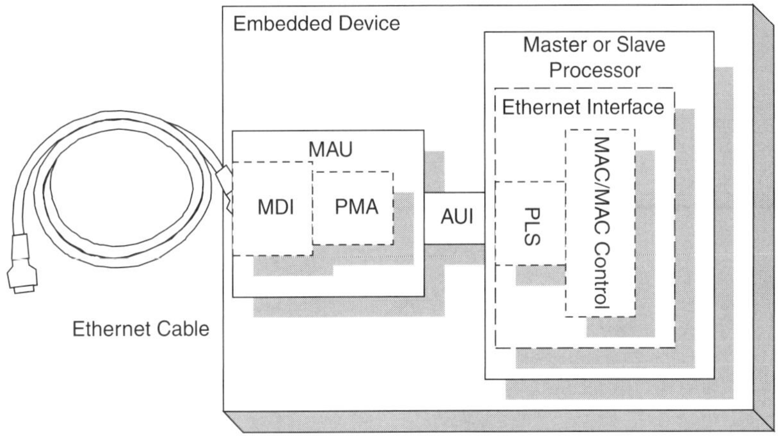

Figure 4.18: Ethernet components diagram.

A transceiver is the physical device that receives and transmits the data bits; in this case it is the *Medium Attachment Unit (MAU)*. The MAU contains not only the MDI but the Physical Medium Attachment (PMA) component as well. It is the PMA which "contains the functions for transmission, reception, and" depending on the transceiver, "collision detection, clock recovery and skew alignment" (p. 25, IEEE 802.3 Spec). Basically, the PMA serializes (breaks down into a bit stream) code groups received for transmission over the transmission medium or deserializes bits received from the transmission medium and converts these bits into code groups.

The transceiver is then connected to an *Attachment Unit Interface (AUI)*, which carries the encoded signals between an MAU and the Ethernet interface in a processor. Specifically, the AUI is defined for up to 10 Mbps Ethernet devices and specifies the connection between the MAU and the *Physical Layer Signaling (PLS)* sublayer (signal characteristics, connectors, cable length, etc.).

The *Ethernet interface* can exist on a master or slave processor and contains the remaining Ethernet hardware and software components. The *Physical Layer Signaling (PLS)* component monitors the transmission medium and provides a carrier sense signal to the *Media Access Control (MAC)* component. It is the MAC that initiates the transmission of data, so it checks the carrier signal before initiating a transmission, to avoid contention with other data over the transmission medium.

Let's start by looking at an embedded board for an example of this type of Ethernet system.

4.2.7 Example 1: Motorola/Freescale MPC823 FADS Board Ethernet System Model

Section 4.9.1 of *The Motorola/Freescale 8xxFADS User's Manual (Rev 1)* details the Ethernet system on the Motorola/Freescale FADS board:

> "4.9.1 Ethernet Port
>
> The MPC8xxFADS has an Ethernet port with a 10-Base-T interface. The communication port on which this resides is determined according to the MPC8xx type whose routing is on the daughter board. The Ethernet port uses an MC68160 EEST 10 Base-T transceiver.
>
> You can also use the Ethernet SCC pins, which are on the expansion connectors of the daughter board and on the communication port expansion connector (P8) of the motherboard. The Ethernet transceiver can be disabled or enabled at any time by writing a 1 or a 0 to the EthEn bit in the BCSR1."

From this information, we know that the board has an RJ-45 jack as the MDI, and the MC68160 enhanced Ethernet serial transceiver (EEST) is the MAU. The second paragraph, as well as Chapter 28 of the *PowerPC MPC823 User's Manual*, tells us more about the AUI and the Ethernet interface on the MPC823 processor.

On the MPC823, a seven-wire interface acts as the AUI. The SCC2 is the Ethernet interface and "performs the full set of IEEE 802.3/Ethernet CSMA/CD media access control and channel interface functions." (See *MPC823 PowerPC User's Manual*, p. 16–312.)

LAN devices that are able to transmit and receive data at a much higher rate than 10 Mbps implement a different combination of Ethernet components. The IEEE 802.3u Fast Ethernet (100 Mbps data rate) and the IEEE 802.3z Gigabit Ethernet (1000 Mbps data rate) systems

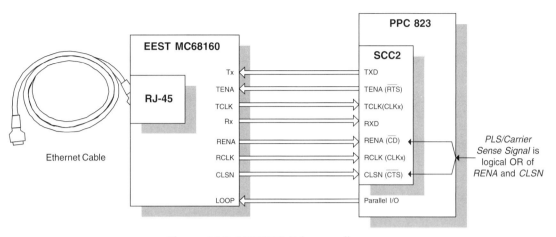

Figure 4.19: MPC823 Ethernet diagram.
Copyright of Freescale Semiconductor, Inc., 2004. Used by permission.

evolved from the original Ethernet system model (described in the previous section) and are based on the system model in Figure 4.20.

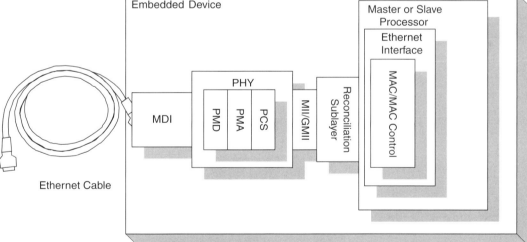

Figure 4.20: Ethernet diagram.

The MDI in this system is connected to the transceiver, not a part of the transceiver (as in the previous system model). The *Physical Layer Device (PHY)* transceiver in this system contains three components: the PMA (same as on the MAU transceiver in the 1/10 Mbps system model), the *Physical Coding Sub layer (PCS)*, and the *Physical Medium Dependent (PMD)*.

The PMD is the interface between the PMA and the transmission medium (through the MDI). The PMD is responsible for receiving serialized bits from the PMD and converting it to the appropriate signals for the transmission medium (optical signals for a fiber optic, etc.). When transmitting to the PMA, the PCS is responsible for encoding the data to be transmitted into the appropriate code group. When receiving the code groups from the PMA, the PCS decodes the code groups into the data format that can be understood and processed by upper Ethernet layers.

The *Media Independent Interface (MII)* and the *Gigabit Media Independent Interface (GMII)* are similar in principle to the AUI, except they carry signals (transparently) between the transceiver and the Reconciliation Sub layer (RS). Furthermore, the MII supports a LAN data rate of up to 100 Mbps, while GMII (an extension of MII) supports data rates of up to 1000 Mps. Finally, the RS maps PLS transmission media signals to two status signals (carrier presence and collision detection) and provides them to the Ethernet interface.

4.2.8 Example 2: Net Silicon ARM7 (6127001) Development Board Ethernet System Model

The *Net+Works 6127001 Development Board Jumper and Component Guide* from NetSilicon has an Ethernet interface section on their ARM based reference board, and from this we can start to understand the Ethernet system on this platform (see Figure 4.21).

> "Ethernet Interface
>
> The 10/100 version of the 3V NET+Works Hardware Development Board provides a full-duplex 10/100 Mbit Ethernet Interface using the Enable 3V PHY chip. The Enable 3V PHY interfaces to the NET+ARM chip using the standard MII interface.
>
> The Enable 3V PHY LEDL (link indicator) signal is connected to the NET+ARM PORTC6 GPIO signal. The PORT6 input can be used to determine the current Ethernet link status (The MII interface can also be used to determine the current Ethernet link status) ...."

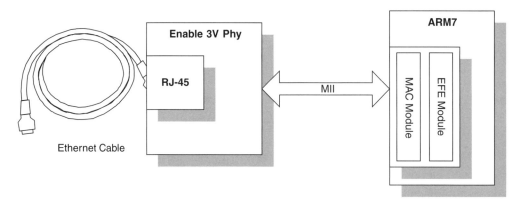

Figure 4.21: Net+ARM Ethernet block diagram.

From this information we can determine that the board has an RJ-45 jack as the MDI, and the Enable 3V PHY is the MAU. Section 5: "Ethernet Controller Interface" of the *NET+Works for NET+ARM Hardware Reference Guide* tells us that the ARM7-based ASIC integrates an Ethernet controller, and that the Ethernet Interface is actually composed of two parts: the Ethernet Front End (EFE) and the Media Access Control (MAC) modules. Finally, Section 1.3 of this manual tells us the Reconciliation Layer (RS) is integrated into the Media Independent Interface (MII).

4.2.9 Example 3: Adastra Neptune x86 Board Ethernet System Model

While both the ARM and PowerPC platforms integrate the Ethernet interface into the main processor (see Figure 4.22), this x86 platform has a separate slave processor for this functionality. According to the *Neptune User's Manual Rev A.2*, the Ethernet controller the ("MAC Am79C791 10/100 Controller") connects to two different transceivers, with each connected to either an AUI or MII for supporting various transmission media.

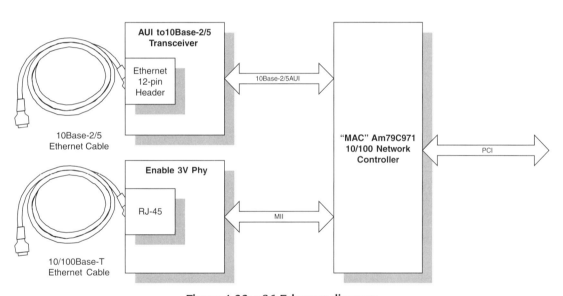

Figure 4.22: x86 Ethernet diagram.

4.3 Interfacing the I/O Components

As discussed at the start of this chapter, I/O hardware is made up of all or some combination of integrated master processor I/O, I/O controllers, a communications interface, a communication port, I/O buses, and a transmission medium (see Figure 4.23).

All these components are interfaced (connected) and communication mechanisms implemented via hardware, software, or both to allow for successful integration and function.

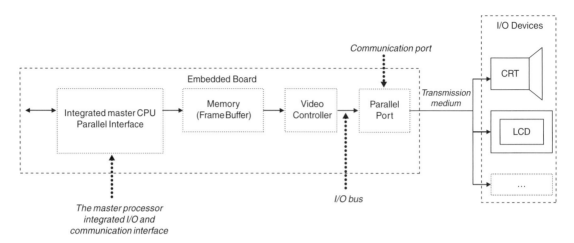

Figure 4.23: Sample I/O subsystem.

4.3.1 Interfacing the I/O Device with the Embedded Board

For off-board I/O devices, such as keyboards, mice, LCDs, printers, and so on, a transmission medium is used to interconnect the I/O device to an embedded board via a *communication port*. Aside from the I/O schemes implemented on the board (serial versus parallel), whether the medium is wireless (Figure 4.24b) or wired (Figure 4.24a) also impacts the overall scheme used to interface the I/O device to the embedded board.

As shown in Figure 4.24a, with a wired transmission medium between the I/O device and embedded board, it is just a matter of plugging in a cable, with the right connector head, to the embedded board. This cable then transmits data over its internal wires. Given an I/O device transmitting data over a wireless medium, such as the remote control in Figure 4.24b, understanding how this interfaces to the embedded board means understanding the nature of infrared wireless communication, since there are no separate ports for transmitting data versus

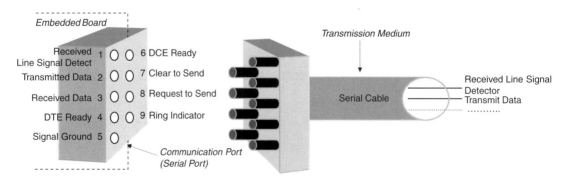

Figure 4.24a: Wired transmission medium. b: Wireless transmission medium.

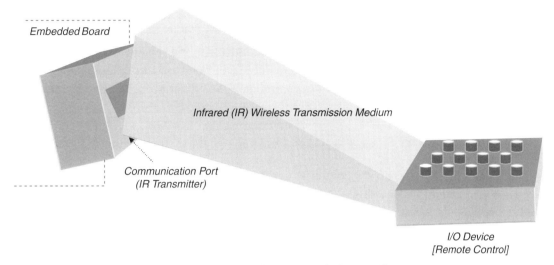

Figure 4.24b: Wireless transmission medium.

control signals. Essentially, the remote control emits electromagnetic waves to be intercepted by the IR receiver on the embedded board.

The communication port would then be interfaced to an I/O controller, a communication interface controller, or the master processor (with an integrated communication interface) via an *I/O bus* on the embedded board (see Figure 4.25). An I/O bus is essentially a collection of wires transmitting the data.

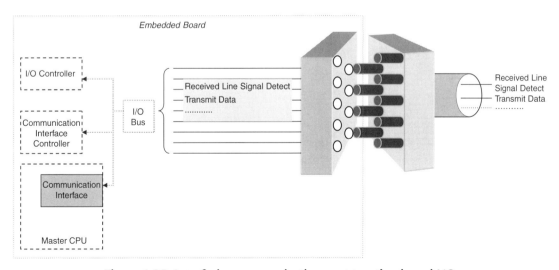

Figure 4.25: Interfacing communication port to other board I/O.

In short, an I/O device can be connected directly to the master processor via *I/O ports* (processor pins) if the I/O devices are located on the board, or it can be connected indirectly using a *communication interface* integrated into the master processor or a separate IC on the board and the *communication port*. The communication interface itself is what is either connected directly to the I/O device or the device's I/O controller. For off-board I/O devices, the relative board I/O components are interconnected via I/O buses.

4.3.2 Interfacing an I/O Controller and the Master CPU

In a subsystem that contains an I/O controller to manage the I/O device, the design of the interface between the I/O controller and master CPU—via a communications interface—is based on four requirements:

- *An ability of the master CPU to initialize and monitor the I/O controller.* I/O controllers can typically be configured via *control registers* and monitored via *status registers.* These registers are all located on the I/O controller. Control registers are data registers that the master processor can modify to configure the I/O controller. Status registers are read-only registers in which the master processor can get information as to the state of the I/O controller. The master CPU uses these status and control registers to communicate and/or control attached I/O devices via the I/O controller.

- *A way for the master processor to request I/O.* The most common mechanisms used by the master processor to request I/O via the I/O controller are *special I/O instructions* (I/O mapped) in the ISA and *memory-mapped* I/O, in which the I/O controller registers have reserved spaces in main memory.

- *A way for the I/O device to contact the master CPU.* I/O controllers that have the ability to contact the master processor via an interrupt are referred to as interrupt driven I/O. Generally, an I/O device initiates an asynchronous interrupt requesting signaling to indicate (for example) control and status registers can be read from or written to. The master CPU then uses its interrupt scheme to determine when an interrupt will be discovered.

- *Some mechanism for both to exchange data.* This refers to how data is actually exchanged between the I/O controller and the master processor. In a *programmed transfer*, the master processor receives data from the I/O controller into its registers, and the CPU then transmits this data to memory. For memory-mapped I/O schemes, DMA (direct memory access) circuitry can be used to bypass the master CPU entirely. DMA has the ability to manage data transmissions or receptions directly to and from main memory and an I/O device. On some systems, DMA is integrated into the master processor, and on others there is a separate DMA controller. Essentially, DMA requests control of the bus from the master processor.

4.4 I/O and Performance

I/O performance is one of the most important issues of an embedded design. I/O can negatively impact performance by *bottlenecking the entire system*. To understand the type of performance hurdles I/O must overcome, it is important to understand that, with the wide variety of I/O devices, each device will have its own unique qualities. Thus, in a proper design, the engineer will have taken these unique qualities on a case-by-case basis into consideration. Some of the most important shared features of I/O that can negatively impact board performance include:

- *The data rates of the I/O devices.* I/O devices on one board can vary in data rates from a handful of characters per second with a keyboard or a mouse to devices that can transmit in Mbytes per second (networking, tape, disk).

- *The speed of the master processor.* Master processors can have clocks rates anywhere from tens of MHz to hundreds of MHz. Given an I/O device with an extremely slow data rate, a master CPU could have executed thousands of times more data in the time period that the I/O needs to process a handful of *bits* of data. With extremely fast I/O, a master processor would not even be able to process anything before the I/O device is ready to move forward.

- *How to synchronize the speed of the master processor to the speeds of I/O.* Given the extreme ranges of performance, a realistic scheme must be implemented that allows for either the I/O or master processor to process data successfully regardless of how different their speeds. Otherwise, with an I/O device processing data much slower than the master processor transmits, for instance, data would be lost by the I/O device. If the device is not ready, it could hang the entire system if there is no mechanism to handle this situation.

- *How I/O and the master processor communicate.* This includes whether there is an intermediate dedicated I/O controller between the master CPU and I/O device that manages I/O for the master processor, thus freeing up the CPU to process data more efficiently. Relative to an I/O controller, it becomes a question whether the communication scheme is interrupt driven, polled, or memory mapped (with dedicated DMA to, again, free up the master CPU). If interrupt-driven, for example, can I/O devices interrupt other I/O, or would devices on the queue have to wait until previous devices finished their turn, no matter how slow.

To improve I/O performance and prevent bottlenecks, board designers need to examine the various I/O and master processor communication schemes to ensure that every device can be managed successfully via one of the available schemes. For example, to synchronize slower I/O devices and the master CPU, status flags or interrupts can be made available for all ICs so that they can communicate their status to each other when processing data. Another example occurs when I/O devices are faster than the master CPU. In this case, some type of interface (i.e., DMA) that allows these devices to bypass the master processor altogether could be an alternative.

The most common units measuring performance relative to I/O include:

- *Throughput* of the various I/O components (the maximum amount of data per unit time that can be processed, in bytes per second). This value can vary for different components. The components with the *lowest* throughput are what drives the performance of the whole system.

- The *execution time* of an I/O component. The amount of time it takes to process all of the data it is provided with.

- The *response time* or delay time of an I/O component. It is the amount of time between a request to process data and the time the actual component begins processing.

To accurately determine the type of performance to measure, the benchmark has to match how the I/O functions within the system. If the board will be accessing and processing several larger stored data files, benchmarks will be needed to measure the *throughput* between memory and secondary/tertiary storage medium. If the access is to files that are very small, then *response time* is the critical performance measure, since execution times would be very fast for small files, and the I/O rate would depend on the number of storage accesses per second, including delays. In the end, the performance measured would need to reflect how the system would actually be used in order for any benchmark to be of use.

4.5 Board Buses

All the other major components that make up an embedded board—the master processor, I/O components, and memory—are interconnected via *buses* on the embedded board. As defined earlier, a bus is simply a collection of wires carrying various data signals, addresses, and control signals (clock signals, requests, acknowledgements, data type, etc.) between all the other major components on the embedded board, which include the I/O subsystems, memory subsystem, and the master processor. On embedded boards, at least one bus interconnects the other major components in the system (see Figure 4.26).

On more complex boards, multiple buses can be integrated on one board (see Figure 4.27). For embedded boards with several buses connecting components that need to inter-communicate, *bridges* on the board connect the various buses and carry information from one bus to another. In Figure 4.27, the PowerManna PCI bridge is one such example. A bridge can automatically provide a transparent mapping of address information when data is transferred from one bus

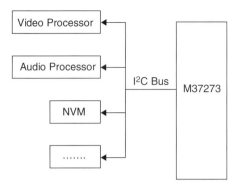

Figure 4.26: General bus structure.

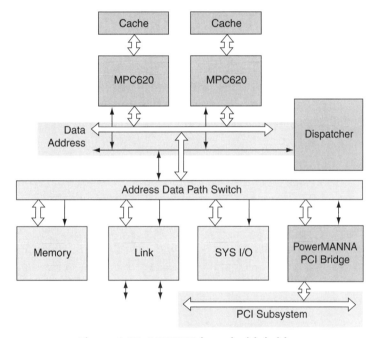

Figure 4.27: MPC620 board with bridge.
Copyright of Freescale Semiconductor, Inc., 2004. Used by permission.

to another, implement different control signal requirements for various buses—acknowledgment cycles, for example—as well as modify the data being transmitted if any transfer protocols differ bus to bus. For instance, if the byte ordering differs, the bridge can handle the byte swapping.

Board buses typically fall under one of three main categories: *system buses, backplane buses* or *I/O buses*. *System buses* (also referred to as "main," "local," or "processor-memory" buses) interconnect external main memory and cache to the master CPU and/or any bridges to the other buses. System buses are typically shorter, higher-speed custom buses. *Backplane buses* are also typically faster buses that interconnect memory, the master processor, and I/O, all on one bus. *I/O buses*, also referred to as "expansion," "external," or "host" buses, in effect act as extensions of the system bus to connect the remaining components to the master CPU, to each other, to the system bus via a bridge, and/or to the embedded system itself, via an I/O communication port. I/O buses are typically standardized buses that can be either shorter, higher-speed buses such as PCI and USB, or longer, slower buses such as SCSI.

The major difference between system buses and I/O buses is the possible presence of IRQ (interrupt request) control signals on an I/O bus. There are a variety of ways I/O and the master processor can communicate, and interrupts are one of the most common methods. An IRQ line allows for I/O devices on a bus to indicate to the master processor that an event has taken place or an operation has been completed by a signal on that IRQ bus line. Different I/O buses can have different impacts on interrupt schemes. An ISA bus, for example, requires that each card that generates interrupts must be assigned its own unique IRQ value (via setting switches or jumpers on the card). The PCI bus, on the other hand, allows two or more I/O cards to share the same IRQ value.

Within each bus category, buses can be further divided into whether the bus is expandable or nonexpandable. An *expandable bus* (PCMCIA, PCI, IDE, SCSI, USB, and so on) is one in which additional components can be plugged into the board on the fly, whereas a *nonexpandable bus* (DIB, VME, I2C are examples) is one in which additional components cannot be simply plugged into the board and then communicate over that bus to the other components.

While systems implementing expandable buses are more flexible because components can be added ad-hoc to the bus and work "out of the box," expandable buses tend to be more expensive to implement. If the board is not initially designed with all of the possible types of components that could be added in the future in mind, performance can be negatively impacted by the addition of too many "draining" or poorly designed components onto the expandable bus.

4.6 Bus Arbitration and Timing

Associated with every bus is some type of *protocol* that defines how devices gain access to the bus (arbitration), the rules attached devices must follow to communicate over the bus (handshaking), and the signals associated with the various bus lines.

Board devices obtain access to a bus using a *bus arbitration* scheme. Bus arbitration is based upon devices being classified as either *master* devices (devices that can initiate a bus transaction) or *slave* devices (devices which can only gain access to a bus in response to a master device's request). The simplest arbitration scheme is for only one device on the board—the master processor—to be allowed to be master, while all other components are slave devices. In this case, no arbitration is necessary when there can only be one master.

For buses that allow for multiple masters, some have an *arbitrator* (separate hardware circuitry) that determines under what circumstances a master gets control of the bus. There are several bus arbitration schemes used for embedded buses, the most common being *dynamic central parallel, centralized serial* (daisy-chain), and *distributed self-selection*.

Dynamic central parallel arbitration (shown in Figure 4.28a) is a scheme in which the arbitrator is centrally located. All bus masters connect to the central *arbitrator*. In this scheme, masters are then granted access to the bus via a *FIFO* (first in, first out—see Figure 4.28b) or *priority-based* system (see Figure 4.28c). The FIFO algorithm implements some type of FIFO

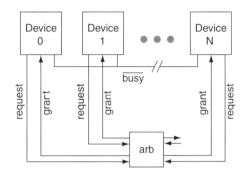

Figure 4.28a: Dynamic central parallel arbitration.

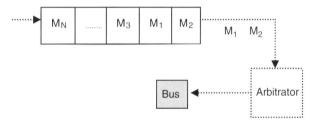

Figure 4.28b: FIFO-based arbitration.

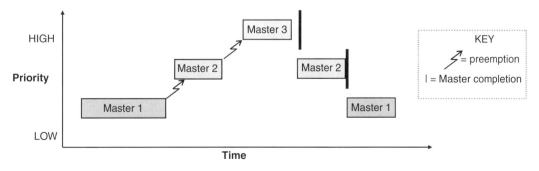

Figure 4.28c: Priority-based arbitration.

queue that stores a list of master devices ready to use the bus in the order of bus requests. Master devices are added at the end of the queue and are allowed access to the bus from the start of the queue. One main drawback is the possibility of the arbitrator not intervening if a single master at the front of the queue maintains control of the bus, never completing and not allowing other masters to access the bus.

The priority arbitration scheme differentiates between masters based upon their relative importance to each other and the system. Basically, every master device is assigned a priority, which acts as an indicator of order of precedence within the system. If the arbitrator implements a *preemption* priority-based scheme, the master with the highest priority always can preempt lower priority master devices when they want access to the bus, meaning a master currently accessing the bus can be forced to relinquish it by the arbitrator if a higher priority master wants the bus. Figure 4.28c shows three master devices (1, 2, 3 where master 1 is the lowest priority device and master 3 is the highest); master 3 preempts master 2, and master 2 preempts master 1 for the bus.

Central-serialized arbitration, also referred to as *daisy-chain arbitration*, is a scheme in which the arbitrator is connected to all masters, and the masters are connected in serial. Regardless of which master makes the request for the bus, the first master in the chain is granted the bus and passes the "bus grant" on to the next master in the chain if/when the bus is no longer needed (see Figure 4.29).

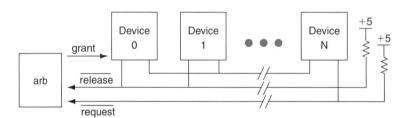

Figure 4.29: Centralized serial/daisy-chain arbitration.

There are also distributed arbitration schemes, which means there is no central arbitrator and no additional circuitry, as shown in Figure 4.30. In these schemes, masters arbitrate themselves by trading priority information to determine if a higher-priority master is making a request for the bus or even by removing all arbitration lines and waiting to see if there is a collision on the bus, which means that the bus is busy with more than one master trying to use it.

Again, depending on the bus, bus arbitrators can grant a bus to a master *atomically* (until that master is finished with its transmission) or allow for *split* transmissions, where the arbitrator can preempt devices in the middle of transactions, switching between masters to allow other masters to have bus access.

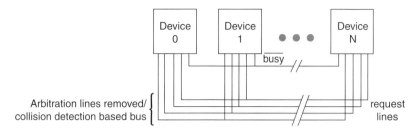

Figure 4.30: Distributed arbitration via self-selection.

Once a master device is granted the bus, only two devices—a master and another device in slave mode—communicate over that bus at any given time. There are only two types of transactions that a bus device can do—READ (receive) and/or WRITE (transmit). These transactions can take place either between two processors (a master and I/O controller, for example) or processor and memory (a master and memory, for example). Within each type of transaction, whether READ or WRITE, there can also be several specific rules that each device needs to follow in order to complete a transaction. These rules can vary widely between the types of devices communicating, as well as from bus to bus. These sets of rules, commonly referred to as the bus handshake, form the basis of any bus protocol.

The basis of any bus handshake is ultimately determined by a bus's *timing scheme*. Buses are based upon one or some combination of synchronous or asynchronous bus timing schemes, which allow for components attached to the bus to synchronize their transmissions. A *synchronous bus* (such as that shown in Figure 4.31) includes a *clock signal* among the other signals it transmits, such as data, address and other control information. Components using a synchronous bus all are run at the same clock rate as the bus and (depending on the bus) data is transmitted either on the rising edge or falling edge of a clock cycle. In order for this scheme to work, components either must be in rather close proximity for a faster clock rate, or the clock rate must be slowed for a longer bus. A bus that is too long with a clock rate that is too fast (or even too many components attached to the bus) will cause a skew in the synchronization of transmissions, because transmissions in such systems won't be in sync with the clock. In short, this means that faster buses typically use a synchronous bus timing scheme.

An *asynchronous bus*, such as the one shown in Figure 4.32, transmits no clock signal, but transmits other (non-clock based) "handshaking" signals instead, such as request and acknowledgment signals. Although the asynchronous scheme is more complex for devices having to coordinate request commands, reply commands, and so on, an asynchronous bus has no problem with the length of the bus or a larger number of components communicating over the bus, because a clock is not the basis for synchronizing communication. An asynchronous bus, however, does need some other "synchronizer" to manage the exchange of information, and to interlock the communication.

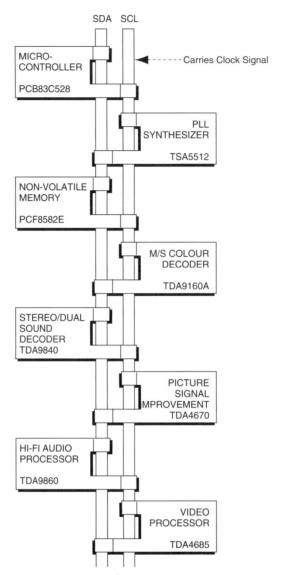

Figure 4.31: I²C bus with SCL clock.

The two most basic protocols that start any bus handshaking are the master indicating or requesting a transaction (a READ or WRITE) and the slave responding to the transaction indication or request (for example, an acknowledgment/ACK or enquiry/ENQ). The basis of these two protocols are *control* signals transmitted either via a dedicated control bus line or over a data line. Whether it's a request for data at a memory location, or the value of an I/O controller's control or status registers, if the slave responds in the affirmative to the master device's transaction request, then either an *address* of the data involved in the transaction is

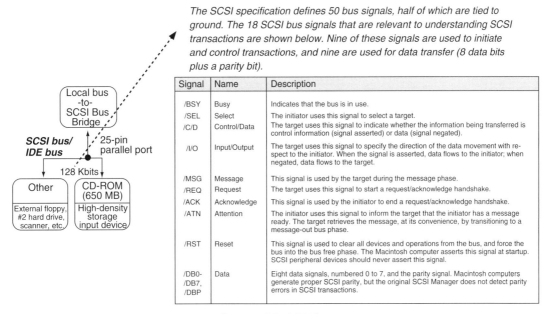

Figure 4.32: SCSI bus.

exchanged via a dedicated address bus line or data line, or this address is transmitted as part of the same transmission as the initial transaction request. If the address is valid, then a data exchange takes place over a data line (plus or minus a variety of acknowledgments over other lines or multiplexed into the same stream). Again, note that handshaking protocols vary with different buses. For example, where one bus requires the transmission of enquiries and/or acknowledgments with every transmission, other buses may simply allow the broadcast of master transmissions to all bus (slave) devices, and only the slave device related to the transaction transmits data back to the sender. Another example of differences between handshaking protocols might be that, instead of a complex exchange of control signal information being required, a clock could be the basis of all handshaking.

Buses can also incorporate a variety of transferring mode schemes, which dictate how the bus transfers the data. The most common schemes are single, where an address transmission precedes every word transmission of data, and blocked, where the address is transmitted only once for multiple words of data. A blocked transferring scheme can increase the bandwidth of a bus (without the added space and time for retransmitting the same address), and is sometimes referred to as burst transfer scheme. It is commonly used in certain types of memory transactions, such as cache transactions. A blocked scheme, however, can negatively impact bus performance in that other devices may have to wait longer to access the bus. Some of the strengths of the single transmission scheme include not requiring slave devices to have buffers to store addresses and the multiple words of data associated with the address, as well as not

Chapter 4

having to handle any problems that could arise with multiple words of data either arriving out of order or not directly associated with an address.

4.6.1 Nonexpandable Bus: I²C Bus Example

The I²C (Inter IC) bus interconnects processors that have incorporated an I²C on-chip interface, allowing direct communication between these processors over the bus. A master/slave relationship between these processors exists at all times, with the master acting as a master transmitter or master receiver. As shown in Figure 4.33, the I²C bus is a two-wire bus with one serial data line (SDA) and one serial clock line (SCL). The processors connected via I²C

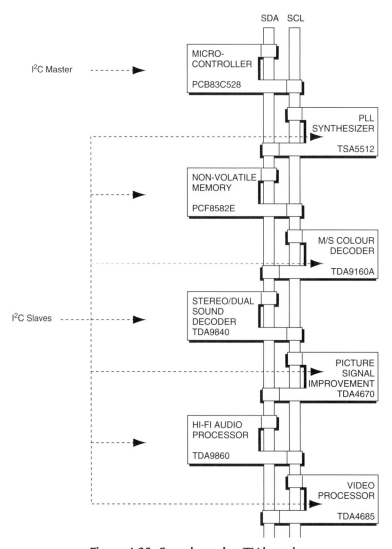

Figure 4.33: Sample analog TV board.

are each addressable by a unique address that is part of the data stream transmitted between devices.

The I²C master initiates data transfer and generates the clock signals to permit the transfer. Basically, the SCL just cycles between HIGH and LOW (see Figure 4.34).

Figure 4.34: SCL cycles.

The master then uses the SDA line (as SCL is cycling) to transmit data to a slave. A session is started and terminated as shown in Figure 4.35, where a "START" is initiated when the master pulls the SDA port (pin) LOW while the SCL signal is HIGH, whereas a "STOP" condition is initiated when the master pulls the SDA port HIGH when SCL is HIGH.

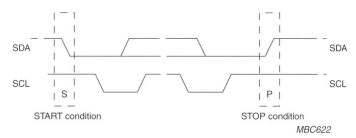

Figure 4.35: I²C START and STOP conditions.

With regard to the transmission of data, the I²C bus is a serial, 8-bit bus. This means that, while there is no limit on the number of bytes that can be transmitted in a session, only one byte (8 bits) of data will be moved at any one time, 1 bit at a time (serially). How this translates into using the SDA and SCL signals is that a data bit is "read" whenever the SCL signal moves from HIGH to LOW, edge to edge. If the SDA signal is HIGH at the point of an edge, then the data bit is read as a "1". If the SDA signal is LOW, the data bit read is a "0". An example of byte "00000001" transfer is shown in Figure 4.36a, while Figure 4.36b shows an example of a complete transfer session.

4.6.2 PCI (Peripheral Component Interconnect) Bus Example: Expandable

The latest PCI specification at the time of writing, *PCI Local Bus Specification Revision 2.1*, defines requirements (mechanical, electrical, timing, protocols, etc.) of a PCI bus implementation. PCI is a synchronous bus, meaning that it synchronizes communication using a clock. The latest standard defines a PCI bus design with at least a 33 MHz clock (up to 66 MHz) and a bus width of at least 32 bits (up to 64 bits), giving a possible minimum

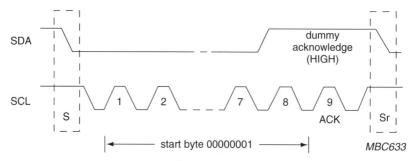

Figure 4.36a: I²C data transfer example.

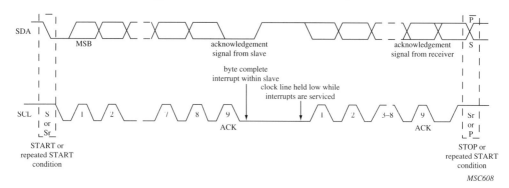

Figure 4.36b: I²C complete transfer diagram.

throughput of approximately 132 Mbytes/sec ((33 MHz * 32 bits) / 8)—and up to 528 Mbytes/sec maximum with 64-bit transfers given a 66-MHz clock. PCI runs at either of these clock speeds, regardless of the clock speeds at which the components attached to it are running.

As shown in Figure 4.37, the PCI bus has two connection interfaces: an internal PCI interface that connects it to the main board (to bridges, processors, etc.) via EIDE channels, and

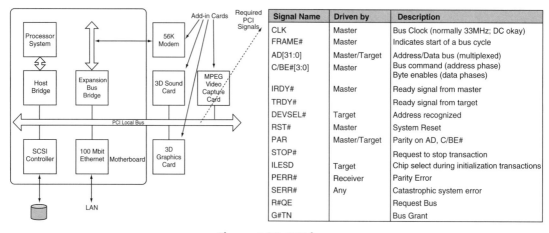

Figure 4.37: PCI bus.

the expansion PCI interface, which consists of the slots into which PCI adaptor cards (audio, video, etc.) plug. The expansion interface is what makes PCI an expandable bus; it allows for hardware to be plugged into the bus, and for the entire system to automatically adjust and operate correctly.

Under the 32-bit implementation, the PCI bus is made up of 49 lines carrying multiplexed data and address signals (32 pins), as well as other control signals implemented via the remaining 17 pins (see table in Figure 4.37).

Because the PCI bus allows for multiple bus masters (*initiators* of a bus transaction), it implements a *dynamic centralized, parallel* arbitration scheme (see Figure 4.38). PCI's arbitration scheme basically uses the REQ# and GNT# signals to facilitate communication between initiators and bus arbitrators. Every master has its own REQ# and GNT# pin, allowing the arbitrator to implement a fair arbitration scheme, as well as determining the next target to be granted the bus while the current initiator is transmitting data.

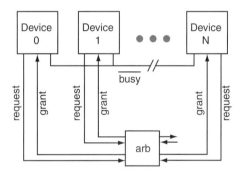

Figure 4.38: PCI arbitration scheme.

In general, a PCI transaction is made up of five steps:

1. An initiator makes a bus request by asserting a REQ# signal to the central arbitrator.

2. The central arbitrator does a bus grant to the initiator by asserting GNT# signal.

3. The address phase which begins when the initiator activates the FRAME# signal, and then sets the C/BE[3:0]# signals to define the type of data transfer (memory or I/O read or write). The initiator then transmits the address via the AD[31:0] signals at the next clock edge.

4. After the transmission of the address, the next clock edge starts the one or more data phases (the transmission of data). Data is also transferred via the AD[31:0] signals. The C/BE[3:0], along with IRDY# and #TRDY signals, indicate if transmitted data is valid.

5. Either the initiator or target can terminate a bus transfer through the deassertion of the #FRAME signal at the last data phase transmission. The STOP# signal also acts to terminate all bus transactions

Figures 4.39a and b demonstrate how PCI signals are used for transmission of information.

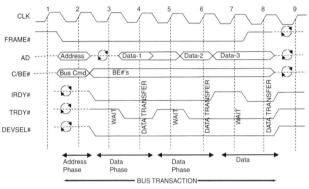

Figure 4.39a: PCI read example.

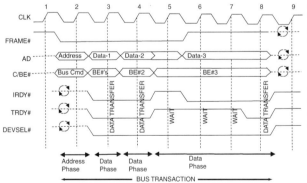

Figure 4.39b: PCI write example.

4.7 Integrating the Bus with Other Board Components

Buses vary in their physical characteristics, and these characteristics are reflected in the components with which the bus interconnects, mainly the pinouts of processors and memory chips which reflect the signals a bus can transmit (shown in Figure 4.40).

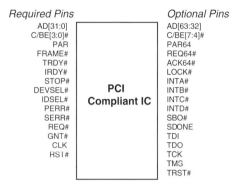

Figure 4.40: PCI-compliant IC.

Within an architecture, there may also be logic that supports bus protocol functionality. As an example, the MPC860 shown in Figure 4.41a includes an integrated I2C bus controller.

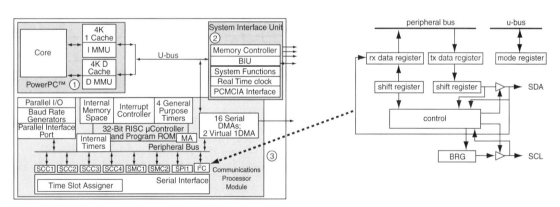

Figure 4.41: (a) I²C on MPC860. Copyright of Freescale Semiconductor, Inc., 2004. Used by permission. (b) I2C on MPC860. Copyright of Freescale Semiconductor, Inc., 2004. Used by permission.

As discussed earlier this section, the I²C bus is a bus with two signals: SDA (serial data) and SCL (serial clock), both of which are shown in the internal block diagram of the PowerPC I²C controller in Figure 4.41b. Because I²C is a synchronous bus, a baud rate generator within the controller supplies a clock signal if the PowerPC is acting as a master, along with two units (receiver and transmitter) covering the processing and management of bus transactions. In this I^{2}C integrated controller, address and data information is transmitted over the bus via the transmit data register and out the shift register. When the MPC860 receives data, data is transmitted into the receive data register via a shift register.

4.8 Bus Performance

A bus's performance is typically measured by its *bandwidth*, the amount of data a bus can transfer for a given length of time. A bus's design—both physical design and its associated protocols—will impact its performance. In terms of protocols, for example, the simpler the handshaking scheme the higher the bandwidth (fewer "send enquiry," "wait for acknowledgment," etc., steps). The actual physical design of the bus (its length, the number of lines, the number of supported devices, and so on) limits or enhances its performance. The shorter the bus, the fewer connected devices, and the more data lines, typically the faster the bus and the higher its bandwidth.

The number of bus lines and how the bus lines are used—for example, whether there are separate lines for each signal or whether multiple signals multiplex over fewer shared lines—are additional factors that impact bus bandwidth. The more bus lines (wires), the more data that can be physically transmitted at any one time, in parallel. Fewer lines mean more data has to share access to these lines for transmission, resulting in less data being transmitted at any one time. Relative to cost, note that an increase in conducting material on the board, in this case the wires of the bus, increases the cost of the board. Note, however, that multiplexing lines will introduce delays on either end of the transmission, because of the logic required on either end of the bus to multiplex and demultiplex signals that are made up of different kinds of information.

Another contributing factor to a bus's bandwidth is the number of data bits a bus can transmit in a given bus *cycle* (transaction); this is the *bus width*. Buses typically have a bandwidth of some binary power of 2—such as 1 (2^0) for buses with a serial bus width, 8 (2^3) bit, 16 (2^4) bit, 32 (2^5) bit, and so on. As an example, given 32 bits of data that needs to be transmitted, if a particular bus has a width of 8 bits, then the data is divided and sent in four separate transmissions; if the bus width is 16 bits, then there are two separate packets to transmit; a 32-bit data bus transmits one packet, and serial means that only 1 bit at any one time can be transmitted. The bus width limits the bandwidth of a bus because it limits the number of data bits that can be transmitted in any one transaction. Delays can occur in each transmission session, because of handshaking (acknowledgment sequences), bus traffic, and different clock frequencies of the communicating components, that put components in the system in delaying situations, such as a *wait state* (a time-out period). These delays increase as the number of data packets that need to be transmitted increases. Thus, the bigger the bus width, the fewer the delays, and the greater the bandwidth (throughput).

For buses with more complex handshaking protocols, the transferring scheme implemented can greatly impact performance. A block transfer scheme allows for greater bandwidth over the single transfer scheme, because of the fewer handshaking exchanges per blocks versus single words, bytes (or whatever) of data. On the flip side, block transfers can add to the latency due to devices waiting longer for bus access, since a block transfer-based transaction lasts longer than a single transfer-based transaction. A common solution for this type of latency is a bus that allows for *split transactions*, where the bus is released during the handshaking, such as while waiting for a reply to acknowledgement. This allows for other transactions to take place, and allows the bus not to have to remain idle waiting for devices of one transaction. However, it does add to the latency of the original transaction by requiring that the bus be acquired more than once for a single transaction.

CHAPTER 5
Memory Systems

David J. Katz
Rick Gentile

5.1 Introduction

To attain maximum performance, an embedded processor should have independent bus structures to fetch data and instructions concurrently. This feature is a fundamental part of what's known as a *Harvard architecture*. Nomenclature aside, it doesn't take a Harvard grad to see that, without independent bus structures, every instruction or data fetch would be in the critical path of execution. Moreover, instructions would need to be fetched in small increments (most likely one at a time) because each data access the processor makes would need to utilize the same bus. In the end, performance would be horrible.

With separate buses for instructions and data, on the other hand, the processor can continue to fetch instructions while data is accessed simultaneously, saving valuable cycles. In addition, with separate buses a processor can pipeline its operations. This leads to increased performance (higher attainable core-clock speeds) because the processor can initiate future operations before it has finished its currently executing instructions.

So it's easy to understand why today's high-performance devices have more than one bus each for data and instructions. In Blackfin processors, for instance, each core can fetch up to 64 bits of instructions and two 32-bit data words in a single core-clock (`CCLK`) cycle. Alternately, it can fetch 64 bits of instructions and one data word in the same cycle as it writes a data word.

There are many excellent references on the Harvard architecture (see Endnotes). However, because it is a straightforward concept, we will instead focus on the memory hierarchy underlying the Harvard architecture.

5.2 Memory Spaces

Embedded processors have hierarchical memory architectures that strive to balance several levels of memory with differing sizes and performance levels. The memory closest to the core processor (known as Level 1, or L1, memory) operates at the full core-clock rate. The use of the term *closest* is literal in that L1 memory is physically close to the core processor on the silicon die so as to achieve the highest operating speeds. L1 memory is most often partitioned

(a) **L1 Instruction Memory**

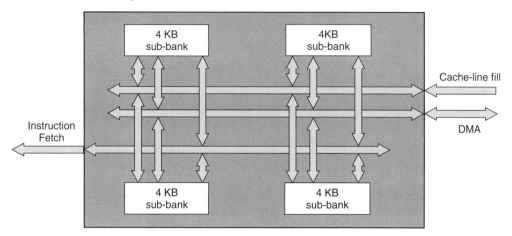

(b) **L1 Data Memory**

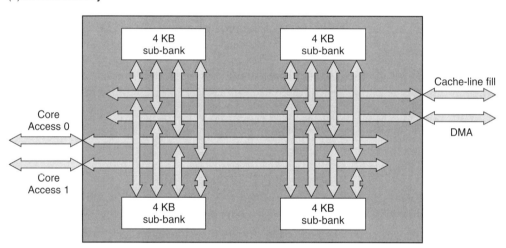

Figure 5.1: L1 memory architecture.

into instruction and data segments, as shown in Figure 5.1, for efficient utilization of memory bus bandwidth. L1 instruction memory supports instruction execution in a single cycle; likewise, L1 data memory runs at the full core-clock rate and supports single-cycle accesses.

Of course, L1 memory is necessarily limited in size. For systems that require larger code sizes, additional on-chip and off-chip memory is available—with increased latency. Larger on-chip memory is called Level 2 (L2) memory, and we refer to external memory as Level 3 (L3) memory. Figure 5.2 shows a summary of how these memory types vary in terms of speed and size. The L1 memory size usually comprises tens of kbytes, whereas the L2 memory on-chip is measured in hundreds of kbytes. What's more, Harvard-style L1 requires us to partition our

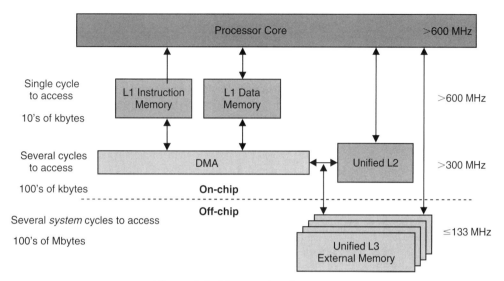

Figure 5.2: Memory-level summary.

code and data into separate places, but L2 and L3 provide a "unified" memory space. By this we mean that the instruction and data fetch units of the processor can both access a common memory space.

In addition, note the operating speed of each memory block. L1 memory runs at the `CCLK` rate. L2 memory does as well, except that accesses typically take multiple `CCLK` cycles. With L3 memory, the fastest access we can achieve is measured in system clock (`SCLK`) cycles, usually much slower than the `CCLK` rate.

On Blackfin processors, L1 and L2 memories are each further divided into sub-banks to allow concurrent core and DMA access in the same cycle. For dual-core devices, the core path to L2 memory is multiplexed between both cores, and the various DMA channels arbitrate for the DMA path into L2 memory. Don't worry too much about DMA right now; we'll focus on it in Chapter 3. For now, it is just important to think of it as a resource that can compete with the processor for access to a given memory space.

As we mentioned earlier, L3 memory is defined as "off-chip memory." In general, multiple internal data paths lead to the external memory interface. For example, one or two core access paths, as well as multiple DMA channels, all can contend for access to L3 memory. When SDRAM is used as external memory, some subset of the processor's external memory interface can be shared with asynchronous memory, such as flash or external SRAM. However, using DDR-SDRAM necessitates a separate asynchronous interface because of the signal integrity and bus loading issues that accompany DDR. Later in this chapter, we will review the most popular L3 memory types.

5.2.1 L1 Instruction Memory

Compared with data fetches, instruction fetches usually occur in larger block sizes. The instructions that run on a processor may range in size in order to achieve the best code density. For instance, the most frequently used Blackfin instructions are encoded in 16 bits, but Blackfin instruction sizes also come in 32-bit and 64-bit variants. The instruction fetch size is 64 bits, which matches the largest instruction size. When the processor accesses instructions from internal memory, it uses 64-bit bus structures to ensure maximum performance.

What happens when code runs from L3 memory that is less than 64 bits wide? In this case, the processor still issues a fetch of 64 bits, but the external memory interface will have to make multiple accesses to fill the request. Take a look at Figure 5.3 to see how instructions might actually align in memory. You'll see that the 64-bit fetch can contain as many as four instructions or as few as one. When the processor reads from memory instructions that are different in size, it must align them to prevent problems accessing those instructions later.

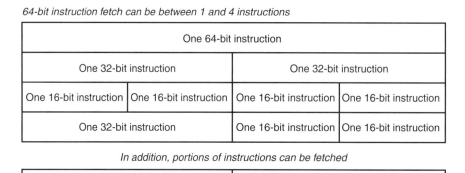

Figure 5.3: Instruction alignment.

5.2.2 Using L1 Instruction Memory for Data Placement

In general, instruction memory is meant to be used only for instructions, because in a Harvard architecture, data can't be directly accessed from this memory. However, due to the code efficiency and low byte count that some applications require, data is sometimes staged in L1 instruction memory. In these cases, the DMA controller moves the data between instruction and data memories. Although this is not standard practice, it can help in situations where you'd otherwise have to add more external memory. In general, the primary ways of accessing instruction memory are via instruction fetches and via the DMA controller. A back-door method is frequently provided as well.

5.2.3 L1 Data Memory

L1 data memory exists as the complement to L1 instruction memory. As you might expect from our L1 instruction memory discussion, the processor can access L1 data memory in a single cycle. As stated earlier, internal memory banks usually are constructed of multiple sub-banks to allow concurrent access between multiple resources.

On the Blackfin processor, L1 data memory banks consist of four 4-kbyte sub-banks, each with multiple ports, as shown in Figure 5.1b. In a given cycle, the core can access data in two of the four sub-banks, whereas the DMA controller can access a third one. Also, when the core accesses two data words on different 32-bit boundaries, it can fetch data (8, 16, or 32 bits in size) from the same sub-bank in a single cycle while the DMA controller accesses another sub-bank in the same cycle. See Figure 5.4 to get a better picture of how using sub-banks can increase performance. In the "unoptimized" diagram, all the buffers are packed into two sub-banks. In the "optimized" diagram, the buffers are spread out to take advantage of all four sub-banks.

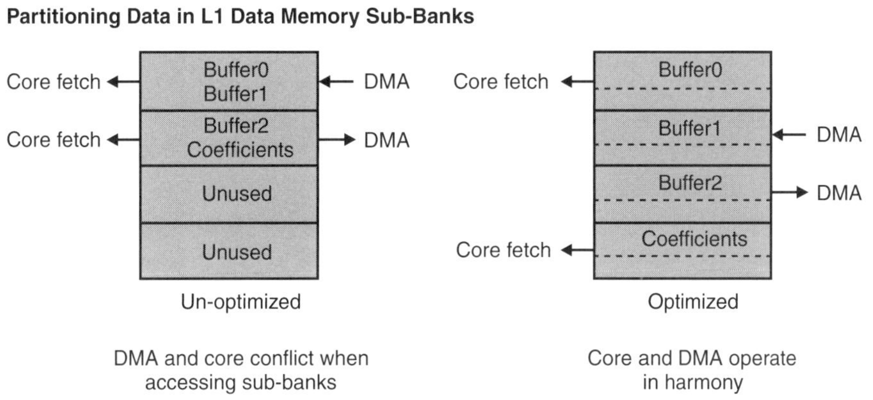

Figure 5.4: Memory bank structure and corresponding bus structure.

We've seen that there are two separate internal buses for data accesses, so up to two fetches can be performed from L1 memory in a single cycle. When fetching to a location outside L1 memory, however, the accesses occur in a pipelined fashion. In this case, the processor has both fetch operations in the same instruction. The second fetch initiates immediately after the first fetch starts. This creates a *head start* condition that improves performance by reducing the cycle count for the entire instruction to execute.

5.3 Cache Overview

By itself, a multilevel memory hierarchy is only moderately useful, because it could force a high-speed processor to run at much slower speeds to accommodate larger applications that only fit in slower external memory. To improve performance, there's always the option of manually moving important code into and out of internal SRAM. But there's also a simpler option: cache.

On the Blackfin processor, portions of L1 data and instruction memory can be configured as either SRAM or cache, whereas other portions are always SRAM. When L1 memory is configured as cache, it is no longer directly accessible for reads or writes as addressable memory.

5.3.1 What Is Cache?

You might be wondering, "Why can't more L1 memory be added to my processor?" After all, if all of a processor's memory ran at the same speed as L1, caches and external memory would not be required. Of course, the reason is that L1 memory is expensive in terms of silicon size, and big L1 memories drive up processor prices.

Enter the cache. Specifically, *cache* is a smaller amount of advanced memory that improves performance of accesses to larger amounts of slower, less expensive memories.

By definition, a cache is a set of memory locations that can store something for fast access when the application actually needs it. In the context of embedded processors, this "something" can be either data or instructions. We can map a relatively small cache to a very large cacheable memory space. Because the cache is much smaller than the overall cacheable memory space, the cacheable addresses alias into locations in cache, as shown in Figure 5.5. The high-level goal of cache is to maximize the percentage of "hits," which are instances when the processor finds what it needs in cache instead of having to wait until it gets fetched from slower memory.

Actually, cache increases performance in two ways. First, code that runs most often will have a higher chance of being in single-cycle memory when the processor needs it. Second, fills done as a result of cache-line misses will help performance on linear code and data accesses because by the time the first new instruction or data block has been brought into the cache, the next set of instructions (or data) is also already on its way into cache. Therefore, any cycles associated with a cache-line miss are spread out across multiple accesses.

Instruction cache almost always helps improve performance, whereas data cache is sometimes beneficial. The only time that instruction cache can cause problems is in the highest-performance systems that tax the limits of processor bus bandwidths.

Each sub-bank of cache consists of *ways*. Each way is made up of *lines*, the fundamental components of cache. Each cache line consists of a collection of consecutive bytes. On Blackfin devices, a cache line is 32 bytes long. Figure 5.6 shows how Blackfin instruction and data caches are organized.

Ways and lines combine to form locations in cache where instructions and data are stored. As we mentioned earlier, memory locations outside the cache alias to specific ways and lines,

4-Way Set-Associative Cache

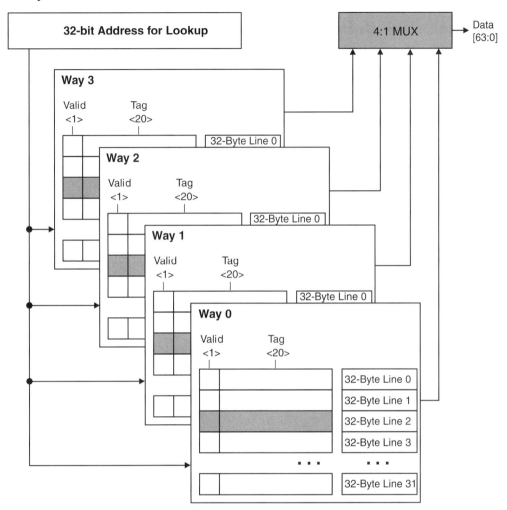

Figure 5.5: Cache concept diagram.

(a) **Instruction Cache**
- One 16 KB bank
- Four 4 KB sub-banks
- Four ways per sub-bank
- Each 1 KB way has 32 lines
- There are 32 bytes per line

(b) **Data Cache**
- Two 16 KB banks
- Four sub-banks per bank
- Two ways per sub-bank
- Each 2 KB way has 64 lines
- There are 32 bytes per line

Figure 5.6: Blackfin cache organization.

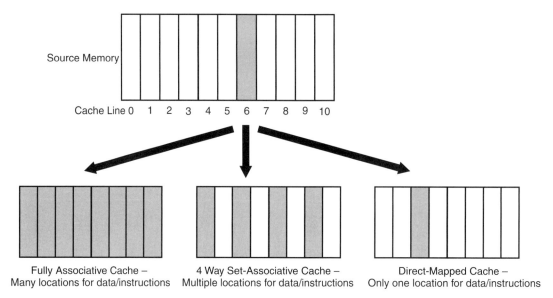

Figure 5.7: Cache architectures.

depending on the type of cache that is implemented. Let's talk now about the three main kinds of cache: direct-mapped, fully associative, and set-associative. Figure 5.7 illustrates the basic differences between types. It is important to understand the variety of cache your processor employs, because this determines how the cacheable memory aliases to the actual cache memory.

5.3.2 Direct-Mapped Cache

When we talk about a cache that is *direct-mapped*, we mean that each memory location maps to a single cache line that it shares with many other memory locations. Only one of the many addresses that share this line can use it at a given time. Although this is the simplest scheme to implement, it provides the lowest percentage of performance increase. Since there is only one site in the cache for any given external memory location, code that has lots of branches will always result in cache-line misses. Direct mapping only helps when code flow is very linear, a situation that does not fit the control nature of the typical embedded application.

The primary problem with this type of cache is that the probability the desired code or data is actually in cache is the lowest of the three cache models we describe. *Thrashing* occurs when a line in cache is constantly being replaced with another line. This is much more common in a direct-mapped cache than in other cache architectures.

5.3.3 Fully Associative Cache

In a *fully associative* cache, any memory location can be cached in any cache line. This is the most complicated (and costly) scheme to implement in silicon, but performance will always be the best. Essentially, this implementation greatly reduces the number of cache-line misses in steady-state operation. Because all addresses map to all sites in the cache, the probability increases that what you want to be in cache will actually be there.

5.3.4 N-Way Set-Associative Cache

The previous two cache designs represent the two ends of the performance spectrum. The final design we will discuss is actually the most common implementation. It is called the *N-way set-associative* cache, where *N* is typically 2 or 4. This scheme represents a compromise between the two previously mentioned types of cache. In this scenario, the cache comprises sets of *N* lines each, and any memory address can be cached in any of those *N* lines within a set. This improves hit ratios over the direct-mapped cache, without the added design complexity and silicon area of the fully associative design. Even so, it achieves performance very close to that of the fully associative model.

In Blackfin processors, the data cache is two-way set-associative, and the instruction cache is four-way set-associative. This mostly has to do with the typical profile of execution and data access patterns in an embedded application. Remember, the number of ways increases the number of locations within the cache to which each address can alias, so it makes sense to have more for instruction cache, where addressing normally spans a larger range.

5.3.5 More Cache Details

As we saw in Figure 5.6, a cache structure is made up of lines and ways. But there's certainly more to cache than this. Let's take a closer look.

Each line also has a "tag array" that the processor uses to find matches in the cache. When an instruction executes from a cached external memory location, the processor first checks the tag array to see if the desired address is in cache. It also checks the "validity bit" to determine whether a cache line is *valid* or *invalid*. If a line is valid, this means that the contents of this cache line contain values that can be used directly. On the other hand, when the line is invalid, its contents can't be used.

As we stated before, a *cache hit* refers to a case when the data (or instruction) the core wants to access is already in the cache. Likewise, a *cache miss* refers to the case when the processor needs information from the cache that is not yet present. When this happens, a *cache-line fill* commences.

As noted earlier, the information in the tag array determines whether a match exists or not. At the simplest level, a cache-line fill is just a series of fetches made from cacheable memory. The difference is that when cache is off, the core fetches only what it needs, and when cache is on, the core may actually fetch more than it needs (or hopefully, what it will need soon!).

So, as an example, let's assume that cache is enabled and the location being accessed has been configured as cacheable. The first time a processor accesses a specific location in, say,

L3 memory, a cache miss will generate a cache-line fill. Once something is brought into cache, it stays there until it is forced out to make room for something else that the core needs. Alternatively, as we will soon see, it is sometimes prudent to manually *invalidate* the cache line.

As we noted earlier, Blackfin processors fetch instructions 64 bits at a time. When instruction cache is enabled and a cache miss occurs, the cache-line fill returns four 64-bit words, beginning with the address of the missed instruction. As Figure 5.8 illustrates, each cache line aligns on a fixed 32-byte boundary. If the instruction is in the last 64-bit word in a cache line, that will be the first value returned. The fill always wraps back around to the beginning of the line and finishes the fetch. From a performance standpoint this is preferable because we wouldn't want the processor to wait for the three unwanted 64-bit fetches to come back before receiving the desired instruction.

Cache Line Replacement

Cache line fill begins with requested word

Target Word	Fetching Order for Next Three Words
WD0	WD0, WD1, WD2, WD3
WD1	WD1, WD2, WD3, WD0
WD2	WD2, WD3, WD0, WD1
WD3	WD3, WD0, WD1, WD2

Figure 5.8: Cache-line boundaries.

When a cache hit occurs, the processor treats the access as though it were in L1 memory—that is, it fetches the data/instruction in a single `CCLK` cycle. We can compute the cache *hit rate* as the percentage of time the processor has a cache hit when it tries to fetch data or instructions. This is an important measure because if the hit rate is too low, performance will suffer. A hit rate over 90% is desirable. You can imagine that, as the hit rate drops, performance starts to approach a system in which everything to which the core needs access resides in external memory.

Actually, when the hit rate is too low, performance will be worse than it is when everything is in external memory and cache is off. This is due to the fact that the cache-line size is larger than the data or instruction being fetched. When cache misses are more common than cache hits, the core will end up waiting for unwanted instructions or data to come into the system. Of course, this situation degrades performance; fortunately, it's difficult to create a case where this happens.

As a rule of thumb, the more associative a cache is, the higher the hit rate an application can achieve. Also, as the cache size grows, the number of cache ways has less impact on hit rate, but performance still does increase.

As more items are brought into cache, the cache itself becomes full with valid data and/or instructions. When the cache has an open line, new fetches that are part of cache-line fills populate lines that are invalid. When all the lines are valid, something has to be replaced to make room for new fetches. How is this replacement policy determined?

A common method is to use a *least recently used (LRU)* algorithm, which simply targets for replacement the data or instruction cache line that has not been accessed for the longest time. This replacement policy yields great performance because applications tend to run small amounts of code more frequently. This is true even when application code approaches Mbytes in size.

5.3.6 Write-Through and Write-Back Data Cache

Data cache carries with it some additional important concepts. There are generally two modes of operation for data cache: write-through and write-back, as shown in Figure 5.9.

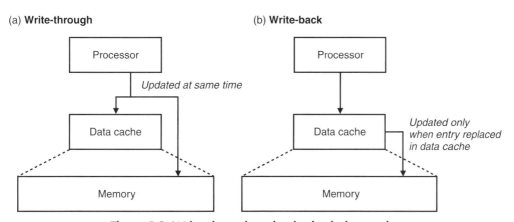

Figure 5.9: Write-through and write-back data cache.

The term *write-through* means that the information is written both to the cache *and* to the source memory at the time it is modified, as shown in Figure 5.9a. This means that if a data element is brought into cache and modified a million times while the processor is executing an algorithm, the data is written to the source memory a million times as well. In this case, the term *source memory* refers to the actual location being cached in L2 or L3 memory. As you can see, write-through mode can result in lots of traffic on the bus between cache and the source memory. This activity can impact performance when other resources are accessing the memory.

Write-through data cache does have one advantage over write-back, however: It helps keep buffers in source memory coherent when more than one resource has access to them. For example, in a dual-core device with shared memory, a cache configured as write-through can help ensure that the shared buffer has the latest data in it. Assume that core A modifies the buffer and then core B needs to make subsequent changes to the same buffer. In this case, core A would notify core B when the initial processing was complete, and core B would have access to the latest data.

The term "write-back" means that the information is written only to the cache, until it is being replaced, as shown in Figure 5.9b. Only then is the modified data written back to source memory. Therefore, if a data value is modified a million times, the result is only written locally to the cache until its cache entry is replaced, at which point source memory will be written a single time. Using our dual-core example again, coherency can still be maintained if core A manually "flushes" the cache when it is done processing the data. The flush operation forces the data in cache to be written out to source memory, even though it is not being replaced in cache, which is normally the only time this data would be written to source memory.

Although these two modes each have merit, write-back mode is usually faster because the processor does not need to write to source memory until absolutely necessary. Which one should you choose? The choice depends on the application, and you should try both ways to see which will give the best performance. It is important to try these options multiple times in your development cycle. This approach will let you see how the system performs once peripherals are integrated into your application. Employing the write-back mode (versus write-through mode) can usually increase performance between 10% and 15%.

These write policy concepts don't apply to instruction memory, because modifying code in cached instruction memory isn't a typical programming model. As a result, instruction cache is not designed with this type of feature.

Before we move on, let's look at one additional write-back mode mechanism. Figure 5.10 illustrates the composition of a cache line. Specifically, an address tag precedes the set of four 64-bit words (in the case of Blackfin devices). The cache array tags possess a *dirty bit* to

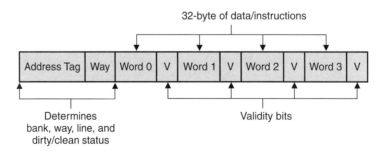

Figure 5.10: Cache array with tags.

mark data that has changed in cache under write-back mode. For example, if we read a value from memory that ends up in cache and the value is subsequently modified by the processor, it is marked as "dirty." The cache uses this bit as a reminder that before the data is completely removed from cache, it needs to be written out to its source memory. Processors often have a *victim buffer* that holds data that was replaced in the cache. Let's consider why and how this helps performance.

When a data miss occurs in write-back mode and the entire cache contains valid data, something must be removed from the cache to hold the data about to be fetched. Recall that when this happens, the cache (assuming an LRU policy) will replace the data least recently used. What if there's "dirty" data in the line that is replaced—that is, data which has changed and needs to be updated in the source memory? The processor is most immediately interested in obtaining data for the currently executing instruction. If it had to wait for the dirty data to be written back to source memory, and then wait again for the new data to be fetched, the core would stall longer than desired. This is where the victim buffer comes in; it holds the data that needs to be written back, while the core gets its new data quickly. Once the cache-line fill completes, the victim buffer empties and the source memory is current.

5.4 External Memory

So far in this chapter, our discussions have centered on internal memory resources. Let's now focus on the many storage options available external to the processor. We generically refer to external memory as "L3" throughout this text, but you will soon see that the choices for L3 memory vary considerably in terms of the way they operate and their primary uses. They all can play an important role in media-based applications. We will start with the highest-performance volatile memories and move to various nonvolatile options. It is important to note here that the synchronous and asynchronous memories described here are directly memory-mapped to a processor's memory space. Some of the other memories we'll discuss later in this chapter, such as NAND flash, are also mapped to an external memory bank, but they have to be indirectly accessed.

5.4.1 Synchronous Memory

We begin our discussion with synchronous memory because it is the highest-performance external memory. It's widely available and provides a very cost-effective way to add large amounts of memory without completely sacrificing performance. We focus on SDRAM technology to provide a good foundation for understanding, but then we'll proceed to an overview of current and near-term follow-ons to SDRAM: DDR-SDRAM 1 and 2.

Both SDRAM and DDR are widely available and very cost-effective because the personal computer industry uses this type of memory in standard DIMM modules.

5.4.1.1 SDRAM

SDRAM is synchronous addressable memory composed of banks, rows, and columns. All reads and writes to this memory are locked to a processor-sourced clock. Once the processor initializes SDRAM, the memory must be continually refreshed to ensure that it retains its state.

Clock rates vary for SDRAM, but the most popular industry specifications are PC100 and PC133, indicating a 100 MHz and 133 MHz maximum clock, respectively. Today's standard SDRAM devices operate at voltages between 2.5 V and 3.3 V, with the PC133 memory available at 3.3 V.

Memory modules are specified in terms of some number of Mbits in addition to a variety of data bus sizes (e.g., ×8, ×16, ×32). This sometimes causes confusion, but it is actually a straightforward nomenclature. For example, the ×16 designation on a 256-Mbit device implies an arrangement of 16 Mbits × 16. Likewise, for a ×32 bus width, a 256-Mbit device would be configured as 8 Mbits × 32. This is important when you are selecting your final memory size. For example, if the external bus on your processor is 32 bits wide and you want 128 Mbytes of total memory, you might connect two 32 Mbits × 16 modules.

At its lowest level, SDRAM is divided into rows and columns. To access an element in an SDRAM chip, the row must first be "opened" to become "active." Next, a column within that row is selected, and the data is transferred from or written to the referenced location. The process of setting up the SDRAM can take several processor system clock cycles. Every access requires that the corresponding row be active.

Once a row is active, it is possible to read data from an entire row without re-opening that row on every access. The address within the SDRAM will automatically increment once a location is accessed. Because the memory uses a high-speed synchronous clock, the fastest transfers occur when performing accesses to contiguous locations, in a burst fashion.

Figure 5.11 shows a typical SDRAM controller (SDC) with the required external pins to interface properly to a memory device. The data access size might be 8, 16, or 32 bits. In addition, the actual addressable memory range may vary, but an SDC can often address hundreds of Mbytes or more.

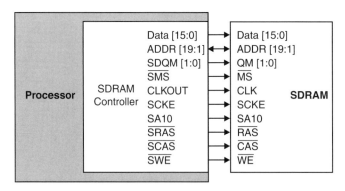

Figure 5.11: Representative SDRAM controller.

SDRAMs are composed of internal banks—most commonly, four equally sized banks. So if you connected 64 Mbytes of SDRAM to a processor, the SDRAM would consist of four 16-Mbyte internal banks. This is important to remember because you'll derive performance benefits from partitioning your application across the internal banks. Another thing to note—the term *bank* is unfortunately used to describe both the internal structure of an SDRAM and an entire SDRAM module (as viewed from the system level). For example, two 64-Mbyte *external* SDRAM banks may connect to a processor, and each 64-Mbyte module may consist of four 16-Mbyte *internal* banks each.

The SDRAM controller uses the external pins shown in Figure 5.11 to issue a series of commands to the SDRAM. Table 5.1 provides a brief description of each of the pins, and Table 5.2 shows how the pins work in concert to send commands to the SDRAM. It generates these commands automatically based on writes and reads by the processor or DMA controller.

Table 5.1: SDRAM pin description.

Pin	Description
ADDR	External address bus
DATA	External data bus
$\overline{SRAS}$	SDRAM row address strobe (connect to SDRAM's $\overline{RAS}$ pin)
$\overline{SCAS}$	SDRAM column address strobe (connect to SDRAM's $\overline{CAS}$ pin)
$\overline{SWE}$	SDRAM write-enable pin (connect to SDRAM's $\overline{WE}$ pin)
SDQM	SDRAM data mask pins (connect to SDRAM's DQM pins)
$\overline{SMS}$	Memory select pin of external memory bank configured for SDRAM
SA10	SDRAM A10 pin (used for SDRAM refreshes; connect to SDRAM's A[10] pin)
SCKE	SDRAM clock-enable pin (connect to SDRAM's CKE pin)
CLKOUT	SDRAM clock pin (connect to SDRAM's CLK pin; Opverates at SCLK frequency)

Table 5.2: SDRAM commands.

Command	$\overline{SMS}$	$\overline{SCAS}$	$\overline{SRAS}$	$\overline{SWE}$	SCKE	SA10
Precharge All	low	high	low	low	high	high
Single Precharge	low	high	low	low	high	low
Bank Activate	low	high	low	high	high	–
Load Mode Register	low	low	low	low	high	–
Load Extended Mode Register	low	low	low	low	high	low
Read	low	low	high	high	high	low
Write	low	low	high	low	high	low
Auto-Refresh	low	low	low	high	high	–
Self-Refresh	low	low	low	high	low	–
NOP (No Operation)	low	high	high	high	high	–
Command Inhibit	high	high	high	high	high	–

Figure 5.12 illustrates an SDRAM transaction in a simplified manner. First, the higher bits of the address are placed on the bus and /RAS is asserted. The lower address bits of the address are then placed on the bus and /CAS is asserted. The number of rows and columns will depend on the device you select for connection.

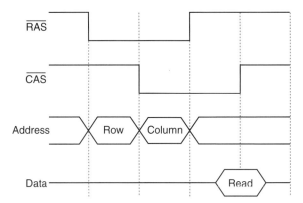

Figure 5.12: Basic SDRAM timing diagram.

It is important to consider the way your processor's SDC multiplexes SDRAM addresses. Consider two possibilities shown in Figure 5.13 as an example. In Figure 5.13b, the SDRAM row addresses are in the higher bit positions. In Figure 5.13a, the bank address lines are in the higher bit positions, which can result in better performance, depending on the application. Why is this the case? The SDRAM can keep four pages open across four internal banks, thus reducing page opening/closing latency penalties. If your system is connected as shown in Figure 5.13b, it would be very hard to partition your code and data to take advantage of this

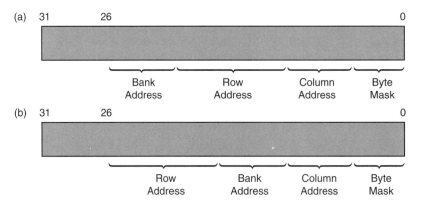

Figure 5.13: Possible SDRAM address muxing.

feature. Specifically, you would have to slice up your data and code and essentially interleave it in memory to make use of all open rows.

Let's now discuss briefly some of the key commands that the SDRAM controller uses to interface with the memory device.

The *Bank Activate* command causes the SDRAM to open an internal bank (specified by the bank address) in a row (specified by the row address). The pins that are used for the bank and row addresses depend on the mappings of Figure 5.13. When the SDC issues the *Bank Activate* command, the SDRAM opens a new row address in the dedicated bank. The memory in the open internal bank and row is referred to as the *open page*. The *Bank Activate* command must be applied before issuing a read or write command.

The *Precharge* command closes a specific internal bank in the active page or all internal banks in the page.

The *Precharge All* command precharges all internal banks at the same time before executing an auto-refresh.

A *Read/Write* command executes if the next read/write access is in the present active page. During the *Read* command, the SDRAM controller drives the column address. The delay between *Bank Activate* and *Read* commands is determined by the t_{RCD} parameter in the SDRAM data sheet. The SDRAM then makes data available after the CAS latency period described below.

In the *Write* command, the SDRAM controller drives the column address. The write data becomes valid in the same cycle. The delay between *Bank Activate* and *Write* commands is determined by the t_{RCD} parameter in the SDRAM data sheet.

Whenever a page miss occurs (an access to a location on a row that is not open), the SDC executes a *Precharge* command followed by a *Bank Activate* command before executing the *Read* or *Write* command. If there is a page hit, the *Read* or *Write* command can be issued immediately without requiring the *Precharge* command.

The *Command Inhibit* and *NOP* commands are similar in that they have no effect on the current operations of the SDRAM. The only difference between the two is that the *NOP* is used when the actual SDRAM bank has been selected.

The *Load Mode* command is used to initialize an SDRAM chip. *Load Extended Mode* is an additional initialization command that's used for mobile SDRAMs.

Auto-Refresh and *Self-Refresh* commands regulate the way the contents of the SDRAM are refreshed periodically. We'll talk more about them shortly.

5.4.1.2 CAS Latency

SDRAM data sheets are swimming in numbers that specify the performance of the device. What do all these parameters really mean?

The *Column Address Strobe (CAS) latency*, abbreviated as CL2 or CL3, is the delay in clock cycles between when the SDRAM detects a *Read* command and when it provides the data at its output pins. This CAS latency is an important selection parameter. A common term used to identify SDRAM devices is either CAS2 or CAS3. These actually represent CL2 or CL3, since they refer to *CAS latency* timings (e.g., two system clocks versus three system clocks). An SDRAM with a CAS latency of two cycles will likely yield better throughput than one with a three-cycle latency. This is based on the fact that for random accesses, a cycle will be saved each time an access is made. You should specify this parameter based on application needs. Does the extra performance of the faster device justify the extra cost? For high-performance systems, the answer is usually "Yes."

The CAS latency of a device must be greater than or equal to its column access time (t_{CAC}) and its frequency of operation (t_{CLK}). That is, the selection of CL must satisfy this equation:

$$CL \times t_{CLK} \geq t_{CAC}$$

For example, if t_{CLK} is 7.5 ns (133 MHz system clock) and t_{CAC} is 15 ns, you can select a CL2 device. If t_{CAC} is 20 ns, you must choose CL3. The PC133 SDRAM specification only allows for CAS latency values of 1, 2, or 3.

Sometimes you will see memory devices described as 3-2-2 or 2-2-2. These numbers represent the CAS latency, RAS-to-CAS delay (t_{RCD}), and Row Precharge (t_{RP}) values, respectively, in clock cycles at 100 MHz. Note that for any other speed of operation, such as 66 MHz or 133 MHz, these numbers would change. For example, let's assume that for a given module, t_{CAC} is 25 ns, t_{RCD} is 20 ns, and t_{RP} is 20 ns. This would indicate 3-2-2 timings at 100 MHz (substituting t_{RCD} or t_{RP} for t_{CAC} in the preceding equation as appropriate), but what would they be at 133 MHz? Since 133 MHz corresponds to a 7.5 ns clock cycle (t_{CLK}), our equation gives timings of 4-3-3, which would be invalid for the SDRAM, since the CAS latency cannot be higher than 3. Therefore, you would not be able to operate this module at 133 MHz.

One more point to understand about the CAS latency figure (CAS2 or CAS3) is that SDRAM suppliers often advertise their top PC100 SDRAM as CAS2. Recall that the CAS latency number is derived from t_{CAC}. Unfortunately, the vendor doesn't provide you with the t_{CAC} value. Imagine a CL2 part with a 20-ns t_{CAC} and a CL3 part with a 21-ns t_{CAC}. At 133 MHz (7.5 ns clock period), both parts would have a CAS latency value of 3. This means that although a CL2 part may be designed to handle a faster system clock speed than a CL3 part, you won't always see a performance difference. Specifically, if you don't plan on running your SDRAM faster than 125 MHz, the device with the lower value of CL doesn't provide any additional benefit over the device with the higher CL value.

5.4.1.3 Refreshing the SDRAM

SDRAM controllers have a refresh counter that determines when to issue refresh commands to the SDRAM. When the SDC refresh counter times out, the SDC precharges all banks of SDRAM and then issues an *Auto-Refresh* command to them. This causes the SDRAM to generate an internal refresh cycle. When the internal refresh completes, all internal SDRAM banks are precharged.

In power-saving modes where SDRAM data needs to be maintained, it is sometimes necessary to place the SDRAM in self-refresh mode. Self-refresh is also useful when you want to buffer the SDRAM while the changes are made to the SDRAM controller configuration. Moreover, the self-refresh mode is useful when sharing the memory bus with another resource. It can prevent conflicts on shared pins until the first processor regains ownership of the bus.

When you place the SDRAM in self-refresh mode, it is the SDRAM's internal timer that initiates auto-refresh cycles periodically, without external control input. Current draw when the SDRAM is in self-refresh mode is on the order of a few milliamps, versus the typical "on" current of 100 mA.

The SDC must issue a series of commands, including the *Self-Refresh* command, to put the SDRAM into this low power mode, and it must issue another series of commands to exit self-refresh mode. Entering self-refresh mode is controlled by software in an application. Any access made by the processor or the DMA controller to the SDRAM address space causes the SDC to remove the SDRAM from self-refresh mode.

It is important to be aware that core or DMA accesses to SDRAM are held off until an in-process refresh cycle completes. This is significant because if the refresh rate is too high, the potential number of accesses to SDRAM decreases, which means that SDRAM throughput declines as well. Programming the refresh rate to a higher value than necessary is a common mistake that programmers make, especially on processors that allow frequency modification on the fly. In other words, they forget to adjust the Refresh Control register to a level commensurate with the newly programmed system clock frequency. As long as the effective refresh rate is not too slow, data will not be lost, but performance will suffer if the rate is too high.

5.4.1.4 Mobile SDRAM

A variant of SDRAM that targets the portable device realm is called *mobile SDRAM*. It comes in a smaller form factor and smaller memory sizes than its traditional cousin, and it can operate down to 2.5 V or 1.8 V, greatly reducing SDRAM power consumption. Mobile SDRAM is also known as LP SDRAM, or low-power SDRAM. It is worth mentioning that mobile SDRAM devices typically specify a supply voltage and an I/O voltage. For the most power-sensitive applications, the supply voltage is at 2.5 V while the I/O supply is 1.8 V.

In addition to a reduced form factor and greatly reduced power budget, mobile SDRAM has three key JEDEC-specified features that set it apart from SDRAM. The first is a temperature compensated self-refresh mode. The second is a partial array self-refresh capability, and the third is a "deep power-down" mode.

The temperature-compensated self-refresh capability allows you to program the mobile SDRAM device to automatically adjust its refresh rate in self-refresh mode to accommodate case temperature changes. Some mobile devices assume you will connect a temperature sensor to the case and adjust the parameters associated with this feature. Others have a built-in temperature sensor. Either way, the goal is to save power by adjusting the frequency of self-refresh to the minimum level necessary to retain the data.

The partial array self-refresh feature allows you to control which banks within a mobile SDRAM are actually refreshed when the device is in self-refresh mode. You can program the device so that 100%, 50%, or 25% of the banks are kept in self-refresh mode. Obviously, any data you need to keep must reside in a bank that is self-refreshed. The other banks are then used for data that does not need to be retained during power-saving modes.

Finally, the deep power-down feature allows you to remove power from the device via a write to the control register of the SDRAM chip to prevent any current draw during self-refresh mode (thus saving hundreds of μA). Of course, all data is lost when you do this, but this mode can be very useful in applications in which the entire board is not powered down (e.g., some components are operational and are powered by the same regulator) but you want to extend the battery life as long as possible.

5.4.1.5 Double Data Rate (DDR) SDRAM/DDR1

SDRAM and mobile SDRAM chips provide the bulk of today's synchronous memory in production. This is quickly changing, however, as an evolved synchronous memory architecture provides increased performance. Double data rate (DDR) SDRAM provides a direct path to double memory bandwidth in an application. In addition, although the industry has more or less skipped 2.5 V SDRAM devices (with the notable exception of mobile SDRAM), standard DDR1 chips are all at 2.5 V.

The traditional SDRAM controller often shares processor pins with various types of asynchronous memory, but the DDR specification is much tighter to allow for much faster operation. As such, a DDR memory module will not share pins with other types of memory. Additionally, DDR memories require a DDR controller to interface with them; a processor's SDRAM controller is incompatible with DDR memory.

As the name suggests, a key difference between SDRAM and DDR is that whereas SDRAM allows synchronous data access on each clock cycle, DDR allows synchronous access on both edges of the clock—hence the term *double data rate*. This results in an immediate increase in peak performance when you use DDR in an application.

As an example, whereas the peak transfer rate of a PC133 SDRAM running at 133 MHz would be 4 bytes × 133 MHz, the maximum peak transfer rate of the equivalent DDR1 module running at the same frequency is 4 bytes × 266 MHz.

Another important advantage DDR has over SDRAM is the size of the "prefetch" it accepts. DDR1 has a prefetch size of 2n, which means that data accesses occur in pairs. When reads are performed, each access results in a fetch of two data words. The same is true for a single write access—that is, two data words must be sent.

This means that the minimum burst size of DDR1 is two external data words. For reads, the DDR controller can choose to ignore either of the words, but the cycles associated with both will still be spent. So you can see that although this feature greatly enhances performance of sequential accesses, it erodes performance on random accesses.

DDR SDRAM also includes a strobe-based data bus. The device that drives the data signals also drives the data strobes. This device is the DDR module for reads and the DDR controller for writes. These strobes allow for higher data rates by eliminating the synchronous clock model to which SDRAM adheres.

When all these DDR feature enhancements are combined, performance really shines. For example, at 400 MHz, DDR1 increases memory bandwidth to 3.2 Gbytes/s, compared to PC133 SDRAM theoretical bandwidth of 133 MHz × 4 bytes, or 532 Mbytes/s.

Just as a low-power version of SDRAM exists, mobile DDR1 also exists, with many similarities to mobile SDRAM. Mobile DDR1 devices run as low as 1.8 V. Another advantage of mobile DDR1 is that there is no minimum clock rate of operation. This compares favorably to standard DDR1, whose approximate 80 MHz minimum operating frequency can cause problems in power-sensitive applications.

5.4.1.6 DDR2 SDRAM

DDR2 SDRAM is the second generation of DDR SDRAM. It offers data rates of up to 6.4 Gbytes/s, lower power consumption, and improvements in packaging. It achieves this higher level of performance and lower power consumption through faster clocks, 1.8 V operation and signaling, and a simplified command set. Like DDR1, DDR2 has a mobile variety as well, targeted for handheld devices.

Table 5.3 shows a summary of the differences between DDR1, mobile DDR1, and DDR2. It should be noted that DDR1 is not compatible with a conventional SDRAM interface, and DDR2 is not compatible with DDR1. However, DDR2 is planned to be forward-compatible with next-generation DDR technology.

5.4.2 Asynchronous Memory

As we have just seen, SDRAM and its successors both are accessed in a synchronous manner. Asynchronous memory, as you can guess, does not operate synchronously to a clock. Each

Table 5.3: Comparison of DDr1, mobile DDR1, and DDr2.

Feature	DDR1	Mobile DDR1	DDR2
Data transfer rate	266, 333, 400 MHz	200, 250, 266, 333 MHz	400, 533, 667, 800 MHz
Operating voltage	2.5 V	1.8 V	1.8 V
Densities	128 Mb–1 Gb	128–512 Mb, 1 Gb (future)	256 Mb–4 Gb
Internal banks	4	4	4 and 8
Prefetch	2	2	4
CAS latency	2, 2.5, 3	2, 3	3, 4, 5, 6
Additive latency	No	No	0, 1, 2, 3, 4, 5
READ latency	CAS latency	CAS latency	Additive latency + CAS latency
WRITE latency	Fixed = 1 cycle	Fixed = 1 cycle	READ latency − 1 cycle
I/O width	×4, ×8, ×16	×4, ×8, ×16, ×32	×4, ×8, ×16
On-die termination	No	No	Selectable
Off-chip driver	No	No	Yes
Burst length	2, 4, 8	2, 4, 8, 16, full page	4, 8
Burst terminate command	Yes	Yes	No
Partial array self-refresh	No	Full, 1/2, 1/4, 1/8, 1/16	No
Temperature-compensated self-refresh	No	Supported	No
Deep power-down	No	Supported	No

access has an associated read and write latency. Burst operations are always available with synchronous memory, but the same can't be said of asynchronous memory.

The asynchronous bus of today has evolved from buses that were popular in the past, such as IBM's *Industry Standard Architecture* (ISA) bus. Asynchronous devices come in many different flavors. The most common include Static RAM (SRAM), which is volatile, and nonvolatile memories like PROM and Flash. SRAM can substitute for the pricier SDRAM when high performance isn't required.

Processors that have an asynchronous memory controller (AMC) typically share its data and address pins with the SDRAM controller. As we mentioned earlier, because DDR has much tighter capacitive loading requirements, an AMC would not share pins with a DDR controller.

A characteristic AMC is shown in Figure 5.14. It contains read- and write-enable pins that can interface to memory devices. Table 5.4 shows a summary of the common AMC pins. Several

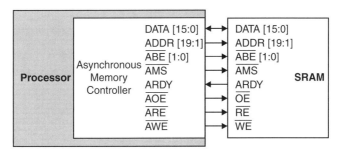

Figure 5.14: A typical asynchronous memory controller (AMC).

Table 5.4: Typical AMC pins.

ADDR	External address bus (outputs)
DATA	External data bus (inputs/outputs)
AMS	Asynchronous memory selects (outputs)
AWE	Asynchronous memory write enable (output)
ARE	Asynchronous memory read enable (output)
AOE	Asynchronous memory read enable (output)
ARDY	Asynchronous memory ready response (input)
ABE[1:0]	Byte enables (outputs)

memory-select lines allow the AMC to choose which of several connected devices it's talking to at a given point in time. The AMC has programmable wait states and setup/hold timing for each connected device, to accommodate a wide array of asynchronous memories.

The AMC is especially useful because, in addition to interfacing with memory devices, it allows connection to many other types of components. For example, FIFOs and FPGAs easily map to an asynchronous bank. Chip sets for USB and Ethernet, as well as bridges to many other popular peripherals, also easily interface to the asynchronous memory interface.

When connecting a nonmemory device to an AMC, it is always best to use a DMA channel to move data into and out of the device, especially when the interface is shared with SDRAM. Access to these types of components usually consumes multiple system clock cycles, whereas an SDRAM access can occur in a single system clock cycle. Be aware that, when the AMC and SDC share the same L3 memory bus, slow accesses on an asynchronous bank could hold off access to a SDRAM bank considerably.

Synchronous random access memory (synchronous SRAM) is also available for higher performance than traditional SRAMs provide, at increased cost. Synchronous SRAM devices are capable of either pipelined or flow-through functionality. These devices take the asynchronous devices one step closer to SDRAM by providing a burst capability.

Whereas synchronous SRAM provides for higher performance than ordinary SRAM, other technologies allow for lower power. Pseudo-SRAM (and a variant called CellularRAM) connect to a processor via an SDRAM-like interface. Additionally, they sport an I/O supply requirement in line with other processor I/O (2.5 V or 3.3 V, for instance) while powering the Vcc supply of the memory itself at 1.8 V. This presents a good compromise that allows processors to take advantage of some power savings even when they don't have 1.8 V-capable I/O.

Figure 5.15 shows a high-level view comparing several representative types of external memory from the dual standpoints of performance and capacity.

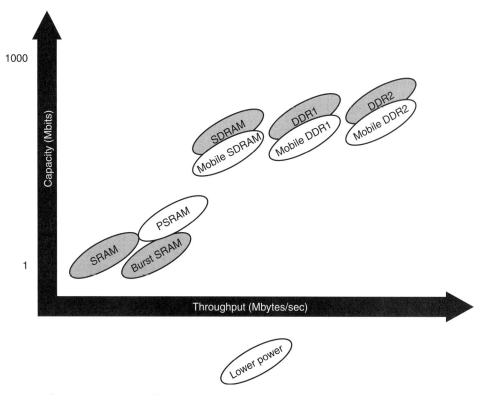

Figure 5.15: External memory comparison of performance and capacity.

5.4.3 Nonvolatile Memories

Nonvolatile memories—memories that retain their contents even when not powered—come in several forms. The simplest, a ROM (read-only memory), is written once (at the factory) but can be read many times. A PROM, or Programmable ROM, also can only be written once, but this can be done in the field, not just at the time of manufacture. An erasable PROM (EPROM) can be reprogrammed after erasing its contents via ultraviolet light exposure. An electrically erasable PROM (EEPROM), commonly used today, needs only electricity for erasure. A Flash

EEPROM takes this technology one step further by allowing data to be stored and erased in blocks, rather than in byte-sized increments. This significantly speeds up access rates to Flash memories compared with regular EEPROM access times. Finally, a burst-mode Flash sharply reduces read access times for sequential bytes by adding a few extra control pins to the standard Flash interface.

5.4.3.1 NAND and NOR Flash Memories

The two main types of nonvolatile Flash memory widely used today are NOR-based and NAND-based. There are many differences between these two technologies, each of which is optimal for certain classes of use. Table 5.5 gives a rundown of their major characteristics.

Table 5.5: Summary of NOR and NAND Flash characteristics.

Trait	NOR Flash	NAND Flash
Capacity/bit density	Low (<64 MB)	High (16–512 MB)
Directly execute code from?	Yes	No
Erase performance	Very slow (5 sec)	Fast (3 ms)
Write performance	Slow	Fast
Read performance	Fast	Medium
Reliability	OK	Low; requires error checking and bad-block checking
Erase cycles	10 K–100 K	100–1000 K
Life span	OK	Excellent (10× NOR)
Interface	ISA-like/MCU-friendly; (Addr + Data + Control), Serial (SPI or I2C)	I/O only (command sequence)
Pin count	High	Low
Access method	Random	Sequential
Ease-of-use	Easy; memory-mapped address + data scheme	Difficult (file system needed)
Cost/bit	High	Low
Primary usage	Low-density, high-speed code access, some data storage	High-density data block storage
Bootable?	Yes	Not generally
Power/Energy Dissipation	Higher (due to long program and erase cycles)	Lower

Fundamentally, NAND and NOR Flash technologies differ in the structure of their respective memory cell arrays. In NOR devices, memory cells are connected in parallel between a bit line and ground, such that selecting any number of memory cells can end up grounding a bit line. Because this is similar to a wired-OR configuration, it is termed *NOR Flash*. This arrangement enables very fast read times and thus makes NOR Flash a good candidate for random access—a trait associated with processor code.

NAND realizes a more efficient cell architecture that results in only about half the space taken by a NOR cell. However, this space-saving layout connects memory cells in series, sharing bit-select lines across several cells. To ground a bit line, an entire group of memory cells must be turned on simultaneously. This makes NAND Flashes a poor choice for random accesses, but it provides excellent cell compactness. What's more, it allows several NAND cells to be programmed simultaneously, providing very fast write performance compared to NOR devices.

NOR Flash holds the edge in read access speed, but NAND is superior in programming and erasing speeds. Erase time is especially crucial because Flash devices must all be erased (each bit set to 1) before programming occurs (selectively setting bits back to 0). Therefore, erasing is an integral part of writing to a Flash device. Moreover, NAND Flashes can tolerate many more erase cycles than NOR Flashes can, usually on the order of tenfold, thus providing much longer life spans than NOR devices.

One intricacy of NAND Flashes is that a file system is necessary to use them. This is because, to maintain high yields, NAND Flashes are shipped with randomly located bad blocks. It is the job of the file system (running on the embedded processor) to find these bad blocks, tag them, and avoid using them. Usually, the memory vendor has already scanned and marked the bad blocks at the factory, and the file system just needs to maintain a table of where these are located, keeping in mind that some additional bad blocks will be formed over the lifetime of the part. The good news is that because each block is independent of all others, the failure of one block has no impact on the operation of others.

As it turns out, NOR Flash can also have bad blocks, but manufacturers typically allocate extra blocks on the chip to substitute for any bad blocks discovered during factory tests. NAND has no such spare blocks because it's assumed that a file system is necessary anyway for the mass storage systems of which NAND devices are a part.

As another consideration, NAND is somewhat prone to bit errors due to periodic programming (one bit error out of every 10 billion bits programmed). Therefore, error-correcting codes (Hamming codes, usually) are employed by the file system to detect and correct bit errors.

As far as processor interface, NOR devices have a couple of options. For one, they hook directly up to the asynchronous memory controller of microprocessors, with the conventional address and data bus lines as well as write/read/select control. Depending on the address space desired, this interface can encompass a large number of pins. As a slower, lower pin-count alternative, serial Flash devices can connect through just a few wires via an SPI or I^2C interface. Here, the address and data are multiplexed on the serial bus to achieve memory access rates only a fraction of those attainable using parallel address/data buses.

Interfacing to NAND devices is more complex and occurs indirectly, requiring a sequence of commands on the 8-bit bus to internal command and address registers. Data is then accessed in pages, usually around 528 bytes in length. Although this indirect interface makes NAND unsuitable for booting, this approach provides a fixed, low pin-count interface that remains static as higher-density devices are substituted. In this respect (and many others), NAND Flash acts like a conventional hard disk drive. This accounts for the similar structure and access characteristics between the two storage technologies.

In fact, NAND Flash devices were specifically intended to replace magnetic hard disk drives in many embedded applications. To their advantage, they are solid-state devices, meaning that they have no moving parts. This leads to more rugged, reliable devices than magnetic disk drives, as well as much less power dissipation.

NAND flash serves as the basis for all of the most popular removable solid-state mass storage cards: CompactFlash, SmartMedia, Secure Digital/Multimedia Card (SD/MMC), Extreme Digital Picture Card (xD), MemoryStick, and the like. With the exception of SmartMedia, all these cards have built-in controllers that simplify access to the NAND memory on the device. These products find wide use in consumer electronics (like cameras, PDAs, cell phones, etc.) and other embedded applications requiring mass storage. Table 5.6 provides a high-level comparison of these storage cards.

Devices like SD, MemoryStick, and CompactFlash also have an I/O layer, which makes them quite attractive as interfaces on embedded processors. They enable a wide range of peripherals through these interfaces, including Bluetooth, 802.11b, Ethernet transceivers, modems, FM radio, and the like.

5.4.3.2 Hard Disk Storage: IDE, ATA, and ATAPI

The terms *IDE* and *ATA* actually refer to the same general interface, but they mean different things, and people often get confused and use the terms interchangeably. For simplicity, we'll just refer to the interface as ATA or ATAPI, as explained below.

IDE is short for *Integrated Drive Electronics*, which refers to a pivotal point in the PC storage market when the disk drive controllers were integrated directly on the hard disk itself, rather than as a separate board inside the PC. This reduced the cost of the interface and increased reliability considerably. IDE drives can support a master and a slave device on each IDE channel.

ATA stands for *AT Attachment*, dating back to the days of the IBM PC/AT. It describes a device interface for hooking up to hard disk drives. ATAPI stands for *ATA Packet Interface*, which specifies a way to hook up to CD-ROM, DVD and tape drives, which communicate by means of packetized commands and data. Although the packet layer of ATAPI adds considerable complexity, it still maintains the same electrical interface as ATA. ATA is such a ubiquitous standard for

Table 5.6: Storage card comparison.

	SmartMedia	MMC	CompactFlash	Secure Digital (SD)	MemoryStick	xD	TransFlash/Micro SD
Developer	Toshiba	Infineon, SanDisk	SanDisk	SanDisk, Panasonic, Toshiba	Sony (later with SanDisk)	Fujifilm, Olympus	SanDisk, Motorola
Variants		MMCplus, HS-MMC (High Speed), RS-MMC (Reduced Size), SecureMMC, MMCmobile		Mini-SD	MemoryStick, ProMemoryStick, DuoMemory Stick Pro Duo		
Volume (mm^3)	1265	1075605 (RS-MMC, MMCmobile)	5141 (Type I) 7790 (Type II)	1613 (SD) 602 (mini-SD)	3010992 (Duo and Pro Duo)	850	165
Weight (g)	2	1.5	11.4	2 (1 for mini-SD)	4 (2 for Duo & Pro Duo)	2	0.4

Interface pins	22	7–13	50	9	10	18	8
Present capacity	128 MB	1 GB	8 GB	2 GB (1 GB for mini-SD)	128 MB 4 GB (Pro and Pro Duo)	1 GB	512 MB
Content security	ID copy protection	Depends on variant	No	CPRM, SD	MagicGate	ID copy protection	CPRM, SD
Max data transfer rate	1 MB/s Write, 3.5 MB/s Read	Up to 52 MB/s (depends on variant)	66 MB/s	10 MB/s	1.8 MB/s Write, 2.5 MB/s Read, 20 MB/s (Pro and Pro Duo)	3 MB/s Write, 5 MB/s Read	1.8 MB/s typical
Comments	No on-board controller, limited capacity	Small form factor, compatible with SD interfaces	IDE-compatible	Mini-SD has adapter to fit SD card slot	Mostly Sony products, supports real-time DVD-quality video transfer	Adapters available to other card types	bridges between embedded and removable memory worlds
I/O capability?	No	No	Yes	Yes	Yes	No	No
Voltage	3.3V/5V	3.3V (1.8V/3.3V for MMCplus and MMCmobile)	3.3V/5V	3.3V	3.3V	3.3V	3.3V
Interface	8-bit I/O	SPI, MMC	IDE	SPI, SD	MemoryStick	NAND Flash	SPI, SD

mass storage (dwarfing SCSI, for instance) that ATA-compliant storage has found its way into the PC-MCIA and CompactFlash worlds as well.

The ATA interface specifies the way commands are passed through to the hard drive controller, interpreted, and processed. The interface consists of eight basic registers: seven are used by the processor's I/O subsystem to set up a read or write access, and the eighth one is used to perform block reads/writes in 512-byte chunks called sectors. Each sector has 256 words, and each access utilizes a 16-bit data bus.

The original scheme for addressing drive data followed a CHS method of accessing a particular cylinder, head, and sector (CHS) of the hard disk. A more intuitive mode known as *logical block addressing (LBA)*, in which each sector on a hard disk has its own unique identification number, was soon added. LBA led the way toward breaking through the 8 GB addressing barrier inherent in the CHS approach (which allows for only 1,024 cylinders, 256 heads, and 63 sectors).

ATA lays down timing requirements for control and data accesses. It offers two modes of data access—Programmed I/O (PIO) and DMA—each offering various levels of performance. PIO modes operate in analogous fashion to direct processor accesses to the hard drive. That is, the processor has to schedule time out of what it's doing to set up and complete a data processing cycle. DMA modes, on the other hand, allow data transfer between the drive and processor memory to occur without core intervention. The term *Ultra DMA* refers to higher-performance levels of DMA that involve double-edge clocking, where data is transferred on both the rising and falling edges of the clock. Ultra DMA usually requires an 80-conductor IDE cable (instead of the standard 40-pin one for parallel ATA drives) for better noise immunity at the higher transfer rates, and it also incorporates cyclical redundancy checking (CRC) for error reduction.

Table 5.7 lists the several different levels of ATA, starting with ATA-1 and extending to ATA/ATAPI-7, which is still under development. At Level 4, ATAPI was introduced, so every level from 4 onward is designated ATA/ATAPI. The levels differ mainly in I/O transfer rates, DMA capability, and reliability. All new drives at a given ATA level are supposed to be backward-compatible to the previous ATA levels.

5.4.3.3 Other Hard Drive Interfaces

SATA

Serial ATA (SATA) is a follow-on to ATA that serializes the parallel ATA data interface, thus reducing electrical noise and increasing performance while allowing for longer, skinnier cabling. SATA uses differential signaling at very high clock rates to achieve data transfer rates of several hundred megabytes per second. SATA sheds the legacy 5 V power supply baggage that parallel ATA carries, and its differential signaling operates in the range of 200–300 mV. More good news: It's software-compatible with parallel ATA. This allows simple

Table 5.7 ATAPI summary.

ATA Version	Max PIO Transfer Rate (MB/s)	Max DMA Transfer Rate (MB/s)	Added Features
ATA-1	8.3	8.3	Original standard for IDE drives
ATA-2	16.7	16.7	Logical Block Addressing, Block Transfers
ATA-3	16.7	16.7	Improved reliability, Drive security, SMART (self-monitoring)
ATA/ATAPI-4	16.7	33.3	Packet Interface extension. Ultra DMA (80-conductor ribbon cable), CRC error checking and correction
ATA/ATAPI-5	16.7	66.7	Faster Ultra DMA transfer modes
ATA/ATAPI-6	16.7	100	Support for drives larger than 137 GB
ATA/ATAPI-7	16.7	133	SATA and DVR support

SATA-to-parallel-ATA converters to translate between the two technologies in the near term, without disturbing the software layer underneath the different physical layers. On the downside, SATA allows only for point-to-point connections between a controller and a device, although external port multipliers are available to effect a point-to-multipoint scheme.

SATA is targeted primarily at the PC market, to improve reliability and increase data transfer speeds. It will slowly make its way into embedded applications, first targeting those systems where the storage density of solid state flash cards isn't adequate, leaving parallel or serial ATA as the only options. But as flash card densities grow into the several-Gbyte range, they will maintain a firm foothold in a wide swath of embedded applications due to the increased reliability of their solid-state nature and their small form factors.

SCSI

SCSI stands for Small Computer Systems Interface, a high-end interconnect that outperforms ATA but is also much more complex. It offers expandability to many more devices than ATA and has a feature set geared toward higher performance. SCSI is more expensive than ATA for the same storage capacity, and it's typically the realm of niche applications, not very popular in the embedded world.

Microdrive

The Microdrive is an actual miniature hard disk in a CompactFlash Type II form factor. Invented by IBM, the product line is now owned and propagated by Hitachi. Whereas CompactFlash is a solid-state memory, Microdrives are modeled after conventional magnetically based hard drives, with tiny spinning platters. For this reason, they are not as rugged and reliable as CompactFlash memory, and they also consume more power. Currently they are available at capacities up to several gigabytes.

USB/Firewire

These drives are nothing more than Parallel ATA disk drives with a USB 2.0 high-speed or Firewire (IEEE 1394) front end, usually used to facilitate access to a PC. However, in the embedded world, where USB and Firewire are gaining traction, these interfaces provide a handy way to add storage to a system.

USB pen drives, also called keychain drives, are not related to these ATA drives. Instead, they are Flash memory devices with a USB front end, similar in all respects to the NAND memory devices described previously.

5.4.3.4 Emerging Nonvolatile Memory Technologies

There are some exciting new technologies on the horizon that will probably be very important in high-performance embedded applications. Two of particular interest are FRAM and MRAM. FRAM, or Ferroelectric RAM, uses electric field orientation to store charge. This gives it almost infinite write capability. By comparison, conventional EEPROMs can only be written on the order of 10,000 times. MRAM, or Magnetoresistive RAM, uses electron spin to store information. It represents the best of several memory domains: it's nonvolatile, it has bit densities rivaling DRAM, and it operates at SRAM-like speeds.

5.5 Direct Memory Access

The processor core is capable of doing multiple operations, including calculations, data fetches, data stores, and pointer increments/decrements, in a single cycle. In addition, the core can orchestrate data transfer between internal and external memory spaces by moving data into and out of the register file.

All this sounds great, but in reality you can only achieve optimum performance in your application if data can move around without constantly bothering the core to perform the transfers.

This is where a direct memory access (DMA) controller comes into play. Processors need DMA capability to relieve the core from these transfers between internal/external memory and peripherals or between memory spaces. There are two main types of DMA controllers. "Cycle-stealing" DMA uses spare (idle) core cycles to perform data transfers. This is not a workable solution for systems with heavy processing loads like multimedia flows. Instead, it is much more efficient to employ a DMA controller that operates independently from the core.

Why is this so important? Well, imagine if a processor's video port has a FIFO that needs to be read every time a data sample is available. In this case, the core has to be interrupted tens of millions of times each second. As though that's not disruptive enough, the core has to perform an equal amount of writes to some destination in memory. For every core processing cycle spent on this task, a corresponding cycle would be lost in the processing loop.

We know from experience that PC-based software designers transitioning to the embedded world are hesitant to rely on a DMA controller for moving data around in an application. This usually stems from their impression that the complexity of the programming model increases exponentially when DMA is factored in. Yes, it is true that a DMA controller adds another dimension to your solution. We will, in fact, explore some intricacies that DMA introduces—such as contention for shared resources and new challenges in maintaining coherency between data buffers. Our goal, however, is to put your mind at ease, to show you how DMA is truly your friend. In this chapter, we'll focus on the DMA controller itself.

5.5.1 DMA Controller Overview

Because you'll typically configure a DMA controller during code initialization, the core should only need to respond to interrupts after data set transfers are complete. You can program the DMA controller to move data in parallel with the core while the core is doing its basic processing tasks—the jobs on which it's supposed to be focused! In an optimized application, the core would never have to move any data but rather only access it in L1 memory. The core wouldn't need to wait for data to arrive, because the DMA engine would have already made it available by the time the core was ready to access it. Figure 5.16 shows a typical interaction between the processor and the DMA controller. The steps allocated to the processor involve setting up the transfer, enabling interrupts, and running code when an interrupt is generated. The dashed lines/arrows between memory and the peripheral indicate operations the DMA controller makes to move data independent of the processor. Finally, the interrupt input back to the processor can be used to signal that data is ready for processing.

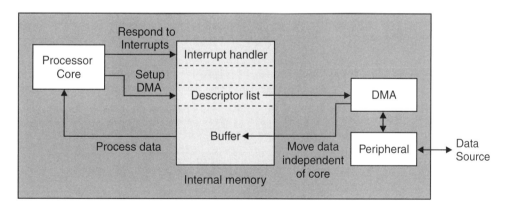

Figure 5.16: DMA controller.

In addition to moving to and from peripherals, data also needs to move from one memory space to another. For example, source video might flow from a video port straight to L3 memory, because the working buffer size is too large to fit into internal memory. We don't want to make the processor fetch pixels from external memory every time we need to perform a calculation, so a memory-to-memory DMA (MemDMA for short) can bring pixels into L1 or L2 memory for more efficient access times. Figure 5.17 shows some typical DMA data flows.

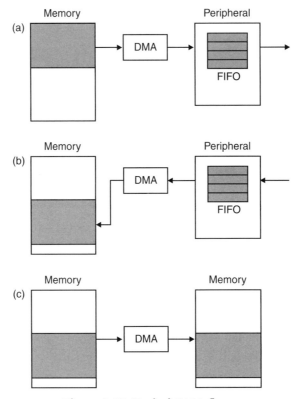

Figure 5.17: Typical DMA flows.

So far we've focused on data movement, but a DMA transfer doesn't always have to involve data. We can use code overlays to improve performance, configuring the DMA controller to move code into L1 instruction memory before execution. The code is usually staged in larger external memory.

In this chapter, we will use the Blackfin processor's DMA controller as a model to illustrate the basic concepts of direct memory access and how it can boost system performance. We will also offer some helpful ways to manage the DMA controller and review examples of "two-dimensional" transfers that can save valuable data passes by markedly reducing the time an application spends traversing a data buffer.

5.5.2 More on the DMA Controller

A DMA controller is a unique peripheral devoted to moving data around a system. Think of it as a controller that connects internal and external memories with each DMA-capable peripheral via a set of dedicated buses. It is a peripheral in the sense that the processor programs it to perform transfers. It is unique in that it interfaces to both memory and selected peripherals. Notably, only peripherals where data flow is significant (kbytes per second or greater) need to be DMA-capable. Good examples of these are video, audio, and network interfaces. Lower-bandwidth peripherals can also be equipped with DMA capability, but it's less an imposition on the core to step in and assist with data transfer on these interfaces.

In general, DMA controllers will include an address bus, a data bus, and control registers. An efficient DMA controller will possess the ability to request access to any resource it needs, without having the processor itself get involved. It must have the capability to generate interrupts. Finally, it has to be able to calculate addresses within the controller.

A processor might contain multiple DMA controllers. Each controller has multiple DMA channels, as well as multiple buses that link directly to the memory banks and peripherals, as shown in Figure 5.18. There are two types of DMA controllers in the Blackfin processor. The first category, usually referred to as a System DMA controller, allows access to any resource (peripherals and memory). Cycle counts for this type of controller are measured in *system clocks* (SCLKs) at frequencies up to 133 MHz. The second type, an Internal Memory DMA (IMDMA) controller, is dedicated to accesses between internal memory locations. Because the accesses are internal (L1 to L1, L1 to L2, or L2 to L2), cycle counts are measured in *core clocks* (CCLKs), which can exceed 600 MHz rates.

Figure 5.18 also shows the Blackfin DMA bus structure, where the DMA External Bus (DEB) connects the DMA controller to external memory, the DMA Core Bus (DCB) connects the

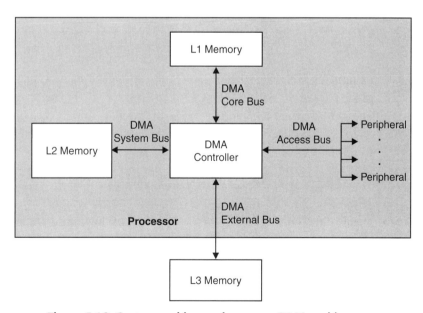

Figure 5.18: System and internal memory DMA architecture.

controller to internal memory, and the DMA Access Bus (DAB) connects to the peripherals. An additional DMA bus set is also available when L2 memory is present, to move data within the processor's internal memory spaces.

Each DMA channel on a Blackfin DMA controller has a programmable priority associated with it. If more than one channel requests the DMA bus in a single cycle, the highest-priority channel wins access. For memory DMA channels, a "round robin" access capability exists. That is, one memory channel can access the bus for a programmable number of cycles before turning the bus over to the next MemDMA channel, which also gets the same number of cycles on the bus. When more than one DMA controller is present on a processor, the channels from one controller can run at the same time as channels on the other controller. This is possible, for example, when a memory-to-memory transfer takes place from L3 to L2 memory while the second controller feeds a peripheral from L1 memory. If both DMA controllers try to access the same resource (L3 memory, for example), arbitration must take place. In this case, one of the controllers can be programmed to a higher priority than the other.

Each DMA controller has a set of FIFOs that act as a buffer between the DMA subsystem and peripherals or memory. For MemDMA, a FIFO exists on both the source and destination sides of the transfer. The FIFO improves performance by providing a place to hold data while busy resources are preventing a transfer from completing.

5.5.3 Programming the DMA Controller

Let's take a look at the options we have in specifying DMA activity. We will start with the simplest model and build up to more flexible models that, in turn, increase in setup complexity.

For any type of DMA transfer, we always need to specify starting source and destination addresses for data. In the case of a peripheral DMA, the peripheral's FIFO serves as either the source or the destination. When the peripheral serves as the source, a memory location (internal or external) serves as the destination address. When the peripheral serves as the destination, a memory location (internal or external) serves as the source address.

In the simplest MemDMA case, we need to tell the DMA controller the source address, the destination address and the number of words to transfer. With a peripheral DMA, we specify either the source or the destination, depending on the direction of the transfer. The word size of each transfer can be 8, 16, or 32 bits. This type of transaction represents a simple one-dimensional (1D) transfer with a unity "stride." As part of this transfer, the DMA controller keeps track of the source and destination addresses as they increment. With a unity stride, as in Figure 5.19a, the address increments by 1 byte for 8-bit transfers, 2 bytes for 16-bit transfers, and 4 bytes for 32-bit transfers.

We can add more flexibility to a one-dimensional DMA simply by changing the stride, as in Figure 5.19b. For example, with nonunity strides, we can skip addresses in multiples of the transfer sizes. That is, specifying a 32-bit transfer and striding by four samples results in an address increment of 16 bytes (four 32-bit words) after each transfer.

Couching this discussion in Blackfin DMA controller lingo, we have now described the operations of the XCOUNT and XMODIFY registers. XCOUNT is the number of *transfers* that need to be made. Note that this is not necessarily the same as the number of bytes to transfer. XMODIFY is the number of bytes to increment the address pointer after the DMA controller moves the first data element. Regardless of the transfer word size, XMODIFY is always expressed in bytes. XMODIFY can also take on the value of 0, which has its own advantage, as we'll see later in this chapter.

Whereas the 1D DMA capability is widely used, the two-dimensional (2D) capability is even more useful, especially in video applications. The 2D feature is a direct extension to what we discussed for 1D DMA. In addition to an XCOUNT and XMODIFY value, we also program corresponding YCOUNT and YMODIFY values. It is easiest to think of the 2D DMA as a nested loop, where the inner loop is specified by XCOUNT and XMODIFY, and the outer loop is

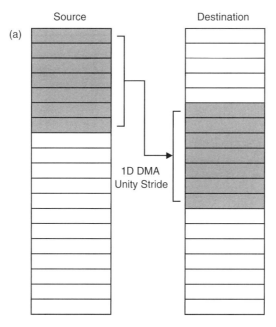

Figure 5.19: 1D DMA examples.
(a) 1D DMA with unity stride.

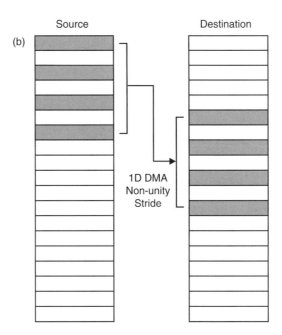

Figure 5.19: (Continued)
(b) 1D DMA with nonunity stride.

specified by `YCOUNT` and `YMODIFY`. A 1D DMA can then be viewed simply as an "inner loop" of the 2D transfer of the form:

```
for y = 1 to YCOUNT STEP YMODIFY /* 2D with outer loop */
for x = 1 to XCOUNT STEP XMODIFY /* 1D inner loop */
         {
            /*Loop goes here */
         }
```

While `XMODIFY` determines the stride value the DMA controller takes every time `XCOUNT` decrements, `YMODIFY` determines the stride taken whenever `YCOUNT` decrements. As is the case with `XCOUNT` and `XMODIFY`, `YCOUNT` is specified in terms of the number of transfers, while `YMODIFY` is specified as a number of bytes. Notably, `YMODIFY` can be negative, which allows the DMA controller to wrap back around to the beginning of the buffer. We'll explore this feature shortly.

For a peripheral DMA, the "memory side" of the transfer can be either 1D or 2D. On the peripheral side, though, it is always a 1D transfer. The only constraint is that the total number of bytes transferred on each side (source and destination) of the DMA must be the same. For example, if we were feeding a peripheral from three 10-byte buffers, the peripheral would have to be set to transfer 30 bytes using any possible combination of supported transfer width and transfer count values available.

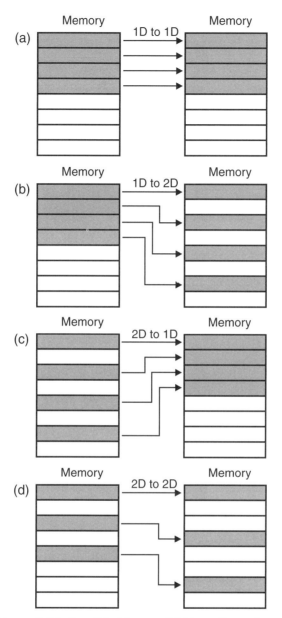

Figure 5.20: Possible Memory DMA configurations.

MemDMA offers a bit more flexibility. For example, we can set up a 1D-to-1D transfer, a 1D-to-2D transfer, a 2D-to-1D transfer, and of course a 2D-to-2D transfer, as shown in Figure 5.20. The only constraint is that the total number of bytes being transferred on each end of the DMA transfer block has to be the same.

Let's now look at some DMA transfer examples:

> **Example 5.1**
>
> Consider a 4-pixel (per line) × 5-line array, with byte-sized pixel values, ordered as shown in Figure 5.21a.
>
>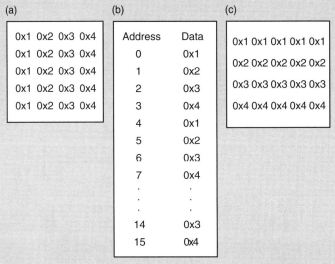
>
> **Figure 5.21: Source and destination arrays for Example 5.1.**

While this data is shown as a matrix, it appears consecutively in memory as shown in Figure 5.21b.

We now want to create the array shown in Figure 5.21c using the DMA controller.

The source and destination DMA register settings for this transfer are:

Source	Destination
XCOUNT = 5	XCOUNT = 20
XMODIFY = 4	XMODIFY = 1
YCOUNT = 4	YCOUNT = 0
YMODIFY = −15	YMODIFY = 0

Source and destination word transfer size = 1 byte per transfer.

Let's walk through the process. In this example, we can use a MemDMA with a 2D-to-1D transfer configuration. Since the source is 2D, it should be clear that the source channel's

XCOUNT and YCOUNT are 5 and 4, respectively, since the array size is 5 lines × 4 pixels/line. Because we will use a 1D transfer to fill the destination buffer, we only need to program XCOUNT and XMODIFY on the destination side. In this case, the value of XCOUNT is set to 20, because that is the number of bytes that will be transferred. The YCOUNT value for the destination side is simply 0, and YMODIFY is also 0. You can see that the count values obey the rule we discussed earlier (e.g., 4 × 5 = 20 bytes).

Now let's talk about the correct values for XMODIFY and YMODIFY for the source buffer. We want to take the first value (0x1) and skip 4 bytes to the next value of 0x1. We will repeat this five times (Source XCOUNT = 5). The value of the source XMODIFY is 4, because that is the number of bytes the controller skips over to get to the next pixel (including the first pixel). XCOUNT decrements by 1 every time a pixel is collected. When the DMA controller reaches the end of the first row, XCOUNT decrements to 0, and YCOUNT decrements by 1. The value of YMODIFY on the source side then needs to bring the address pointer back to the second element in the array (0x2). At the instant this happens, the address pointer is still pointing to the last element in the first row (0x1). Counting back from that point in the array to the second pixel in the first row, we traverse back by 15 elements. Therefore, the source YMODIFY = −15.

If the core carried out this transfer without the aid of a DMA controller, it would consume valuable cycles to read and write each pixel. Additionally, it would have to keep track of the addresses on the source and destination sides, tracking the stride values with each transfer.

Here's a more complex example involving a 2D-to-2D transfer:

Example 5.2

Let's assume now we start with the array that has a border of 0xFF values, shown in Figure 3.7.

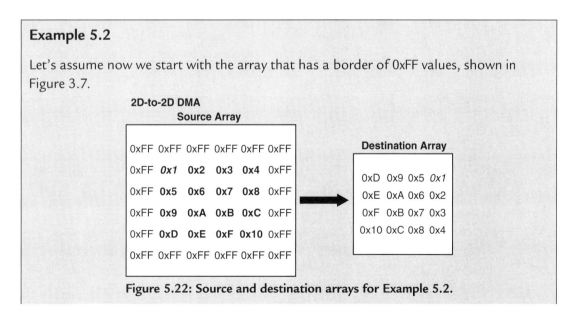

Figure 5.22: Source and destination arrays for Example 5.2.

We want to keep only the inner square of the source matrix (shown in bold), but we also want to rotate the matrix 90 degrees, as shown in Figure 5.22.

The register settings below will produce the transformation shown in this example, and now we will explain why.

Source	Destination
XCOUNT = 4	XCOUNT = 4
XMODIFY = 1	XMODIFY = 4
YCOUNT = 4	YCOUNT = 4
YMODIFY = 3	YMODIFY = −13

As a first step, we need to determine how to access data in the source array. As the DMA controller reads each byte from the source array, the destination builds the output array 1 byte at a time.

How do we get started? Well, let's look at the first byte that we want to move in the input array. It is shown in italics as 0x1. This will help us select the start address of the source buffer. We then want to sequentially read the next three bytes before we skip over the "border" bytes. The transfer size is assumed to be 1 byte for this example.

Because the controller reads 4 bytes in a row before skipping over some bytes to move to the next line in the array, the source XCOUNT is 4. Because the controller increments the address by 1 as it collects 0x2, 0x3, and 0x4, the source XMODIFY = 1. When the controller finishes the first line, the source YCOUNT decrements by 1. Since we are transferring four lines, the source YCOUNT = 4. Finally, the source YMODIFY = 3, because as we discussed earlier, the address pointer does not increment by XMODIFY after XCOUNT goes from 1 to 0. Setting YMODIFY = 3 ensures the next fetch will be 0x5.

On the destination side of the transfer, we will again program the location of the *0x1* byte as the initial destination address. Since the second byte fetched from the source address was 0x2, the controller will need to write this value to the destination address next. As you can in see in the destination array in Figure 5.22, the destination address has to first be incremented by 4, which defines the destination XMODIFY value. Since the destination array is 4 × 4 in size, the values of both the destination XCOUNT and YCOUNT are 4. The only value left is the destination YMODIFY. To calculate this value, we must compute how many bytes the destination address moves back in the array. After the destination YCOUNT decrements for the first time, the destination address is pointing to the value 0x4. The resulting destination YMODIFY value of −13 will ensure that a value of 0x5 is written to the desired location in the destination buffer.

For some applications, it is desirable to split data between both cores. The DMA controller can be configured to spool data to different memory spaces for the most efficient processing.

> ### Example 5.3
>
> Consider when the processor is connected to a dual-channel sensor that multiplexes alternating video samples into a single output stream. In this example, each channel transfers four 8-bit samples packed as a 32-bit word. The samples are arranged such that a "packed" sample from Channel 2 follows a "packed" sample from Channel 1, and so on, as shown in Figure 5.22. Here the peripheral serves as the source of the DMA, and L2 memory serves as the destination. We want to spread the data out in L2 memory to take advantage of its internal bank structures, as this will consequently allow the processor and the DMA controller access to different banks simultaneously.
>
>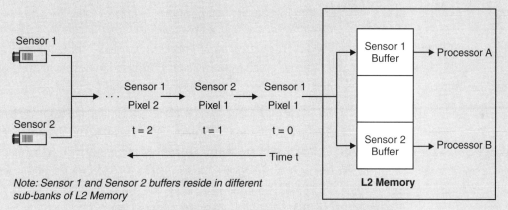
>
> Note: Sensor 1 and Sensor 2 buffers reside in different sub-banks of L2 Memory
>
> **Figure 5.23: Multiplexed stream from two sensors.**
>
> Because a sample is sent from each sensor, we set the destination XCOUNT to 2 (one word each from Sensor 1 and Sensor 2). The value of XMODIFY is set to the separation distance of the sensor buffers, in bytes. The controller will then write the first 4 bytes to the beginning of Sensor 1 buffer, skip XMODIFY bytes, and write the first 4 bytes of Sensor 2 buffer. The value of YCOUNT is based on the number of transfers required for each line. For a QVGA-sized image, that would be 320 pixels per line × 2 bytes per pixel / 4 bytes per transfer, or 160 transfers per line. The value of YMODIFY depends on the separation of the two buffers. In this example, it would be negative (buffer separation + number of line transfers − 1, which already accounts for the fact that the pointer doesn't increment when XCOUNT goes to 0).

Earlier, we mentioned that it's useful in some applications to set XMODIFY to 0. A short example will illustrate this concept.

> **Example 5.4**
>
> Consider the case where we want to zero-fill a large section—say, 1024 bytes—of L3 memory. To do so, we can first create a 32-bit buffer in internal memory that contains all zeros, and then perform core writes to the block of external memory, but then the core would not be available to do more useful tasks.
>
> So why not use a simple 1D DMA instead? In this case, if we assume a 32-bit word transfer size, the `XCOUNT` values for the source and destination are (1024 bytes/4 bytes per transfer), or simply 256 transfers. The `XMODIFY` value for the destination will be 4 bytes. The source value of `XMODIFY` can be set to 0 to ensure that the address of the source pointer stays on the same 32-bit word in the source buffer, meaning that only a single 32-bit "zero word" is needed in L1 memory. This will cause the source side of the DMA to continually fetch the value of 0x0000 from the same L1 location, which is subsequently written to the buffer in external memory.

The previous examples show how the DMA controller can move data around without bothering the core to calculate the source and destination addresses. Everything we have shown so far can be accomplished by programming the DMA controller at system initialization.

The next example will provide some insight into implications of transfer sizes in a DMA operation. The DMA bus structure consists of individual buses that are either 16- or 32-bits wide. When 8-bit data is not packed into 16-bit or 32-bit words (by either the memory or peripheral subsystems), some portion of the bus in question goes unused. Example 5.5 considers the scenario where a video port sends 8-bit YCbCr data straight into L2 memory. (Don't worry if you are not too familiar with the term YCbCr—you will be after reading Chapter 6!).

> **Example 5.5**
>
> Assume we have Field 1 of a 4:2:2 YCbCr video buffer in L2 memory as shown in Figure 5.24a. We would like to separate the data into discrete Y, Cb and Cr buffers in L3 memory where we can fit the entire field of data, since L2 memory can't hold the entire field for large image sizes. The peripheral sends data to L2 memory in the same order in which the camera sends it. Because there is no re-ordering of the data on the first pass into L2 memory, the word transfer size should be maximized (e.g., to 32 bits). This ensures that the best performance is achieved when the data enters the processor.

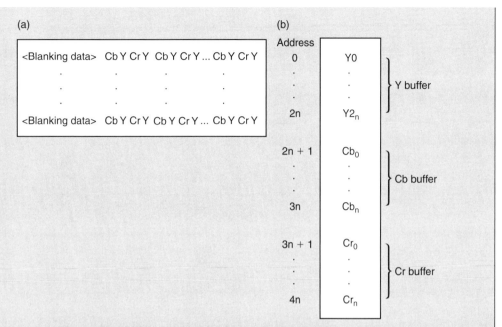

Figure 5.24: Source and destination buffers for Example 5.5.

How should we separate the buffers? One viable option is to set up three 2D-to-1D DMAs for each line—one each for Y, Cb, and Cr pixel components. Because the data that needs to be separated is spread out in the array, 8-bit transfers must be used. Since there are twice as many values of Y as there are of Cr and Cb, the `XCOUNT` for the source and destination would be twice that of the Cb buffer, and twice that of the Cr buffer as well. On the source side, `XCOUNT` would be the number of Y values in each line, and `YCOUNT` would be the number of lines in the source buffer. This is typically some subset of a video field size. The source `XMODIFY` = 2, which is the number of bytes to increment the address to reach the next Y value. For Cb or Cr transfers, the source `XMODIFY` = 4. `YMODIFY` is simply the number of bytes in the horizontal blanking data that precedes each line.

The destination parameters for the Y buffer in L3 memory are much simpler. Since the destination side of the transfer is one-dimensional, only `XCOUNT` and `XMODIFY` are needed. The value of `XCOUNT` on the destination side is equal to the product of the source `XCOUNT` and `YCOUNT` values. The `XMODIFY` value is simply 1.

This example is important because transfers to L3 memory are not efficient when they are made in byte-sized increments. It is much more efficient to move data into external memory at the maximum transfer size (typically 16 or 32 bits). As such, in this case it is better to create new data buffers from one L2 buffer using the technique we just

> described. Once the separate buffers are created in L2 memory as shown in Figure 5.24b, three 1D DMAs can transfer them to L3 memory. As you can see, in this case we have created an extra pass of the data (Peripheral to L2, L2 to L3, versus Peripheral to L2 to L3). On the surface, you may think this is something to avoid, because normally we try to reduce data movement passes.
>
> In reality, however, bandwidth of external memory is often more valuable than that of internal memory. The reason the extra pass is more efficient is that the final transfer to L3 memory can be accomplished using 32-bit transfers, which is far more efficient than using 8-bit transfers. When doing four times as many 8-bit transfers, the number of times the DMA bus has to change directions, as well as the number of actual transfers on the bus, eats into total available bandwidth. You may also recall that the IMDMA controller is available to make the intermediate pass in L2 memory, and thus the transfers can be made at the CCLK rate.

5.5.4 DMA Classifications

There are two main classes of DMA transfer configuration: Register mode and Descriptor mode. Regardless of the class of DMA, the same type of information depicted in Table 5.8 makes its way into the DMA controller. When the DMA runs in Register mode, the DMA controller simply uses the values contained in the DMA channel's registers. In the case of Descriptor mode, the DMA controller looks in memory for its configuration values.

Table 5.8: DMA registers.

Next descriptor pointer (lower 16 bits)	Address of next descriptor
Next descriptor pointer (upper 16 bits)	Address of next descriptor
Start address (lower 16 bits)	Start address (source or destination)
Start address (upper 16 bits)	Start address (source or destination)
DMA configuration	Control information (enable, interrupt selection, 1D vs. 2D)
X_Count	Number of transfers in inner loop
X_Modify	Number of bytes between each transfer in inner loop
Y_Count	Number of transfers in outer loop
Y_Modify	Number of bytes between end of inner loop and start of outer loop

5.5.5 Register-Based DMA

In a register-based DMA, the processor directly programs DMA control registers to initiate a transfer. Register-based DMA provides the best DMA controller performance because registers don't need to keep reloading from descriptors in memory, and the core does not have to maintain descriptors.

Register-based DMA consists of two submodes: *Autobuffer mode* and *Stop mode*. In Autobuffer DMA, when one transfer block completes, the control registers automatically reload to their original setup values and the same DMA process restarts, with zero overhead.

As we see in Figure 5.25, if we set up an Autobuffer DMA to transfer some number of words from a peripheral to a buffer in L1 data memory, the DMA controller would reload the initial parameters immediately upon completion of the 1024th word transfer. This creates a "circular buffer" because after a value is written to the last location in the buffer, the next value will be written to the first location in the buffer.

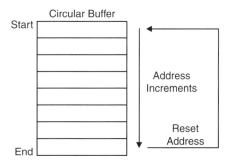

Figure 5.25: Implementing a circular buffer.

Autobuffer DMA especially suits performance-sensitive applications with continuous data streams. The DMA controller can read in the stream independent of other processor activities and then interrupt the core when each transfer completes. While it's possible to stop Autobuffer mode gracefully, if a DMA process needs to be started and stopped regularly, it doesn't make sense to use this mode.

Let's take a look at an Autobuffer example in Example 5.6.

> ### Example 5.6
>
> Consider an application where the processor operates on 512 audio samples at a time, and the codec sends new data at the audio clock rate. Autobuffer DMA is the perfect choice in this scenario, because the data transfer occurs at such periodic intervals.
>
> Drawing on this same model, let's assume we want to "double-buffer" the incoming audio data. That is, we want the DMA controller to fill one buffer while we operate on the other. The processor must finish working on a particular data buffer before the DMA controller wraps around to the beginning of it, as shown in Figure 5.26. Using Autobuffer mode, configuration is simple.

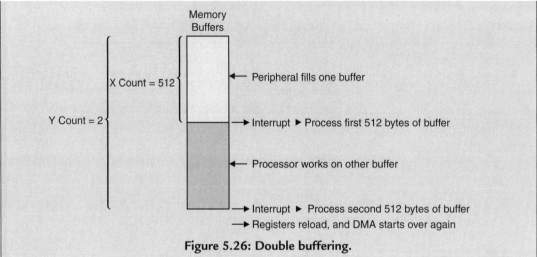

Figure 5.26: Double buffering.

The total count of the Autobuffer DMA must comprise the size of two data buffers via a 2D DMA. In this example, each data buffer size corresponds to the size of the inner loop on a 2D DMA. The number of buffers corresponds to the outer loop. Therefore, we keep XCOUNT = 512. Assuming the audio data element size is 4 bytes, we program the word transfer size to 32 bits and set XMODIFY = 4. Since we want two buffers, we set YCOUNT = 2. If we want the two buffers to be back-to-back in memory, we must set YMODIFY = 1. However, for the reasons we've discussed, in many cases it's smarter to separate the buffers. This way, we avoid conflicts between the processor and the DMA controller in accessing the same sub-banks of memory. To separate the buffers, YMODIFY can be increased to provide the proper separation.

In a 2D DMA transfer, we have the option of generating an interrupt when XCOUNT expires and/or when YCOUNT expires. Translated to this example, we can set the DMA interrupt to trigger every time XCOUNT decrements to 0 (i.e., at the end of each set of 512 transfers). Again, it is easy to think of this in terms of receiving an interrupt at the end of each inner loop.

Stop mode works identically to Autobuffer DMA, except registers don't reload after DMA completes, so the entire DMA transfer takes place only once. Stop mode is most useful for one-time transfers that happen based on some event—for example, moving data blocks from one location to another in a nonperiodic fashion, as is the case for buffer initialization. This mode is also useful when you need to synchronize events. For example, if one task has to complete before the next transfer is initiated, Stop mode can guarantee this sequencing.

5.5.6 Descriptor-Based DMA

DMA transfers that are descriptor-based require a set of parameters stored within memory to initiate a DMA sequence. The descriptor contains all of the same parameters normally programmed into the DMA control register set. However, descriptors also allow the chaining together of multiple DMA sequences. In descriptor-based DMA operations, we can program a DMA channel to automatically set up and start another DMA transfer after the current sequence completes. The descriptor-based model provides the most flexibility in managing a system's DMA transfers.

Blackfin processors offer two main descriptor models—a *Descriptor Array* scheme and a *Descriptor List* method. The goal of these two models is to allow a tradeoff between flexibility and performance. Let's take a look at how this is done.

In the Descriptor Array mode, descriptors reside in consecutive memory locations. The DMA controller still fetches descriptors from memory, but because the next descriptor immediately follows the current descriptor, the two words that describe where to look for the next descriptor (and their corresponding descriptor fetches) aren't necessary. Because the descriptor does not contain this Next Descriptor Pointer entry, the DMA controller expects a group of descriptors to follow one another in memory like an array.

A Descriptor List is used when the individual descriptors are not located "back-to-back" in memory. There are actually multiple sub-modes here, again to allow a tradeoff between performance and flexibility. In a "small descriptor" model, descriptors include a single 16-bit field that specifies the lower portion of the Next Descriptor Pointer field; the upper portion is programmed separately via a register and doesn't change. This, of course, confines descriptors to a specific 64 K ($=2^{16}$) page in memory. When the descriptors need to be located across this boundary, a "large" model is available that provides 32 bits for the Next Descriptor Pointer entry.

Regardless of the descriptor mode, using more descriptor values requires more descriptor fetches. This is why Blackfin processors specify a "flex descriptor model" that tailors the descriptor length to include only what's needed for a particular transfer, as shown in Figure 5.27. For example, if 2D DMA is not needed, the `YMODIFY` and `YCOUNT` registers do no need to be part of the descriptor block.

5.5.6.1 Descriptor Management

So what's the best way to manage a descriptor list? Well, the answer is application-dependent, but it is important to understand what alternatives exist.

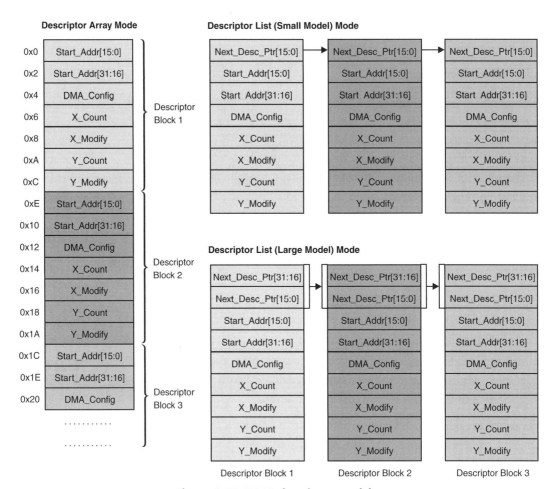

Figure 5.27: DMA descriptor models.

The first option we will describe behaves very much like an Autobuffer DMA. It involves setting up multiple descriptors that are chained together as shown in Figure 5.28a. The term "chained" implies that one descriptor points to the next descriptor, which is loaded automatically once the data transfer specified by the first descriptor block completes. To complete the chain, the last descriptor points back to the first descriptor, and the process repeats. One reason to use this technique rather than the Autobuffer mode is that descriptors allow more flexibility in the size and direction of the transfers. In our YCbCr example (Example 5.5), the Y buffer is twice as large as the other buffers. This can be easily described via descriptors and would be much harder to implement with an Autobuffer scheme.

The second option involves the processor manually managing the descriptor list. Recall that a descriptor is really a structure in memory. Each descriptor contains a configuration word, and each configuration word contains an "Enable" bit which can regulate when a transfer starts. Let's

(a) **Linked List of Descriptors**

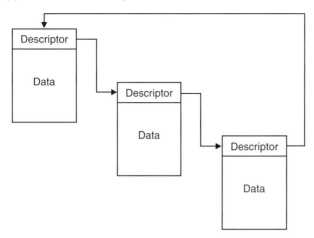

(b) **"Throttled" Descriptor Management**

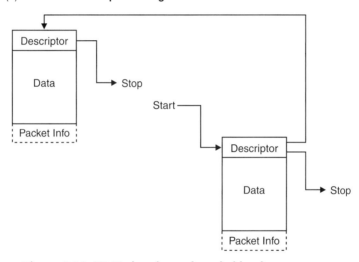

Figure 5.28: DMA descriptor throttled by the processor.

assume we have four buffers that have to move data over some given task interval. If we need to have the processor start each transfer specifically when the processor is ready, we can set up all of the descriptors in advance, but with the "Enable" bits cleared. When the processor determines the time is right to start a descriptor, it simply updates the descriptor in memory and then writes to a DMA register to start the stalled DMA channel. Figure 5.28b shows an example of this flow.

When is this type of transfer useful? EMP applications often require us to synchronize an input stream to an output stream. For example, we may receive video samples into memory at a rate that is different from the rate at which we display output video. This will happen in real systems even when you attempt to make the streams run at exactly the same clock rate.

In cases where synchronization is an issue, the processor can manually regulate the DMA descriptors corresponding to the output buffer. Before the next descriptor is enabled, the processor can synchronize the stream by adjusting the current output descriptor via a semaphore mechanism. For now, you can simply consider semaphores tools that guarantee only one entity at a time accesses a shared resource.

When using internal DMA descriptor chains or DMA-based streams between processors, it can also be useful to add an extra word at the end of the transferred data block that helps identify the packet being sent, including information on how to handle the data and, possibly, a time stamp. The dashed area of Figure 5.28b shows an example of this scheme.

Most sophisticated applications have a "DMA Manager" function implemented in software. This may be provided as part of an operating system or real-time kernel, but it can also run without either of these. In both cases, an application submits DMA descriptor requests to the DMA Queue Manager, whose responsibility it is to handle each request. Usually, an address pointer to a "callback" function is part of the system as well. This function carries out the work you want the processor to perform when a data buffer is ready, without needlessly making the core linger in a high-priority interrupt service routine.

There are two general methods for managing a descriptor queue using interrupts. The first is based on interrupting upon the completion of every descriptor. Use this method only if you can guarantee that each interrupt event will be serviced separately, with no interrupt overrun. The second involves interrupting only on completion of the work transfer specified by the last descriptor of a work block. A work block is a collection of one or more descriptors.

To maintain synchronization of the descriptor queue, you need to maintain in software a count of descriptors added to the queue, while the interrupt handler maintains a count of completed descriptors removed from the queue. The counts are then equal only when the DMA channel pauses after having processed all the descriptors.

5.5.7 Advanced DMA Features

5.5.7.1 System Performance Tuning

To effectively use DMA in a multimedia system, there must be enough DMA channels to support the processor's peripheral set fully, with more than one pair of Memory DMA streams. This is an important point, because there are bound to be raw media streams incoming to external memory (via high-speed peripherals), while at the same time data blocks will be moving back and forth between external memory and L1 memory for core processing. What's more, DMA engines that allow direct data transfer between peripherals and external memory, rather than requiring a stopover in L1 memory, can save extra data passes in numerically intensive algorithms.

As data rates and performance demands increase, it becomes critical to have "system performance tuning" controls at your disposal. For example, the DMA controller might be optimized to transfer a data word on every clock cycle. When there are multiple transfers ongoing in the same direction (e.g., all from internal memory to external memory), this is usually the most efficient way to operate the controller because it prevents idle time on the DMA bus.

But in cases involving multiple bidirectional video and audio streams, "direction control" becomes obligatory in order to prevent one stream from usurping the bus entirely. For instance, if the DMA controller always granted the DMA bus to any peripheral that was ready to transfer a data word, overall throughput would degrade when using SDRAM. In situations where data transfers switch direction on nearly every cycle, the latency associated with turn-around time on the SDRAM bus will lower throughput significantly. As a result, DMA controllers that have a channel-programmable burst size hold a clear advantage over those with a fixed transfer size. Because each DMA channel can connect a peripheral to either internal or external memory, it is also important to be able to automatically service a peripheral that may issue an urgent request for the bus.

Other important DMA features include the ability to prioritize DMA channels to meet current peripheral task requirements, as well as the capacity to configure the corresponding DMA interrupts to match these priority levels. These functions help insure that data buffers do not overflow due to DMA activity on other peripherals, and they provide the programmer with extra degrees of freedom in optimizing the entire system based on the data traffic on each DMA channel.

5.5.7.2 External DMA

Let's close out this chapter by spending a few minutes discussing how to DMA data between the processor and a memory-mapped external device. When a device is memory-mapped to an asynchronous memory bank, a MemDMA channel can move data into and out of the external chip via the DMA FIFOs we described earlier. If the destination for this data is another external memory bank in SDRAM, for example, the bus turns around when a few samples have entered the DMA FIFO, and these samples are then written back out over the same external bus, to another memory bank. This process repeats for the duration of the transfer period.

Normally, these Memory DMA transfers are performed at maximum speed. Once a MemDMA starts, data transfers continuously until the data count expires or the DMA channel is halted. This works well when the transfer is being made as a memory-to-memory transfer, but if one of the ends of the transfer is a memory-mapped device, this can cause the processor to service the transactions constantly, or impede the memory-mapped device from making transfers effectively.

When the data source and/or destination is external to the processor, a separate "Handshake DMA" mode can help throttle the MemDMA transfer, as well as improve performance by removing the processor from having to be involved in every transfer. In this mode, the

Memory DMA does not transfer data automatically when it is enabled. Rather, it waits for an external trigger from another device. Once a trigger event is detected, a user-specified portion of data is transferred, and then the Mem-DMA channel halts and waits for the next trigger.

The handshake mode can be used to control the timing of memory-to-memory transfers. In addition, it enables the Memory DMA to operate efficiently with asynchronous FIFO-style devices connected to the external memory bus. In the Blackfin processor, the external interface acknowledges a Handshake DMA request by performing a programmable number of read or write operations. It is up to the device connected to the designated external pins to de-assert or assert the "DMA request" signal.

The Handshake DMA configuration registers control how many data transfers are performed upon every DMA request. When set to 1, the peripheral times every individual data transfer. If greater than 1, the external peripheral must possess sufficient buffer size to provide or consume the number of words programmed. Once the handshake transfer commences, no flow control can hold off the DMA from transferring the entire data block.

In the next chapter, we will discuss "speculative fetches." These are fetches that are started but not finished. Normally, speculative fetches can cause problems for external FIFOs, because the FIFO can't tell the difference between an aborted access and a real access, and it increments its read/write pointers in either case. Handshake DMA, however, eliminates this issue, because all DMA accesses that start always finish.

Endnotes

Application Note 23710, Rev A: "Understanding Burst Mode Flash Memory Devices," *Spansion*, March 23, 2000.

"DDR2 SDRAM," www.elpida.com/en/ddr2/advantage.html.

"DDR2—Why Consider It?" Micron Corporation, 2005, http://download.micron.com/pdf/flyers/ddr_to_ddr2.pdf.

Dipert, Brian, "Pick a Card: Card Formats," *EDN*, July 8, 2004.

Heath, Steve, *Embedded Systems Design*, Elsevier (Newnes), second edition, 2003.

Hennessy, J., and Patterson, D., *Computer Architecture: A Quantitative Approach*, Morgan Kaufmann, third edition, 2002.

"Innovative Mobile Products Require Early Access to Optimized Solutions," Position Paper by Samsung Semiconductor, Inc.

Inoue, Atsushi, and Wong, Doug, "NAND Flash Applications Design Guide," Toshiba America Electronic Components, Inc., Revision 1.0, April 2003.

MacNamee, C., and Rinne, K., "Semiconductor Memories," www.ul.ie/~rinne/et4508/ET4508_L9.pdf.

"Mobile-RAM" Application Note, V1.1, Feb. 2002, Infineon Technologies.

Pietikainen, V., and Vuori, J., Memories ITKC11 "Mobiilit sovellusalustat (Mobile Application Platforms)," http://tisu.mit.jyu.fi/embedded/itkc11/itkc11.htm.

Tal, Arie, "Two Technologies Compared: NOR vs. NAND White Paper," *M-Systems*, Revision 1.1, July 2003.

Technical Note: TN-46-05. "General DDR SDRAM Functionality," Micron Corporation.

Wong, Doug, "NAND Flash Performance White Paper," Toshiba America Electronic Components, Inc., Revision 1.0, March 2004.

CHAPTER 6
Timing Analysis in Embedded Systems

Ken Arnold

6.1 Introduction

Just as in comedy, timing is essential to the success of a microcomputer design. Often it is quite possible to get one system functioning by simply interconnecting the various components. But it is *significantly* more difficult to be able to guarantee that many systems will work under the entire range of possible conditions that they may be exposed to. There are many designs in production right now that have a number of unidentified failures due to the lack of a worst-case analysis of the design. When timing or loading problems show up in a design, they usually appear as intermittent failures or as sensitivity to power supply fluctuations, temperature changes, and so on.

A *worst-case design* takes into account all available information regarding the components to be used with respect to variations in performance. Even when all parameters are at their most adverse values, the worst-case design can still be proved to meet the specifications. These variants may be due to changing manufacturing conditions, temperature, voltage, and other variables. Without performing a detailed analysis, there is no way of knowing if the design will work reliably under all operating conditions. It is much better to design reliability and simplicity of manufacturing into a product using worst-case design rules than to attempt to correct a problem after the design has been implemented. With the emphasis that must be given to the quality of the final product, a designer is obligated to perform a detailed examination of the timing in a system. As is the case in most quality improvements, these efforts result in direct cost and saving time. *This is clearly one of the places where the designer can have the greatest impact on overall product quality.*

6.2 Timing Diagram Notation Conventions

Timing notation is illustrated in Figure 6.1. The timing notation used in manufacturers' data sheets may vary from this notation but is usually very similar. It is also important to notice that although the diagrams are reasonably standard, there is a wide variation in the selection of symbols for each timing parameter.

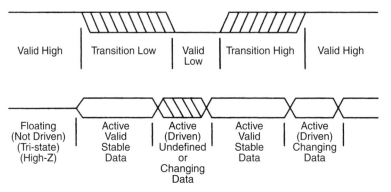

Figure 6.1: Timing diagram notation as used in this book.

The purpose of timing analysis is to determine the sequence of events in each of the bus cycles so that we can delimit, among other things, the time available for each of the components to respond to changes. This time is compared to the requirements as specified in the manufacturers' data sheets to determine whether they are compatible and by what margin.

The most important timing specifications for interfacing components to a bus-oriented design are:

- Rise/fall time
- Propagation delay time
- Setup time
- Hold time
- Tri-state enable and disable delays
- Pulse width
- Clock frequency

There are two general classes of logic: combinatorial and sequential. *Combinatorial logic* has no memory and its output is some logical function of its current inputs, after some delay. Examples of combinatorial logic include gates, buffers, inverters, multiplexers, and decoders. *Sequential logic* has memory, which means that its outputs are a function of both current and past inputs. Examples of sequential logic are flip-flops, registers, microprocessors, and counters. There are two types of sequential logic. *Synchronous logic* is synchronized to change only when there is a clock transition. In contrast, *asynchronous logic* does not use a clock signal. Almost all the logic used in a microcomputer design will either be unclocked asynchronous logic (gates, decoders) or clocked synchronous logic (counter, latch or microprocessor). Some types of devices are available in either form. Each of the timing specifications in the following discussion is described using simple logic devices as they are typically used in embedded computer designs.

6.2.1 Rise and Fall Times

The *rise time* of a signal is usually defined as the time required for a logic signal voltage to change from 20% to 80% of its final value. The *fall time* is from 80% to 20%, as shown in Figure 6.2. These times are also commonly defined by some manufacturers as the transitions between the 10% and 90% levels.

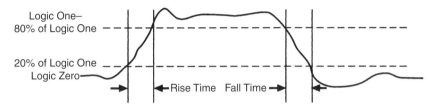

Figure 6.2: Rise and fall times of a signal.

6.2.2 Propagation Delays

The *propagation delay* is the time it takes for a change at the input of a device to cause a change at the output. All devices—even wires—exhibit some propagation delay. Some devices do not have symmetrical delays for positive and negative transitions. In Figure 6.3, the propagation times for a high to low transition are shorter than for a low to high transition. This *asymmetrical delay* is common for TTL and open collector and open drain outputs because they are better at sinking current than sourcing it. Thus, the load capacitance is charged more slowly when the current is being supplied from the weaker "high side" or pull-up device. Propagation delays are usually measured from the 50% amplitude points, as shown in Figure 6.3.

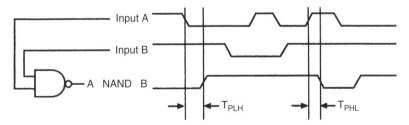

Figure 6.3: Propagation delay.

6.2.3 Setup and Hold Time

In Figure 6.4, a standard D type flip-flop (e.g., a 74xx74 device) is shown along with a sample timing diagram that illustrates the operation and key timing parameters of a flip-flop. This type of flip-flop samples the D input whenever the clock (CK) line goes high, and after a delay, the

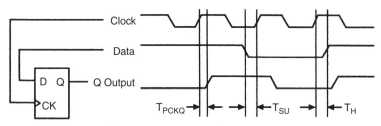

Figure 6.4: Setup and hold time.

output remains in the same state until the next rising edge on the clock line. The triangle on the clock input indicates that it is a rising edge sensitive input, meaning that it will only have an effect when there is a rising edge on the clock pin. A falling edge sensitive input would have a bubble outside the block where the clock enters the flip-flop. In order to be able to guarantee that the flip-flop will operate correctly, the D input must be stable during the setup and hold time.

Figure 6.4 also shows the propagation delay from clock to Q out (T_{PCKQ}), the setup time (T_{SU}), and the hold time (T_H). *Setup time* is the amount of time a sampled input signal must be valid and stable prior to a clock signal transition. *Hold time* is the amount of time that a sampled signal must be held valid and stable after a clock signal transition occurs. If these conditions are not met, the Q output may become invalid or even oscillate. This condition is referred to as *metastability*. The times of these and most other signals are frequently measured with respect to the 50% amplitude points of the clock signal rather than the valid logic one and zero levels. An analogy for the flip-flop as a sampling device is that of an instant camera: The clock is the shutter, the D input is the lens, and the output is the film image. The input is sampled when the shutter is open, and if the subject moves with the shutter open, the picture will be blurred. For the flip-flop, the "shutter open" time, referred to as the *window of uncertainty*, is shown in Figure 6.5, along with some possible results.

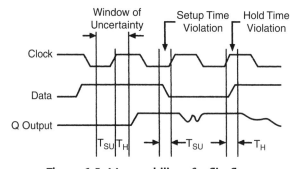

Figure 6.5: Metastability of a flip-flop.

Metastability of a storage device such as a flip-flop or register is caused by the change of an input signal too close to the edge of the clock signal. In other words, if the setup or hold time requirements are not met, the output of the device is unpredictable and may even be unstable. The output may operate normally, take an invalid level, or oscillate (which could also explain why indecisive people take bad photos!).

6.2.4 Tri-State Bus Interfacing

When multiple devices are capable of driving the same line, the possibility exists that two or more of them will try to drive it in opposite directions at the same time. When tri-state devices fight like this it is called *bus contention.* Figure 6.6 illustrates this condition. Although the data is unpredictable during this period, there are far worse things that can happen as a result of this condition. Since most tri-state devices have the ability to drive many loads, they are also capable of sourcing and sinking large currents. When two of these devices are in contention, very large currents with peaks in the tens or hundreds of amperes can flow for time periods on the order of nanoseconds.

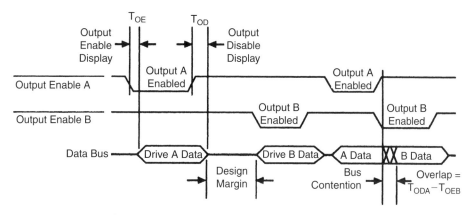

Figure 6.6: Tri-state bus timing and contention.

The large current spikes that occur during contention may stress the devices and significantly reduce their reliability. A far more frequent problem, however, is the temporary drop or glitch in the local power supply wires that can cause any other nearby devices to change state. As you can imagine, this can create havoc in sequential logic, particularly for micros. Based on past experience with Murphy's Law, these glitches generally seem to change the current instruction to "jump immediate to format hard disk routine," thereby erasing all your data. In a properly designed system, there is a "dead time" when no device is driving the bus to act as a safety margin between the times that two devices are enabled to drive their outputs. The problems arise when the output enable time of a device which is just turning on is less than the output disable time of a device which is turning off.

6.2.5 Pulse Width and Clock Frequency

The *width* of a positive going pulse is the period beginning from its *positive transition* (rising edge or leading edge) to its *negative transition* (falling or trailing edge). Figure 6.7 illustrates these concepts. Pulse widths are important in defining the operation of control signals such as the memory read or write signals and clocks. Clock signals used for modern microprocessors usually, but do not always, have equal high and low pulse width requirements. The period (T) of a signal is the sum of the rise time, high time, fall time, and low time. The frequency of a processor clock ($f = 1/T$) may have a lower limit as well as an upper limit. The standard NMOS 8051 family of parts has a lower frequency limit of 1.2 MHz. That means that the processor cannot be operated at a lower frequency. The reason is that the processor's internal design requires a constant clock to correctly maintain its state. Other processors (such as the 80C51 series CMOS devices) can tolerate having their clock stopped completely, since they have been designed to maintain their internal states indefinitely, as long as power is applied.

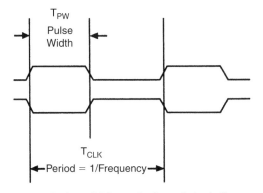

Figure 6.7: Pulse width, period, and clock frequency.

6.3 Fan-Out and Loading Analysis: DC and AC

Another important part of worst-case design is a realistic model of the signal loading for each of the circuit's outputs. If insufficient drive is available, buffer circuits must be added or the number of loads must be reduced to guarantee correct operation. *Fan-out* is the number of equivalent inputs that can be safely driven by one output. A fan-out of 10 indicates that one device output can drive 10 inputs. The fan-out is determined from:

- The source, type, and number of loads
- DC characteristics sources and load
- AC characteristics of the loads vs. the source test conditions

DC characteristics of the output and inputs consist of:

- The maximum current that can be produced by an output
- Maximum currents required to drive an input

The maximum output currents are specified as:

- I_{OLmin}. Minimum output low (sink) current for a valid zero output voltage.
- I_{OHmin}. Minimum output high (source) current for a valid one output voltage.

Note that a low output is sinking currents that are coming out of the inputs that are being driven. Likewise, a high output is sourcing current that goes into the inputs that are being driven.

Maximum currents required to drive an input are specified as:

- I_{ILmax}. Maximum input low current for a valid zero input voltage.
- I_{IHmax}. Maximum input high current for a valid one input voltage.

Another important convention has to do with the sign of the current flowing in or out of a device pin. In most cases, current flowing into a device pin is given a *positive* sign (as shown in Figure 6.8), whereas current flowing out of a pin is given a *negative* sign (as shown in Figure 6.9). In both Figures 6.8 and 6.9, the device on the left is the driving device, which tries to force its output to the desired logic state. In the logic one state, the output sources current (-50 microampere), and the receiving device absorbs that current ($+50$ microampere). In our example, the available output current is exactly equal to the input current used by the load, resulting in a DC fan-out of 1.

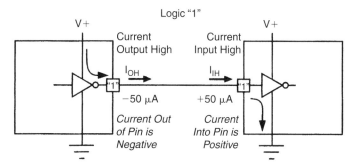

Figure 6.8: Current sign for logic high.

Unfortunately, *this convention is not always followed consistently*, so it is up to you to recognize the current direction from the context of the situation in which it appears. Generally, the current direction can be determined by keeping these images in mind, especially since many data sheets do not specify the sign for the input and output currents.

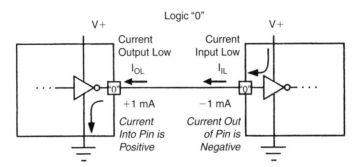

Figure 6.9: Current sign for logic low.

The other type of fan-out limitation is the ability of an output to drive the capacitance of the loads and stray wiring capacitance, also known as *AC fan-out*. The AC fan-out is determined by the specified test load for the driving chip and the load presented by the actual load capacitance. The capacitive load is the parallel combination of all the input capacitances of the gate inputs attached to the signal, plus the wiring capacitance. Since the capacitors in parallel are equivalent to a single capacitor equal to the sum of the individual capacitances, we simply add up all the load capacitor values and compare this to the output's specified test load. The driving device's specified load capacitance, C_L, is the test load capacitance used by the manufacturer for specifying the AC or timing characteristics of the device. Most often, this specification is listed in the test conditions or notes for the timing specifications of the chip. As long as the sum of the load capacitances, including the stray wiring capacitance, is less than the specified test load for the driving device, all the timing specifications will be valid as specified in the timing section of the data sheet. If the driving device is overloaded (actual C_L is greater than specified C_L), then the timing specifications of the device need to be de-rated (slowed down), since additional capacitance will increase the rise and fall times of the signal line in question. Methods for estimating the amount that an overloaded output can withstand are described later.

AC characteristics of the outputs and the inputs consist of:

- C_L. The load capacitance that an output is specified to drive is listed in the timing specifications for the driving device under the name "test conditions," which is usually in the notes at the bottom of the specification sheet.

- C_{in}. Maximum input capacitance of a driven input load.

- C_{stray}. Wiring and stray capacitance can be approximated to be in the range of 1 to 2 picofarads per inch of wiring on a typical PC board.

As long as the inequality below is satisfied, the signal will meet the timing specifications for the driving device. If the actual load is greater, it will delay:

$$\text{Driving device spec } C_L > \text{actual Cload} = C_{in1} + C_{in2} + \cdots + C_{wiring}$$

The AC fan-out is limited by the parallel combination of the logic inputs' capacitance, C_{in}, and the stray or wiring capacitance. Capacitors in parallel are additive, so the load presented to an output is the sum of the input capacitances of the logic inputs plus the wiring capacitance. Logic input capacitance is often difficult to find, since it might not be listed in the component data sheet but rather in another section of the data book describing the characteristics common to all members of a given logic family. Typical logic input capacitance ranges from 1 to 5 pF (picofarads or 10^{-12} F) but may be outside this range. The maximum load capacitance that a device is specified to drive (C_L) is usually defined in the test conditions for the timing specifications of an integrated circuit, because it is the timing which is most affected by capacitance. Load capacitance is usually specified in the range of 50 to 150 pF. Wiring capacitance is often in the range of 1 to 2 pF per inch of wire for a nominal printed circuit trace. Actual values can vary quite a bit, depending on the physical dimensions of the trace, proximity to surrounding signals, and distance from a ground plane, as well as the dielectric constant of the circuit board material.

6.3.1 Calculating Wiring Capacitance

The standard formula for determining capacitance is:

$$C = (\varepsilon * A)/d$$

where A is the area of two closely spaced parallel plates, d is the distance between the plates, and ε represents the permittivity of the material. (*Permittivity* is the measure of how easily a material can carry electric lines of force.)

For the purposes of this section, we can define the area, A, as the trace length multiplied by the trace width. Wiring capacitance is determined as a capacitance per unit length for a given trace width and distance from the ground or power plane.

Let's examine a typical situation. For an eight-layer PC board with 8 mil traces and innermost layer ground/power planes, what is the capacitance per inch of trace on each of the signal layers?

Here are the terms we'll use in the equations to solve this problem and their values:

- Trace width (w) = 8 mils (one mil equals 10^{-3} inch)
- Trace length (l) = 1000 mils
- Area (A) = w times l
- Total board thickness (T) = 0.062 inch
- Number of layers (N) = 8

- Number of layers separating power and ground plane $(n) = 1$
- Fringe effect and inter-trace stray capacitance adjustment factor $(f) = 1.7$
- Permittivity of air $(e) = 8.859 * 10^{-12} * (coul^2/(newton*m^2))$
- Relative permittivity of glass-epoxy dielectric (er) used in this example $= 6$

We start by determining the thickness of each dielectric layer, represented by t:

$$t = T/(N - 1) = 8.857 \text{ mils}$$

Next we need to determine the distance between the trace and ground/power plane, represented by d. This is found by the formula $d = nt$, which in this case makes for a simple calculation!

The capacitance as a function of the number of layers distance (Cd) is found by the formula:

$$Cd = (\varepsilon * \varepsilon r * A * f)/d$$

Using this formula,

$$C(1 * d) = 2.073 \text{ pF (layer closest to ground/power plane)}$$

$$C(2 * d) = 1.037 \text{ pF (layer next closest to ground/power plane)}$$

$$C(3 * d) = 0.691 \text{ pF (layer farthest from ground/power plane)}$$

To find the average capacitance per inch (Cavg), then:

$$Cavg = (C(1 * d) + C(2 * d) + C(3 * d))/3 = 1.267 \text{ pF}$$

From this example, it is apparent that the stray wiring capacitance can vary significantly depending on which layer of a multilayer PC board a particular trace is located. Since a signal may travel on different layers between source and destination, exact values might be difficult to determine.

When performing a worst-case analysis of a given design, it is most effective to calculate the total load capacitance based on the sum of the loads' input capacitances, plus an estimate of the nominal wiring capacitance using 1 or 2 picofarads per inch of wiring using a rough guess for the length of the trace.

In a typical design, we might pick the diagonal distance from one corner of the board to the other and multiply by 1 or 2 picofarads. If the total load capacitance is less than the driving device's specified test load capacitance, the device will perform as specified. If not or if it's very close, we might want to make a more accurate estimate or avoid the problem by using a driving device that has a larger specified test load capacitance. Other alternatives include using two outputs *from the same chip* in parallel to double the drive capacity or splitting the loads into two separate groups and driving them independently from two different sources.

As digital IC technology has improved, allowing signals to be processed at ever-increasing rates, the other non-ideal effects of the devices that could be ignored at lower speeds become more important. At very high speeds, these secondary effects become much more important. A wire ceases to be equivalent to a 0 ohm connection with zero time delay. For the newer high-speed logic devices, the speed of the signal traveling down the wire, distributed resistance, and inductance, as well as capacitance, may become very important. When the time it takes a signal to propagate down a wire is of the same order as the rise and fall time of the signal, it behaves as a *transmission line* rather than an ideal wire. Transmission-line effects are briefly described later in this chapter.

6.3.2 Fan-Out When CMOS Drives LSTTL

A common design problem involves the determination of the number of LSTTL loads a CMOS output can drive. In this section, we will use the parameters shown in Tables 6.1–6.4 to create an example to determine the number of LSTTL loads a CMOS gate can drive.

Table 6.1: LSTTL gate DC parameters.

Symbol	Parameter	Min	Typ	Max	Units	Conditions
V_{IL}	Input low voltage	−0.3		0.8	V	
V_{IH}	Input high voltage	2.4		Vcc+0.3	V	
I_{IL}	Input low current		−120	−360	µA	
I_{IH}	Input high current		30	50	µA	
C_{IN}	Input capacitance			10	pF	

Table 6.2: Absolute maximum operating conditions.

Symbol	Parameter	Min	Typ	Max	Units	Conditions
V_{OL}	Output low voltage		0.2	0.4	V	@ I_{OL} max
V_{OH}	Output high voltage	2.8	3.5		V	@ I_{OH} max
I_{OL}	Output low current	3.2	8		mA	@ V_{OL} max
I_{OH}	Output high current	−600	−1000		µA	@ V_{OH} min
Note: Test conditions $R_L = 1K$, $C_L = 100\ pF$.						

Table 6.3: CMOS gate DC parameters.

Symbol	Parameter	Min	Typ	Max	Units	Conditions
V_{IL}	Input low voltage			2.0	V	
V_{IH}	Input high voltage	3.0			V	
I_I	Input leakage current			~0	μA	
C_{IN}	Input capacitance			25	pF	

Table 6.4: Absolute maximum operating conditions.

Symbol	Parameter	Min	Typ	Max	Units	Conditions
V_{OL}	Output low voltage			0.4	V	@ I_{OL} max
V_{OH}	Output high voltage	4.5			V	@ I_{OH} max
I_{OL}	Output low current	3.6			mA	@ V_{OL} max
I_{OH}	Output high current	600			μA	@ V_{OH} min

Note: Test conditions $R_L = 5K$, $C_L = 150\ pF$.

For Logic one:

CMOS $I_{OH} = 600$ microamperes (μA)

LSTTL $I_{IH} = 50\ \mu A$ so $600\mu A/50\mu A = 12$ loads

For Logic zero:

CMOS $I_{OL} = 3.6$ milliamperes (mA)

LSTTL $I_{IL} = 360\ \mu A$ so $3.6\ mA/360\mu A = 10$ loads

Thus, considering the DC specifications only, the maximum number of loads driven is 10, since the zero state is the worst-case condition. The AC parameters would not be the limiting factor in this case because the CMOS output is specified with a C_L of 150 pF, and each LS input is only 10 pF. Thus, 10 loads would present 100 pF plus stray wiring capacitance of less than 50 pF would present an AC load less than the 150 pF CMOS output load-handling capability.

How many additional CMOS loads could be added? There are two levels of answer for this problem. First, from a DC point of view all the CMOS I_{OL} output sink current is used up, so from this point of view, no loads could be added. However, there is negligible current in a CMOS input, so it is not the practical limit. In fact, the errors in the DC computations above are in excess of the amount required to drive a CMOS input, so in reality the DC current is not a problem. The real limitation is the capacitive loading. Even if you assume that the loading from the TTL inputs and wiring can be ignored, the CMOS input capacitance will limit the loading. For the output to conform to the specs, the test load was specified as 150 pF (C_L). With 10 LSTTL loads of 10 pF each, the C_L on the CMOS gate output would be $10 * 10 = 100$ pF. Since the CMOS gate timing is specified at $C_L = 150$ pF, there is only $150-100 = 50$ pF left over to drive the additional CMOS loads. Since the CMOS C_{in} is 25 pF, the number of additional gates that can be driven is:

$$50 \text{ pF}/25 \text{ pF} = (\text{remaining } C_L)/(C_{in} \text{ of additional CMOS inputs}) = 2$$

Practically speaking, the wiring capacitance on a PC board will generally be in the 2–3 pF per inch range, so allowing 25 pF for wiring capacitance would permit one CMOS load in addition to the 10 LSTTL loads from above.

What if the CMOS output were to drive only CMOS loads? The input capacitance of the CMOS gate is 25 pF, so even if *all* loads were CMOS, it can only drive $C_L/C_{in} = 150 \text{ pF}/25 \text{ pF} = 6$ CMOS loads and still meet its test condition limits. Since we must also allow for the wiring capacitance, we should limit this device to five loads, leaving 25 pF for the wiring capacitance. The additional load capacitance from more than five devices would likely result in timing performance that would be poorer than that specified in the data sheet. Excessive capacitance can also make ground bounce worse, which is the change in on-chip ground voltage due to rapid current spikes caused by charging load capacitance, developing a voltage across the lead inductance of the driving IC.

6.3.3 Transmission-Line Effects

When you're using high-speed logic and the rise and fall times are of the same order as the propagation of the signal, transmission-line effects become significant. When a signal transition propagates down a wire, it will be reflected back if the signal is not absorbed at the destination end. At lower speeds the effect can be ignored, but with the fastest processors now in use, most designers will need to consider whether the effects will have a negative impact on their designs and take appropriate action if necessary.

Several characteristics of digital transmission lines must be addressed, including the following:

- Signal transition time vs. clock rate
- Mutual inductance and capacitance (crosstalk)
- Physical layout effects
- Impedance estimates
- Strip line vs. micro strip
- Effects of unmatched impedances
- Termination and other alternatives
- Series termination vs. parallel termination
- DC vs. AC termination techniques

The techniques for high-speed design are beyond the scope of this text but are covered in detail in an excellent text on the subject, *High-Speed Digital Design*: *A Handbook of Black Magic*, by Howard W. Johnson and Martin Graham. In contrast with the subtitle, this subject is easily understood by applying some very basic physics.

A transmission line is a conductor long enough that the signal at the far end of the line is significantly different from the near end, due to the time it takes the signal to propagate from one end to the other.

In this book, we will assume that the interconnections between the devices are *not* long enough to require transmission-line analysis. To verify that this is the case we can use a simple estimate. The rough estimate we will make is based on the idea that a wire does not have to be analyzed as a transmission line if the signal takes longer to rise or fall than it takes to get from one end of the wire to another. In other words, if the signal doesn't have to travel too far, both ends of the wire are at approximately the same voltage. To come up with a numerical value to determine whether a signal must be treated as a transmission line, we can use a simple calculation:

$$I = T_r/D$$

where:

I = Length of rising or falling edge in inches (in)
T_r = Rise time in picoseconds (pS)
D = Delay in picoseconds per inch (pS/in)

For traces on a standard printed circuit board, the value for D will be in the range of 100 to 200 pS/in. Depending on how much distortion you're willing to live with, the critical trace length will be between one-sixth and one-quarter of the length of a trace corresponding to the signal's transition. For a trace that is shorter than one-sixth the length of the signal's rising or falling edge, the circuit seldom needs to be considered to be a transmission line. Traces that are much longer than one-quarter the length of the fastest edge will start to behave as transmission lines, exhibiting reflections of the signal when the transition gets to the far end of the trace and is reflected back to the near end. Once the trace is about half of the length it takes for a logic transition to propagate, the problems become quite pronounced.

Let's look at an example. A logic device on a standard glass-epoxy printed circuit board has a 2 nS rise time. This signal has a rising edge that is:

$$(2 \text{ nS})/(150 \text{ pS/in}) = \sim 13 \text{ inches long}$$

That means a trace that is one-sixth that length, or about 2 inches or less, does not have to be considered as a transmission line. If the trace is much longer than two inches, it will begin to show significant distortions on the rising and falling edges due to the fact that there is a different signal voltage at each end of the trace at the same instant, resulting in reflections of the signal from the ends of the trace.

This is one of the most important reasons for using logic that is fast enough and not too much faster than required to meet the timing requirements. Although it might seem tempting to buy the fastest device available to reduce the delays in a device which does not meet the timing requirements, doing so can result in many more difficult problems to solve.

6.3.4 Ground Bounce

Another effect of high-speed signal transitions is called *ground bounce*. Ground bounce occurs when a large peak current flows through the ground pin of a chip when one or more logic outputs change state and discharge their load capacitances through the chip's ground pin. The parasitic inductance of the ground pin might not seem very significant, but in the nanohenry (10^{-9} H) range, fast transients can cause large voltages to appear across the ground pin. This occurs most often when multiple bus signal outputs from one chip change state at the same time. The rapid, parallel current pulses which result from charging or discharging stray bus capacitance must be carried through the ground or power pins, which have inductance.

The voltage across an inductor is equal to the inductance times the rate of change of current through the inductor, or:

$$V = L * di/dt$$

where:

V = Instantaneous voltage across the inductor (volts)

L = Inductance (henry)

di/dt = Rate of change of current (amperes/sec)

current $i = Q/t$ (amperes = coulombs per second)

The charge on a capacitor is $Q = CV$ (coulombs = farads * volts)

$$V = L * C * (\text{delta } V)/(\text{delta } t)^2$$

approximately, or:

$$V = L * C * (V_{oh} - V_{ol})/(T_r)^2$$

using the output voltage and rise time.

Because of the high-speed (nS) and large (amperes) peak currents, even the small nanohenry inductance can induce a voltage transient on the order of volts. (The instantaneous voltage across an inductor is $V = L * di/dt$.) For typical high-speed signals, nanohenries * amperes/nanoseconds = volts! This effect is minimized by the use of minimum circuit interconnect trace lengths, wider ground traces, power and ground planes, and small, surface mounted IC packages that have very short leads.

For example, a CMOS output driving a 100 pF load with a rise time of 2 nS would induce a voltage across a typical 1 nH inductance of the chip's ground lead:

$$V = 1\,\text{nH} * 100\,\text{pF} * (4.5 - 0.5\,\text{V})/(2\,\text{nS})^2 = 0.1\,\text{V}$$

Although a voltage of 0.1 volt or 100 millivolts may not seem like much, remember that a part with many outputs, such as a processor, will sometimes switch many outputs at the same time, and *the current that flows through those pins all has to flow through a single ground pin*. An 8-bit output will cause 0.8 volt pulse or ground bounce. If the processor drives an 8-bit data bus and a 16-bit address bus low at the same time, this would result in a 2.4 V bounce! The ground bounce voltage across the ground lead inductance results in a different ground voltage reference for the chip while the chip's ground is bouncing. Needless to say, this ground bounce can cause a logic level to change during the brief pulse, which can cause trouble with circuits, such as clock signals, which are edge sensitive. This is why high-speed logic devices may have multiple, short ground pins and may only be available in small, surface-mounted packages. To make things even worse, if two devices overlap slightly in time driving the bus, very large current transients may briefly generate even larger currents that in turn generate larger ground bounce pulses. This can disturb several chips on the board at the same time.

The power supply leads are also subject to bounce for exactly the same reasons, and even though the power supply is not used as a logic voltage reference, the resulting drop in the local power supply voltage to the chip can result in errors.

Exact ground lead inductances may prove difficult or impossible to measure, but there is always some inductance in the ground lead, and the longer the lead, the greater the inductance. The example above illustrates another reason that it makes sense to avoid logic that is faster then necessary and to use very short ground and power wires. In fact, high-speed PC boards should use separate inner layers of a multilayer board to provide large ground and power planes, allowing the chips' power and ground leads to be connected using very short wires.

The magnitude of the bounce depends on the number and direction of logic transitions, so the noise is also data dependent. This is an apparently intermittent hardware design fault with symptoms that act like a software bug, since it might only happen at certain points in executing a program, with certain data values.

The example also shows why it is so important to maintain sufficient tolerance to noise in the logic. This noise tolerance is referred to as *noise margin*, which is covered in the next section. Noise margin analysis is especially important in a high-speed logic design, to prevent transient logic errors, which are extremely difficult to track down. This is another example of how a proper analysis and worst-case design can save a lot of time and money while delivering much higher quality and, ultimately, reliability. In the next section, the noise margin analysis process is described in detail.

6.4 Logic Family IC Characteristics and Interfacing

The three most common logic families are:

- *TTL.* Transistor-transistor logic (also known as *bipolar logic*).
- *NMOS. n*-channel metal oxide semiconductor field effect transistor logic.
- *CMOS.* Complementary (*n*- and *p*- channel) MOS logic.

All three logic families have versions with TTL compatible inputs, once the most common type, followed by later NMOS and CMOS. Because of its lower power density and relatively high circuit density, however, CMOS has become the most common form of logic, particularly in high-density and low-power battery-operated systems. TTL logic uses bipolar transistors requiring input drive currents on the order of hundreds of microamperes to a few milliamperes, depending on the version. Input voltage ranges for TTL-level compatible logic are generally 0 to 0.8 V for logic zero and 2.4 to 5 V for logic one. Output voltages are from 0 to 0.4 V for logic zero and 2.8 to 5 V for logic one. The 0.4 V difference is called the *noise margin* voltage because additive noise at or below this level will not change zeros to ones or vice versa. The *logic threshold voltage* (V_T) or "0/1 decision point" for TTL logic is typically around 1.5 V. It may range anywhere between 0.8 and 2.0 V depending on supply voltage and temperature and varies from one device to another. For TTL circuits, the noise margin is at least 0.4 V. Figure 6.10 shows the concepts of noise margin and logic threshold voltages.

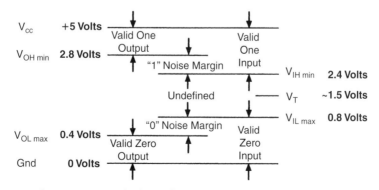

Figure 6.10: Typical TTL logic voltages and noise margin.

Interconnecting different logic families, such as CMOS and TTL, requires the designer to assure the compatibility of the logic signal voltage levels and adapt the circuit as necessary to maintain appropriate noise margins. The equivalent resistance or impedance of the signal network also has an impact on the noise in a specific circuit. High-impedance inputs are more prone to noise than are low-impedance inputs. The interface design process is illustrated by an example at the end of this chapter.

TTL logic is capable of sinking high currents and is used for driving very fast, large, heavily loaded buses. Both active and passive pull-up output devices are used with TTL. The active pull-up, referred to as a *totem-pole output*, uses one transistor to source current and one to sink it. The passive pull-up uses a transistor to sink current and a resistor connected to V+ as a current source. If a pull-up resistor is not connected to the gate's output pin and the collector is connected only to the output pin, it is referred to as an *open collector output*. In both cases, the output current sinking capabilities are greater than current source capacity. Many devices can sink a few milliamperes but can only source hundreds of picoamperes. Figure 6.11 shows both totem pole and open collector outputs.

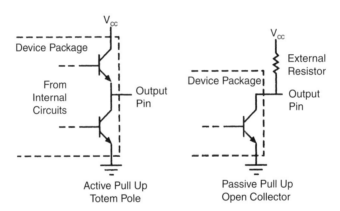

Figure 6.11: TTL outputs: totem pole and open collector.

TTL and CMOS logic are available in several versions, each identified by a distinctive prefix in the part number. Some of the more common versions and their prefixes are:

- *74xx*. Standard TTL.
- *74LSxx*. Low-power Schottky clamped TTL.
- *74ALSxx*. Advanced LS TTL.
- *74Fxx*. (Fast) high-speed TTL.
- *74HCxx*. High-speed CMOS with CMOS compatible inputs (Vt = ~Vcc/2).
- *74HCTxx*. High-speed CMOS with TTL compatible inputs (Vt = ~1.5 V).

- *74FCTxx.* High-speed CMOS with TTL compatible inputs (Vt = ~1.5 V).

- *74ACTxx.* Advanced high-speed CMOS with TTL compatible inputs.

- *74BCTxx.* Very high-speed CMOS/Bipolar with TTL compatible inputs.

Schottky logic (74ALSxx 74LSxx and 74Sxx) incorporates a low V_f (forward voltage drop) Schottky diode across the collector-base junction of a transistor to prevent it from saturating. This increases the speed for turning the transistor off. TTL is generally used where low cost, output drive, and high speed are important and there is no objection to the relatively high power consumption and resulting heat.

NMOS logic was used for moderate complexity logic ICs such as more mature microprocessors. Most NMOS logic ICs have TTL compatible voltage specs and operate at a lower power and speed than TTL. The power consumed by NMOS lies between TTL and CMOS, as does its speed. The input current is nearly zero since the MOSFETs have extremely high input resistance. Unfortunately, they do have fairly large input capacitance, limiting the circuit speed. The output configurations are similar to TTL except the transistors are *n*-channel field effect transistors (FETs) rather than bipolar NPN. Both active totem pole and passive (open drain) outputs are used in microprocessor and microcontrollers. Because of the constant operating current drain, these devices tend to be limited in size and complexity.

CMOS logic has a significant advantage since it does not use any significant amount of power when it is static (not changing state). Most of the power used in an operating device is due to the charge and discharge of internal capacitance and the current transient when both N and P devices arc partially on. As a result, power consumption is a function of clock rate for CMOS devices. Some processors are even designed to take advantage of this fact by incorporating "sleep" or low-power modes, stopping some or all of the clock operations when nothing important is going on. This is frequently required for battery-operated systems to maintain a reasonable battery life. Another advantage is the standard CMOS logic threshold is one-half the supply voltage and the output voltages tend to be very close to Vcc and ground voltage, resulting in higher noise margins than those of TTL devices. This is particularly important for CMOS devices that operate at reduced power supply voltage. CMOS devices that operate at 3 V or less are available.

Because CMOS logic is inherently symmetrical, the rise and fall times tend to be nearly equal. The symmetry also results in equal source and sink capabilities. The inherent increase in noise margin makes CMOS less susceptible to noise than TTL and NMOS. Figure 6.12 illustrates this concept. CMOS devices operating at voltages other than 5 V, such as 3.3 V, will have a threshold voltage corresponding to Vcc/2. Some versions of CMOS logic operate with a reduced noise margin to have TTL-compatible input voltages. This is accomplished by artificially lowering the input threshold voltage to 1.5 V, the same as used for TTL. These TTL input threshold

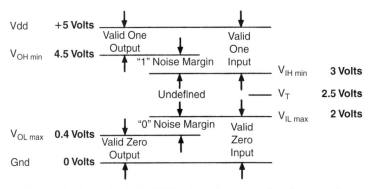

Figure 6.12: Typical CMOS logic voltages and noise margin.

compatible circuits have a *T* in their number (74HCT, 74BCT, etc.), indicating they have TTL compatible inputs. A series of high-speed logic compatible with the TTL logic family in function and input voltage is the 74HCTxx (high-speed CMOS TTL compatible) series. The advantage of the T series CMOS devices is they can be driven directly by devices having TTL output voltage levels. The T series of CMOS devices has the disadvantage that the noise margin is less than it is for true CMOS compatible inputs due to the shifted threshold voltage. The 74HCxx series is pure CMOS with a threshold voltage of one-half the supply voltage (2.5 V for a 5 Vcc) and correspondingly higher noise margins. As a result, a standard TTL output VOHmin of 2.8 volts is not enough to guarantee a logic one value for a 74HCxx gate input.

6.4.1 Interfacing TTL Compatible Signals to 5 V CMOS

Interfacing a CMOS output to a TTL input is a direct connection as long as the CMOS output is capable of sinking the TTL device's input low current. Interfacing a TTL output to a standard CMOS input requires the use of at least a pull-up resistor. A resistor on the TTL output to Vcc will ensure that the output voltage is pulled high enough to guarantee the logic one output signal is interpreted as a logic one by the CMOS input. Another useful technique when using 5 V logic to drive CMOS circuits is to use a higher-voltage open collector or open drain output with a pull-up resistor connected to the higher supply voltage. This level-shifting technique can also be used for driving other high-voltage circuits such as high-voltage outputs. In either case, the objective is to guarantee that there is sufficient noise margin to guarantee a valid logic one when the TTL compatible output drives a CMOS input.

It is important to note that when a TTL output is pulled above its normal output high voltage, it will not source any significant current. This is because the TTL output source is equivalent to a high resistance in series with a voltage source that is effectively limited to around 3 V, due to internal design constraints. As the output voltage increases until it equals the internal voltage, the output can no longer source any current. When the voltage is increased beyond the internal circuitry (up to a limit of Vcc), the internal circuitry is equivalent to a reverse biased diode, so only leakage currents in the sub-microampere range will flow into the output device. As a result, the effect of a TTL output on external circuits is negligible when the pin is pulled high by an external resistor.

Also, a 5 V TTL compatible output is often compatible with a 3 V CMOS device input, since the CMOS threshold (Vcc/2 = 1.5 volt) is the same as a 5 volt TTL gate (TTL Vt = 1.5 V). Most of the 3 V CMOS devices are designed to withstand a 5 V input signal, so it is often possible to interface 5 V TTL outputs directly to 3 V CMOS inputs. However, if the 3 V CMOS inputs are not designed to handle 5 V inputs, the CMOS device could be destroyed with an input signal greater than 3 V, so it is important to verify this. A 3 V CMOS device output will be close to 3 V, so it can drive a 5 V TTL compatible input directly.

A 3 V CMOS output would probably be marginal driving a 5 V CMOS input (Vt = Vcc/2 = 2.5 volt), leaving less than 0.5 V CMOS output generally cannot withstand a pull-up resistor to 5 V, it is necessary to add a level shifting IC to convert 3 V logic levels to 5 V.

Level shifters are available for converting logic levels from one family to another, including 3 V to and from 5 V, or 5 V TTL to +/– V ECL (*emitter-coupled logic*), and 5 V levels to +/–12 V RS-232 signals. There are also special ICs for driving output loads requiring either a high voltage or high current output, such as a light, motor or relay. Most microcontrollers have very weak output drive capability, so external driver ICs may be necessary. These would typically be needed to drive LEDs, a vacuum fluorescent display, or a motor. Solid-state relays even allow large AC loads to be controlled by a micro. Likewise, there are other devices (i.e., optical isolators), allowing high voltages (like 110 V AC inputs) to be safely converted to logic levels for input to a microcontroller. Devices that use potentially hazardous high voltages should be isolated from the rest of the circuitry for reasons of safety. It might be possible to connect such devices directly to our circuits, but they would allow us to come into contact with potentially fatal voltages. The standard 50 or 60 cycle AC power supply used almost everywhere has the unfortunate characteristic that it is very nearly the optimal voltage to guarantee that a human heart will stop functioning due to muscle fibrillation. Customer death by electrocution is sure to result in the next of kin hiring an attorney to relieve you of all your assets … unless, of course, they're *your* next of kin! There are many isolation devices available, most of which use the same basic approach.

The isolation can be accomplished using optical or magnetic means, which can provide a barrier to transient voltages that can be on the order of thousands of volts. The barrier is transparent and so allows light to pass, but it is made of a good insulator to prevent electrical current from flowing across the boundary. Figure 6.13 shows a simple optical isolation circuit.

This isolation approach can be used to input high voltages to a microcontroller safely by connecting the LED to a high-voltage source in series with a resistor and protective diode to limit the LED's current and prevent the LED from being exposed to the potentially destructive

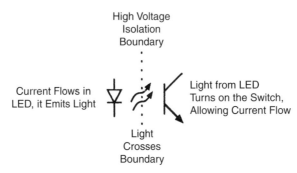

Figure 6.13: Optical isolation allows connection to hazardous voltages.

reverse voltage. The output transistor will then be turned on whenever the LED is turned on by one half of the AC power cycle. This is useful for time-of-day clock functions, since the AC power mains frequency is maintained very accurately by the power utilities over a period of time. The output switch can be connected to the processor counter or interrupt input, allowing the processor to keep track of time and synchronize its operation with the AC power cycle.

High voltage outputs can also be controlled safely by using the micro's output to turn on the LED that turns the output switch on. In this case, another type of switch such as a silicon-controlled rectifier (SCR) or TRIAC (an AC version of the SCR) is used rather than a transistor. SCR and TRIAC switches can be obtained to handle relatively large AC loads, such as lamps and motors. These devices are often referred to as *solid-state relays (SSR)*, since they are equivalent to an electromechanical relay except that they are implemented with solid-state semiconductor devices instead of using a coil to move a switch. Both isolated inputs and outputs are available in complete modules that have all the necessary circuits to monitor and control high voltage and power devices, using optical isolation for safety. They have microcontroller-compatible I/O on one side that is optically isolated from the high-power outputs on the other side.

Very often, even when safety is not an issue, microcontroller chips simply cannot handle the voltages or currents required to interface with other devices. In some cases it is required when connecting one logic family to another, incompatible family, such as emitter-coupled logic (ECL) levels or RS-232 interfaces utilizing negative voltages.

Sometimes a plain, old-fashioned electromechanical relay is a better solution, since relays usually have contact resistances that are far lower than can be found in a semiconductor switch. In some cases, a simple transistor or MOSFET switch can be used to control a load operating at voltages which are greater than the logic supply, such as motors, solenoid actuators, and relays that might require 12 or more volts to operate.

The circuitry required to interface between logic levels and high-level circuits is described in detail elsewhere, including an excellent book titled *The Art of Electronics*, by Horowitz and

Hill. If you don't already have this very handy book—and you have to do much electronic design or interfacing—you should definitely obtain a copy.

The real world is an analog place, and interfacing between the discrete, digital world of computers and the real world demands significant attention. The interface between low-level analog signals and logic is handled in another chapter of this book.

At this point, it is time to look at some simple examples so that we can see exactly how a worst-case analysis should be performed. The next section illustrates part of the worst-case analysis for a real laboratory instrument that is still used in the healthcare industry. This product's poor reliability was seriously inconvenient for the medical staff and patients who depended on it, and if it had led to an incorrect diagnosis, a truly fatal error! It is in these types of applications that worst-case design is most important, and the cost of unreliable hardware in the field almost always greatly exceeds the cost of avoiding the problem by using proper design and analysis techniques. Now let's turn our attention to the analysis of the worst-case noise margin for an 8051-based design example.

6.5 Design Example: Noise Margin Analysis Spreadsheet

The spreadsheet in Table 6.5 shows the results of a noise margin on a design that was already in production at the time of the analysis. The product's users had complained about intermittent glitches, and the author was consulted to determine the source of the problem. After a quick look at a few of the noise margin values, it became obvious that there were deficiencies in the design in that area. A portion of the spreadsheet used in that analysis is shown in Table 6.5, with problems shown in ***bold italic underline*** font.

The first column of Table 6.5 is the signal name, followed by the pin number and chip that is the source of the signal, followed by the source's worst-case output voltages, Volmax and Vohmin. The next columns list the loads on the signals and their respective worst-case input voltages Vilmax and Vihmin. The noise margins are shown in the last two columns, Vil–Vol for the logic zero case and Voh–Vih for the logic one case. As shown, the logic zero noise margins are all probably acceptable, since the lowest value is 0.3 V. The logic one noise margin is zero or negative for most of the devices listed, which is completely unacceptable. Any noise on the power supply, ground, or the signal lines themselves can easily cause a logic input to interpret the wrong logic state, causing an error. An interesting thing to observe is that none of them were very far out of spec, and the instrument worked perfectly most of the time. These problems can be virtually impossible to find in the field. Hooking up a test instrument like a scope or logic analyzer to the problem signals often makes the problem go away due to changing the ground currents and impedances of the circuit. The specs that cause the problem in this case are the high Vih specs of the loads, especially the SRAM chip. The example design in the spreadsheet represents a relatively common problem with devices

Table 6.5: 8051 Noise Margin Analysis Sample.

Output						Input				Noise	Margin
Signal	Pin(s)	Source	Volmax	Vohmin	Load(s)	Signal	Vilmax	Vihmin		logic zero	logic one
PSEN/ (P3.7)	29	8051	0.40	2.00	EPROM	OE/	0.80	2.00		0.40	_0.00_
RD/ (P3.7)	17	8051	0.40	2.00	SRAM	OE/	0.80	2.20		0.40	_0.20_
			0.40	2.00	82C55	RD/	0.80	2.00		0.40	_0.00_
WR/ (P3.6)	16	8051	0.40	2.00	SRAM	WR/	0.80	2.20		0.40	_−0.20_
			0.40	2.00	82C55	WR/	0.80	2.00		0.40	_0.00_
A15(P2.7)	28	8051	0.40	2.00	74LS138A		0.80	2.00		0.40	_0.00_
A8..14 (P2.0–P2.6)	21–27	8051	0.40	2.00	SRAM	A8..14	0.80	2.20		0.40	_−0.20_
			0.40	2.00	EPROM	A8..14	0.80	2.00		0.40	_0.00_
			0.40	2.00	GAL	A8..14	0.80	2.00		0.40	_0.00_
ALE	30	8051	0.40	2.00	74LS373LE		0.80	2.00		0.40	_0.00_
AD0..7 (P0.0–P0.7)	39–32	8051	0.40	2.00	74LS373	A0..7	0.80	2.00		0.40	_0.00_
			0.40	2.00	SRAM	D0..7	0.80	2.20		0.40	_−0.20_
			0.40	2.00	82C55	D0..7	0.80	2.00		0.40	_0.00_
		SRAM	0.40	2.20	8051	D0..7	0.80	2.40		0.40	_−0.20_
		EPROM	0.45	2.40	8051	D0..7	0.80	2.40		0.35	_0.00_
		82C55	0.40	3.50	8051	D0..7	0.80	2.40		0.40	1.10
RAM enable		16V8	0.50	2.40	SRAM	/CE	0.80	2.20		0.30	0.20
EPROM enable		16V8	0.50	2.40	EPROM	/CE	0.80	2.00		0.30	0.40

that are advertised as "compatible" with other logic families. The solution to the problem is very simple and inexpensive: the addition of pull-up resistors to the signals that have zero or negative noise margin in the logic one state. This also impacts the output low current that must be handled by the signal source chip outputs, so it must be taken into account in the load analysis, and pull-up resistors should be chosen accordingly.

It is important to note that there are four sources listed for AD0 .. 7, since there are four devices that drive the data bus. Only the data paths that are used need to be evaluated vs. loading analysis, where unused paths load the bus. The load analysis for another similar design is shown in Table 6.6, which tabulates the capabilities of the various driving devices and the loads that are presented to them. The first three columns (signal, pin, and source) identify the signal source; the next three (IOL, IOH, and CL) list the corresponding source's output drive current and capacitive load values. The next two columns (load, and signal) identify the load's signal names. The Qty column is the number of loads in the case of multiple signals connected to the same output or the number of inches of wire in the case of the wire capacitance. The next three columns (IIL, IIH, and Cin) define the load characteristic of a single input's input current and input capacitance. For the interconnect wiring, Cin is the estimated stray wiring capacitance per inch of the printed circuit trace. The last three columns show the extended totals and grand totals for each signal, followed by the design margin, which should be a positive number. In this case there is only one problem, due to excessive capacitive loading of the SRAM when it drives the data bus, AD0 .. 7.

The output capacitive load specs are usually found as notes within the AC section of the chip specification listing the various timing parameters. This is because the capacitive loading affects the rise and fall time of the signal, so the capacitance value is really used as a test condition for the timing measurements. Input capacitance may be difficult to find in the specification sheet, it might be in a different "family" specification sheet or handbook, or might not be specified at all. When it is not specified, a reasonable estimate can be made by substituting values for similar parts in the same type of package.

The SRAM output is specified with a Cload value of 50 pF, which is relatively low value. By using a very low load capacitance, the SRAM's timing specs look good due to shorter than normal rise and fall times, since the chip is not driving a realistic load. This is a good example of a manufacturer's "specsmanship." They are intentionally playing games with the test conditions to make their device appear to be better than it is. That way when someone looks at their timing specs, the shorter rise and fall times make their chip appear to be faster than another equivalent chip that is specified with a larger capacitive load value when the chips are actually identical. Unfortunately, this practice is all too common, so the designer must view the claims on the cover of a data sheet very critically. If it looks too good to be true, then it probably is!

Table 6.6: Load analysis for a similar design.

Source						Load			Unit			Load		Total		
Signal	Pin#	Source	uA IOL	uA IOH	pF CL	Load	Signal	Qty	uA IIL	uA IIH	pF Cin	pF Cin	uA IIL	uA IIH	pF Cin
PSEN/	29	8051	3200	−60	100	EPROM	OE/	1	−1	1	12	12	−1	1	12
						wire cap		2			2	2			4
									Total				3199	59	16
									Margin						84
RD/	17	8051	1600	−60	80	SRAM	OE/	1	−1	1	7	7	−1	1	7
(P3.7)						82C55	RD/	1	−1	1	10	10	−1	1	10
						wire cap		3		2	2	2		2	6
									Total				1598	58	23
									Margin						57
WR/	16	8051	1600	−60	80	SRAM	WR/	1	−1	1	7	7	−1	1	7
(P3.6)						82C55	WR/	1	−1	1	10	10	−1	1	10
						wire cap		3		2	2	2		2	6
									Total				1598	58	23
									Margin						57
A15	28	8051	1600	−60	80	74LS138	A	1	−200	20	10	10	−200	20	10
(P2.7)						wire cap		2		2	2	2			4
									Total				−200	20	14
									Margin				1400	40	66
A8..14	21-7	8051	1600	−60	80	SRAM	A8..14	1	−1	1	7	7	−1	1	7
(P2.0–P2.6)						EPROM	A8..14	1	−1	1	12	12	−1	1	12
						wire cap		3		2	2	2		2	6
									Total				−2	2	25
									Margin				1598	58	55
ALE	30	8051	3200	−60	100	74LS373	LE	1	−400	20	10	10	−400	20	10
						wire cap		2		2	2	2			4
									Total				−400	20	14
									Margin				2800	40	86

AD0..7 (P0.0–P0.7)	39–2	8051	3200	−800	100	74LS373	A0..7	1	−400	20	10	−400	20	10
						SRAM	D0..7	1	−1	1	7	−1	1	7
						EPROM	D0..7	1	−1	1	12	−1	1	12
						82C55	D0..7	1	−10	10	20	−10	10	20
						wire cap		5			2			10
									Total			−412	32	59
									Margin			2788	768	41
	SRAM		1600	−600	50	74LS373	A0..7	1	−400	20	10	−400	20	10
						8051	D0..7	1	−1	1	20	−1	1	20
						EPROM	D0..7	1	−1	1	12	−1	1	12
						82C55	D0..7	1	−10	10	20	−10	10	20
						wire cap		5			2			10
									Total			−412	32	72
									Margin			1188	568	−22
	EPROM		1600	−600	100	74LS373	A0..7	1	−400	20	10	−400	20	10
						8051	D0..7	1	−1	1	7	−1	1	7
						SRAM	D0..7	1	−1	1	12	−1	1	12
						82C55	D0..7	1	−10	10	20	−10	10	20
						wire cap		5			2			10
									Total			−412	32	59
									Margin			1188	568	41
	82C55		1600	−60	80	74LS373	A0..7	1	−400	20	10	−400	20	10
						8051	D0..7	1	−1	1	20	−1	1	20
						EPROM	D0..7	1	−1	1	12	−1	1	12
						SRAM	D0..7	1	−1	1	7	−1	1	7
						wire cap		5			2			10
									Total			−403	23	59
									Margin			1197	37	21

When an output like this is operated with actual capacitive load greater than the test conditions, the related timing specs for the device must be de-rated due to the degraded rise and fall times that will occur. As long as the load capacitance is no more than twice the spec value, this will be sufficient. The excess C load will increase the stress on the driver. If the overload is much greater than two times normal, the device can be overstressed due to the relatively large currents that will flow into the load capacitance on transitions when the C is charged and discharged through the driving output. As long as the output is not overloaded too much, the resulting increase in the rise/fall time can be estimated, resulting in a de-rated timing spec. All we have to do is calculate the additional rise time and add that to the timing values specified in the data sheet. To do that, we need to evaluate the output circuit's performance. This can be accomplished by noting that the output current drives the load capacitance from a logic low to high or vice versa. For our purposes, we will assume that the interconnect does not behave like a transmission line, which is most often the case for garden variety microcontroller components. If the chips used have a fast rise time and trace length greater than about one-sixth the edge length of the pulse, it is necessary to analyze the circuit as a transmission line. In this case we will look at the simpler problem.

By assuming a constant current charging the capacitance, the voltage will ramp linearly from one logic level to the other. To make a rough estimate, we can use the source's output current and load capacitance to determine the signal slew rate and the difference between the high and low logic levels to determine the delay. Figure 6.14 illustrates this idea.

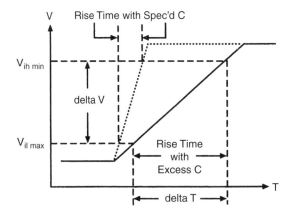

Figure 6.14: De-rating delay for excess CL.

Let's next look at a simple example showing how to de-rate the timing based on the approximation technique just described.

First we make the assumption that the signal timing measurements in the data sheet are made under the specified test conditions, usually with the output loaded by R_L and C_L in parallel to ground. The output delay specifications in the data sheet include the internal delay as well as the rise time. The output drive current charges C_L within the specified time. The circuit can be divided into two parts: the specified load, and the additional output current available to drive the excess load C. So the additional delay (delta T) we are looking for depends upon the leftover drive current (delta I) which is available to charge the excess load capacitance (delta C). The equation for this is:

$$\text{Delta T} = (\text{delta V} * \text{delta C})/(\text{delta I})$$

Let's look at a typical example. An SRAM is specified with a 50 nS access time, but the outputs are overloaded with respect to the C_L spec in the data sheet. What access time spec should be used for the actual conditions specified below?

- The output is specified to drive CL = 50 pF, but the actual load is 100 pF.
- The output is specified to drive 20 mA into the load, but the load is only 10 mA.
- The driven device has input voltage specs Vilmax = 0.4 V, Vihmin = 3.4 V.

Spec values	Actual Values	Difference
C_L = 50 pF	100 pF	50 pF = delta C
Io = 20 mA	10 mA 10	mA = delta I

$$\text{Voltage: Vih} - \text{Vil} = 3.4 - 0.4 = 3\,\text{V} = \text{delta V}$$

$$\text{Delta T} = (\text{delta V} * \text{delta C})/(\text{delta I})$$

$$\text{Delta T} = (3\,\text{V} * 50\,\text{pF})/(10\,\text{mA}) = 15\,\text{nS}$$

So in this case 15 nS should be added to all the output delay specs for the driving device. The access time used should be:

$$\text{Taa(actual)} = \text{Taa(spec)} + (\text{delta T}) = 50\,\text{nS} + 15\,\text{nS} = 65\,\text{nS}$$

Since the output current from most devices is larger at the beginning of the transition and smaller near the end of the transition, the approximation is only a rough guide. Also, the delta V calculation is conservative, since the input threshold voltage is typically halfway between the Vih and Vil values.

So, the estimate as shown will usually be conservative compared to actual performance. All of the above must be used with caution and is only an approximation of the additional delay caused by excess C_L, so it is wise to allow additional margin in the timing for any de-rated specs.

Here's another typical example. An LSTTL gate is to be used to drive one LSTTL load and a CMOS processor clock input, as shown in Figure 6.15. An interface must be made which will guarantee the CMOS input voltage requirement will be met with the same noise margin as a standard LSTTL input. The LSTTL and CMOS gates have the specs as defined below:

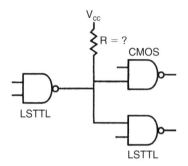

Figure 6.15: TTL-to-CMOS interface example.

LSTTL Gate DC Parameters.

Symbol	Parameter	Min	Typ	Max	Units	Conditions
V_{IL}	Input low voltage	−0.3		0.8	V	
V_{IH}	Input high voltage	2.4		Vcc+0.3	V	
I_{IL}	Input low current		−120	−360	μA	
I_{IH}	Input high current		30	60	μA	

Absolute Maximum Operating Conditions.

Symbol	Parameter	Min	Typ	Max	Units	Conditions
V_{OL}	Output low voltage		0.2	0.4	V	@ I_{OL} max
V_{OH}	Output high voltage	2.8	3.5		V	@ I_{OH} max
I_{OL}	Output low current	3.2	8		mA	@ V_{OL} max
I_{OH}	Output high current	−600	−1000		μA	@ V_{OH} min

Note: Test conditions $R_L = 1K$, $C_L = 100\ pF$.

CMOS Gate DC Parameters.

Symbol	Parameter	Min	Typ	Max	Units	Conditions
V_{IL}	Input low voltage			2.0	V	
V_{IH}	Input high voltage	3.0			V	
I_I	Input leakage current			<1	μA	

Here is how we would determine the answer. Since the LSTTL V_{OL} is 0.4 V and the CMOS V_{IL} is 2.0 V, the CMOS input low voltage is compatible with the LSTTL low output voltage.

Absolute Maximum Operating Conditions.

Symbol	Parameter	Min	Typ	Max	Units	Conditions
V_{OL}	Input low voltage			0.4	V	@ I_{OL} max
V_{OH}	Output high voltage	4.5			V	@ I_{OH} max
I_{OL}	Output low current	3.2			mA	@ V_{OL} max
I_{OH}	Output high current	600			μ	@ V_{OH} max
C_{in}	Input capacitance			20	pF	

Note: Test conditions $R_L = 5K$, $C_L = 150\ pF$.

However, the LSTTL output high voltage of $V_{OH} = 2.8$ V is not sufficient to meet the CMOS input high $V_{Ihmin} = 3.0$ V. A pull-up resistor is required to allow the LSTTL output to go to a higher voltage, $V_{IH} + V_{noise\ margin} = 3.0 + 0.4 = 3.4$ V. There is no exact solution, but the range of resistors meeting the requirements can be determined.

The lowest resistor value that will work is the value which will source enough current so the LSTTL output is just able to sink the resistor current plus the additional LSTTL load when the signal is low and still meets the maximum output low voltage specification. Negligible DC current is flowing from the CMOS input. The voltage across the resistor is Vcc – V_{OLmax}. for the LSTTL input, or $5 - 0.4 = 4.6$ V. The current required is $I = I_{ILmax} + I_{RPU}$ where I_{ILmax} is the current coming from the LSTTL input load and I_{RPU} is the current flowing through the pull-up resistor. The current the LSTTL output must sink is the sum of the I_{IL} of the LSTTL load and the current through the pull-up resistor.

The equation is:

$$I_{OLmin} >= I_{ILmax} + I_{RPU} = 360\ \mu A + (Vcc - V_{OLmax})/R_{min}$$

Solving for R_{min}:

$$R_{min} >= (5 - 0.4\ volts)/(3.2\ mA - 360\ \mu A) = 4.6\ V/2.84\ mA = 1.62\ kilohms$$

$$R_{min}\ is\ 1.62\ Kilohms$$

This value is also greater than specified as a test load of 1 kilohms.

The maximum acceptable value, Rmax, is determined by the minimum output high voltage that will guarantee a CMOS high input plus noise margin. The resistor must be able to supply the LSTTL maximum input high current and not have too large a voltage drop across it. This will determine the upper limit for the resistor value.

Specifically, the resistor voltage is:

$$Vcc - (CMOS\ V_{IH\ min} + V_{noise\ margin}) = 5 - (3.0 + 0.4) = 1.6\ volts$$

This voltage is maintained while sourcing the LSTTL $I_{IH\ max}$ of 60 μA.

Solving for R_{max}:

$$R_{max} <= 1.6\ V/60\ \mu A = 26.7\ \text{kilohms maximum}$$

Thus, the acceptable range for the pull up resistor is:

$$1.62\ \text{kilohms} <= R_{pu} <= 26.7\ \text{kilohms}$$

An acceptable standard value such as 10 kilohms would be appropriate.

Another limit relates to the rise time of the signal under load, due to the R-C time constant of the pull-up resistor charging the load capacitance, C_L. From the example above, let's see what the effect of this time constant is on the selection of the resistor value.

The maximum R value can be approximated by the equation:

$$R = T/C_L$$

where T is the rise time and C_L is the total load capacitance.

Ignoring the Ioh current of the LSTTL driver, if the circuit above had an allowable rise time $T = 50\ nS$ and $C_L = 20\ pF$, then the maximum R value would be:

$$R_{max} = 50\ nS/20\ pF = 2.5$$

kilohms maximum to maintain the 50 nS rise time.

So a better choice might be a standard 2.2 kilohm pull-up resistor. Since the driver will supply some current to charge the load capacitance, this is a fairly conservative value. We would also have to allow for the additional rise time as part of the timing analysis for the low-to-high transition.

6.6 Worst-Case Timing Analysis Example

Let's suppose an LSTTL gate is used to enable the D input of a flip-flop frequency divider, as shown in Figure 6.16. Figure 6.17 shows a functional timing diagram for the circuit in Figure 6.16, and Figure 6.18 illustrates a specification timing diagram for the same circuit.

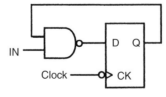

Figure 6.16: Example of worst-case timing.

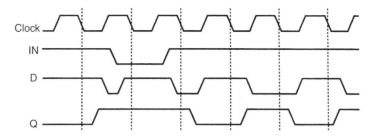

Figure 6.17: Functional timing diagram for Figure 6.16.

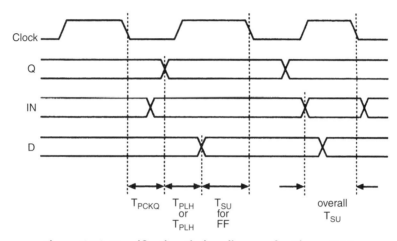

Figure 6.18: Specification timing diagram for Figure 6.16.

Flip-Flop Timing Specs

Symbol	Min	Typ	Max	Units
T_{SU}	10			nS
T_H	1			nS
T_{PCKQ}			15	nS
T_{PWCK}	10			nS
F_{CLK}			50	MHz

Gate Timing Specs

Symbol	Min	Typ	Max	Units
T_{PHL}	1 2	5	nS	
T_{PLH}	2 4	6	nS	
Test Conditions $R_L = 1K$, $C_L = 100\ pF$.				

The timing of the input signals must conform to the combined specs of both devices, as defined below:

For the circuit shown in Figure 6.16 and the accompanying specifications, what is the maximum guaranteed clock rate?

From the timing figures on the previous page, note that the minimum clock cycle time is defined by the sum of the following times: the time it takes for the transition from the active edge of the clock for the signal at D to propagate through the flip-flop through the NAND gate and the time the signal must be stable before the next clock. The maximum propagation times and minimum setup times are used as they are the most severe requirements.

$$T_{PCKQ} + T_{PLH} + T_{SU} = 15 + 6 + 10 = 31 \text{ nS}$$

$$f = 1/t = 1/31 \text{nS} = 32.26 \text{ MHz}$$

Now let's determine the setup and hold time requirements for the overall circuit. The overall setup time is lengthened by the delay of the NAND gate; therefore, the system setup time is the sum of the flip-flop setup time and the worst-case propagation delay.

$$T_{SU}(\text{system}) = T_{PLH} + T_{SU}(\text{flip-flop}) = 16 \text{ nS minimum}$$

For the overall system hold time, the hold time of the flip-flop is offset by the minimum delay through the NAND gate, since this is the minimum amount of time that can be counted on to delay a changing D input to the flip-flop.

$$T_H(\text{system}) = T_H(\text{flip-flop}) - T_{PHL}(\text{min}) = 1 - 1 = 0 \text{ nS}$$

The delay in the D signal path reduced the hold time requirement from 1 nS to 0 nS, meaning that the input can change at the same time as the clock edge or later. This is actually an improvement on the performance of the flip-flop by itself, which requires that the D line be held stable for 1 nS after the clock edge.

Endnotes

Horowitz, Paul and Winfield Hill, *The Art of Electronics.* Cambridge, UK: Cambridge University Press, 1989.

Johnson, Howard W., and Martin Graham, *High-Speed Digital Design: A Handbook of Black Magic.* Upper Saddle River: NJ: Prentice Hall, 1993.

CHAPTER 7
Choosing a Microcontroller and Other Design Decisions

Lewin Edwards

7.1 Introduction

The start of a complex embedded project, particularly in a small organization without engineers who can be dedicated full-time to component procurement, can be extremely stressful. Until a first-round prototype is built and tested (and often even after this stage), it is usual for hardware requirements to be at least slightly vague, particularly vis-à-vis the exact breakdown of which functions are expected to be integrated into the microcontroller and which will be off-chip. As the design engineer, some of your goals are obviously ease of firmware and hardware development, low bill-of-materials cost, and reliability of sourcing. You will probably start with a list of hardware requirements and match those up against selection matrices from different vendors to find a part that has as many of your features as possible on-chip.

At this point, what you really want is a vendor-neutral parametric search engine for which you can select the performance and peripherals you want and obtain a list of suggestions collated from *everybody's* catalogs. Unfortunately, most of the search facilities available online leave much to be desired. Many manufacturers don't have full parametric search engines available, and those that do obviously only list their own parts. Third-party search engines do exist, but they are usually premium services for which you will have to pay—and again, they only list products from manufacturers with whom they have a relationship. Also, the total startup cost of development—evaluation boards, tools, etc.—is an important factor to us (for some readers, perhaps even more important than the unit cost of the microcontroller), and this cost will not be listed by parametric search engines. Finally, as with any other search facility, it can be difficult to match your needs with the list of keywords provided in the search engine.

This is one occasion when there is no substitute for peer support. Even if you think you've found a perfect match already, it's well worth searching Usenet archives (groups.google.com) for discussions on similar applications to your own. A carefully phrased question may lead to even more useful suggestions. Even if you are intimately familiar with every IC vendor that impinges on your industry, you might miss a new product announcement and thereby not know to check manufacturer X's catalog. Sometimes the only clue you need to lead you to the right part is the information that manufacturer X makes 32-bit microcontrollers! Furthermore, other engineers who have worked with the part may be able to point you to low-cost, third-party evaluation platforms or off-the-shelf appliances that can be used as demo boards, and they will be better positioned than anyone else to give you relatively unbiased opinions on real-world difficulties of using a specific device.

In the early days, it is also doubly hard to make an optimal price/performance choice, because the selection sheets generally won't show pricing. For any part that can't be bought anonymously off the shelf (and unfortunately the majority of 32-bit microcontrollers fall into this category), most chip vendors expect you to establish a relationship with their distributors. This can waste a lot of time in profitless face-to-face meetings. My own experiences with local reps and distributors in the United States have been *very* patchy, and I have often found that their knowledge of the 32-bit parts on their line card is limited to whatever bullet points the manufacturer printed on the sales literature. The distributors want accurate annual usage forecasts before they will give you sensible pricing, and they obviously have little or no incentive to deal with small-volume purchasers like students or hobbyists. Political difficulties related to sales commissions also arise when you are designing the product in one country but intend to manufacture it in another. Furthermore, the distributors and reps will be most likely to quiz you on your other requirements and try vigorously to sell you other parts from their line card. Although this possibly has some marginal convenience benefits if you intend to source and manufacture locally, it certainly isn't the ideal way of minimizing the bill-of-materials cost of your product.

It's all too easy to become trapped in an endless circle trying to seek an optimal solution to all these problems, so you shouldn't attempt it. Recognize from the outset that this is a classic "traveling salesman" problem (perhaps even in the literal mathematical sense) and that your goal is merely to find an *acceptable* solution in time to finish your project and send it to the factory (or submit it to your professor, if you're a student). Your goal is not to find the best possible solution. If your team has enough personnel to dedicate a lot of person-hours to sourcing components, you will probably be able to find a better solution than the one-person "team" scouring catalogs on a time limit, but a suboptimal one-person solution can always be refined later if the project goes into production in quantities that justify it. As in any other industry, our goal is to develop a product that works properly and is ready to manufacture in a timely fashion.

With that said, I employ the following useful heuristics to filter my short list for 32-bit microcontroller selection:

- The device should be available for anonymous online or catalog ordering in single-piece quantity from at least one major distributor. (In the U.S., the big names commonly mentioned are Digi-Key, Newark, and Avnet Marshall. Digi-Key and Newark in particular have very broad inventories and generally allow purchases in small quantity. Avnet Marshall seems to cater more to manufacturing rather than prototype runs; they typically have 25- or even 250-piece minimum orders on parts.)

- Full data sheets for the device should be available without requiring a nondisclosure agreement or committing to any kind of purchase.

- A low-cost development board should be available for the part—either the manufacturer-recommended board, a third-party board, or even some appliance based around the chip, as long as sufficient documentation exists to enable use of the appliance as a test bed for your own code. You should also ask the manufacturer and distributor if loaner boards are available; if you can borrow a board for a month or two, it will be enough to get at least bootstrap code up and running and establish a basic level of familiarity with the microcontroller. You can then move to your own hardware and return the evaluation board.

- There should be a direct technical contact available at the chip vendor, at least for emergency issues; it should not be necessary to route all questions through distribution. (Note that I'm *not* advising you to abuse such a privilege—if you have a direct manufacturer contact, it's best to contact him or her only when absolutely necessary. But there are times when a complex problem will take weeks to solve when there are several layers in the communication chain, versus only a day or two if you can communicate directly with the *cognoscenti* at the chip manufacturer. As a small customer, the less you use this resource, the better chance you will have that your next urgent question will be answered speedily.)

- The device should have been shipping to OEMs for at least three to six months.

- The core should be supported by the GNU toolchain.

- There should be at least one currently shipping commercial product that uses the device, and the larger the market for this device, the better. All too often, parts that are consumed only by small niche markets are discontinued in favor of parts with more general applicability.

These are not absolutely binding rules (in particular, the last one can be hard to obey for a brand-new part), but they provide a good way of thinning a short list of any undesirable parts that are going to cause logistical problems later. The first criterion above is especially important to note because it can give you some idea of the part's longevity. One little-mentioned fact of the microcontroller industry is that very few high-end parts are designed only for the marketplace in general; many of the "standard" 32-bit parts and ASSPs started life as proprietary ASICs developed under contract for some specific electronics manufacturer. These contracts typically have large guaranteed order quantities and forward-planned production schedules. However, once that manufacturer's exclusivity expires, the chip vendor is free to sell it to other people, if it conceivably has any generally applicable function.

The first step in this process is usually to offer the part quietly to other existing customers or to carefully selected others, without a highly visible product announcement or other publicity. This small group of privileged customers will, again, work on large volume pre-orders with long-term schedules. If a chip goes on from this stage into retail distribution channels (such as Digi-Key and other stores catering to small orders) it is a very good sign because it usually means one of two things:

1. The chip vendor is seeking to gain market share in the field addressed by this part and is pushing it heavily (also implying that excellent support will be available both from the manufacturer and other users), or

2. The product is so wildly successful that the chip vendor is producing reasonable quantities of it in advance of any firm order, in expectation of future unscheduled orders.

In either case, the part is in wide-scale production, and it is a fairly safe bet to design it into your product. You can be reasonably certain that the part will not be discontinued in the immediate future.

7.2 Choosing the Right Core

Unfortunately, even with the greatest care in choosing parts that appear to be supported for the long term, there are never any guarantees. Parts are discontinued or superseded all the time for marketing reasons that are sometimes not obvious and far from predictable. For that matter, sometimes your requirements change slightly and your previous choice of microcontroller is suddenly no longer suitable. This is particularly annoying when a design change of this sort is a result of entirely external forces. I have been involved in several projects where the microcontroller has been changed just before production, or even after production starts, simply because of sudden supply shortages of other parts.

Obviously, the more careful you are in choosing a part that *exactly* meets your requirements, the more disruptive it is likely to be to have to substitute a different part. A large customer might be able to guarantee the chip vendor enough volume for them to continue occasional production runs or even perhaps migrate an old part to a new process and continue general production. Since we're going to be a tiny customer, we won't have this luxury.

The only truly effective preparation for this inevitability is to anticipate it and pick a microcontroller based around a popular core to minimize the workload of porting to a new processor when circumstances demand it. Generally speaking, there are six very widely used 32-bit cores on the market at the moment: Motorola 680x0, Intel x86, PowerPC, MIPS, SuperH, and ARM. Numerous less popular or proprietary architectures also exist, of course; many of these are associated with specific applications such as laser printers or DVD players.

> **Note:** that we mention only general-purpose microprocessor cores here. DSPs are a separate world beyond the scope of this chapter.

At the risk of antagonizing its userbase, I recommend against choosing the 680x0 series for a new design. Use of this core appears to be in decline, and it is perhaps actually close to the end of its life; the principal consumer use at this time is in PalmOS® devices. These PDAs are now migrating toward ARM, and even Motorola has introduced an ARM-cored processor as its new flagship PDA part. The entry-level laser printer market, which formerly consumed a lot of MC68000 and MC68008 parts, has largely been dominated by cheap devices that lack a rasterizer (they rely on the driver software running on the attached PC), so they only require simple servo control on the printer mainboard.

Architectures based around the high-end x86 family (and code-compatible parts from AMD, National Semiconductor, Via Technologies, etc.) have some immediate advantages:

- You can use almost any PC-compatible operating system and free software development tools.

- Installing operating systems is simple; in most cases there are automated installers that will probe your hardware combination and automatically install appropriate kernels, drivers, etc. Compare this to the norm with embedded systems, where you will need to look at the board, work out the hardware configuration yourself, and sysgen the kernel and driver set on external hardware, probably using a cross-compiler.

- It is simple to interface literally thousands of peripheral components for almost any imaginable function. Because these components are produced for the consumer market, with its enormous volumes and bloodthirsty price competition, peripheral components are cheap and fairly easy to acquire.

- Driver support exists (within the framework of most off-the-shelf operating systems) for almost any piece of hardware you could want to attach to your system.

- Highly integrated mainboards are available with many possible combinations of peripherals, in a wide variety of form factors.

- Migrating to a slightly different hardware platform due to shortages of support parts or evolving customer needs is relatively simple; in many cases, it simply involves recompiling and reinstalling the operating system and preparing a new master disk image for duplication.

Having extolled the obvious virtues of these parts, we must also point out some of the downsides:

- x86 parts are very expensive, in production quantities, compared to RISC alternatives of comparable performance. This may affect your ability to commercialize your device.

> **Note:** The first statement above needs qualification. Although the x86 CPU is quite expensive, you might find that a given system configuration is cheaper when built around an x86 than a RISC processor such as PowerPC because of the significant economies of scale in producing large volumes of the x86 board.

- There are relatively few x86 variants that are true "system on chip" devices, so you are likely to need quite a bit of external hardware in addition to the microprocessor itself. Often, to obtain one specific function, you will need to add a complex multifunction part because the single function you want isn't available as a discrete component. Again, this brings up your system complexity and total bill-of-materials cost.

- x86 has significant power consumption, heat, and size disadvantages. (The Transmeta Crusoe x86-compatible device combats these disadvantages, but it is currently rather expensive and not very many vendors have products based around this microprocessor.)

- Modern x86 parts and their support chips are very high-speed devices in dense packages. It is virtually impossible to hand-prototype your own design based around these parts; unless you want to spend many thousands of dollars on equipment, at the very least you will have to contract out some assembly work.

- PC peripheral ICs often have very short production life spans; twelve to eighteen months is not uncommon, so ongoing sourcing may be an issue.

- Code to cold-boot a "bare" PC platform is usually very complicated because you have to replace numerous layers—motherboard BIOS, expansion card BIOS, and various OS layers. The CPU architecture is also complex.

- Although I personally don't consider this to be a serious downside, it bears pointing out that JTAG-based or other hardware debugging systems aren't usually available on commercial single-board x86 computers.

I recommend x86 as the platform of choice if you are either building just a few of your appliance or if you are prototyping something and want to pull together a lot of miscellaneous hardware features without spending a great deal of time debugging the hardware design. It's also a good choice for an initial production run that you can ship to early adopters while you are developing a cheaper second-round customized hardware design. There are other special situations where you might find x86 to be a good choice, but these are the major ones.

Of course, you aren't restricted to using Intel parts; for instance, one x86-compatible part that is fairly popular in embedded applications is the Geode series from National Semiconductor (based on intellectual property acquired from Cyrix). This part was designed for Internet appliances and can be found in several such devices on the market today. There are also numerous single-board computers built around Geode chips, with various peripheral functions according to the intended application. Geode was also used as the reference platform to develop and showcase the new Microsoft Smart Display device, so the product family is likely to be supported for quite a while.

Using x86 also doesn't mean that your device needs to have a large PC motherboard and expansion cards inside it. Unless your needs are highly specialized (and perhaps even if they are), it is probable that you will be able to find a single-board computer with most or all of your required hardware already integrated. These boards range in size from "biscuit PCs" with the same footprint as a 5.25″ disk drive down to a fairly new standard (consisting of a user-designed baseboard holding an off-the-shelf module containing the CPU and some peripherals) usually referred to as ETX. Embedded computer boards like this typically have PC/104 expansion buses (a condensed, stackable version of ISA using 100 mil headers) or Mini-PCI. Some of the larger boards will have regular PCI slots, but these start to make the overall system unavoidably rather bulky, approaching the size of a normal slim-line PC.

Note that PC-compatible SBC pricing falls into two widely separated categories: industrial and commercial. Industrial SBCs are *extremely* expensive—at least twice the cost of commercial versions. Commercial SBCs, though substantially more expensive than consumer-grade PC hardware of the same nominal specifications, are a much better choice for the budget-constrained purchaser. Many SBC vendors specialize in industrial automation only, so if the prices you are being quoted seem unrealistically high, you should investigate other vendors before concluding that x86 is too expensive for your project.

Moving onto the RISC platforms, MIPS, SuperH and PowerPC are good candidates for many applications, and in particular the SuperH family is large and contains a wide variety of useful devices, though MIPS seems to be a more widely licensed core in third-party ASICs and ASSPs. PowerPC seems to be found mainly in applications requiring very high performance. In evaluating all these parts for various projects, I have found them to be fairly difficult to develop with on a shoestring budget; evaluation hardware is usually costly, and most variants of these parts are not readily available to buyers who are unable to demonstrate a need for large quantities. However, all these cores are likely to remain available and well supported for the foreseeable future, so they are all viable choices as long as you can obtain development systems and parts.

At least in the case of SuperH and MIPS, your cheapest path to a prototype based on these parts is generally to repurpose some existing piece of hardware such as a PDA; for PowerPC, I would suggest buying a commercial single-board industrial control computer based around the chip of interest. Be warned that this is likely to be expensive; PowerPC boards don't have the same kind of mass-market pricing as x86-compatible boards and you can expect to pay between two and three times as much for a PowerPC SBC as for a comparable x86-based board.

Bearing the above discussion in mind, unless some of the Intel arguments apply to your case, my primary recommendation for a 32-bit embedded platform is ARM. This architecture has many important advantages (some of these are also applicable to the other RISC platforms mentioned above, of course):

- It is a mature, well-understood architecture with a solid engineering history and many refinements. The large number of current licensees and now-shipping parts makes ARM a very safe bet for future availability.

- The cores are small and have excellent power consumption vs. performance characteristics.

- Many features—coprocessors, external bus widths, memory-management unit, cache size, etc.—are tunable by the chip designer, meaning that a core variant can be found to meet almost any performance/size/power requirement.

- There are a huge number of attractively priced standard, custom, and semi-custom parts on the market with a wide variety of integrated peripherals.

- Since ARM provides reference designs for many different peripherals as well as the core itself, there are often similarities in peripheral control on different ARM implementations, even from different vendors. To take a trivial example, code to send data out of a serial port can usually be ported from one ARM variant to another with little effort.

- Partly due to the above factors, there is a huge amount of freely available intellectual property—reference designs, ready-ported operating systems, etc.—already extant for this core.

The cliché is that "ARM is the 32-bit 8051," meaning that it is the universal 32-bit microcontroller core known to everybody and used everywhere. This is barely an exaggeration; ARM is to the embedded world what x86 is to the desktop PC world.

It's important to keep your priority—low *overall* development cost—in sight at all times during the selection process. For example, I almost always reject parts that are only available in BGA packages, because it is practically impossible to hand-build prototypes around these devices, and it's costly to hire an external contract assembly house to build your initial development boards. You'll also need to consider the price and availability of evaluation hardware for the devices you're comparing, as well as the complexity of building a working hardware platform of your own. For example, a chip that requires complex analog support circuitry and careful PCB layout will be very difficult to work with in a hand-prototype environment. For such a chip, you would quite likely be better off investing in an expensive known-good evaluation board before attempting to build your own PCB. Diving straight into the deep end by designing your own board around such a part is likely to be costly because of the need for several respins of your board to resolve layout-related and other analog issues.

7.3 Building Custom Peripherals with FPGAs

While you are evaluating different chips for your application, you are likely to find yourself tempted by specialized system-on-chip devices offered by various manufacturers. These chips will have interesting peripherals specific to various applications—for example, dedicated motion compensation and colorspace conversion hardware for digital video playback or discrete cosine transform (DCT) engines for image compression, typical in devices intended for the digital camera market. Unfortunately, these are usually precisely the sorts of devices that are unobtainable to the hobbyist or small-scale developer. They are usually only available with solid, up-front quantity commitments, and often nondisclosure agreements are also required. In some cases, just to view the data sheet for a part, you will need to pay large fees to join some kind of specialized industry cartel. (DVD/DVB playback hardware can be like this, for instance, because of the numerous patents in the field and vested copyright interests at stake.)

Because of this annoying fact, one of the most useful money-saving skills you can acquire is experience working with synthesizable hardware design language (HDL) code on CPLDs and FPGAs. Using such devices, you can design your own custom peripherals, optimized for your specific application, and avoid the trouble of trying to source a rare ASSP. FPGAs are available off-the-shelf in many different packages and complexities, and in many cases the manufacturers supply free development tools.

In fact, there are now products available, such as Altera's Nios® and Excalibur™ devices, which consist of a high-performance RISC core "wrapped" in an FPGA, all on the one chip. Nios is a proprietary microcontroller core; Excalibur is built around a high-performance ARM922T core. With a part like this, you can effectively create your own custom ASIC; it is an extremely powerful tool and it seems likely that we can expect to see many more such devices in the future. ARM and other vendors also supply some cores in soft form, so you could in theory build your own entirely customized system-on-chip using a generic FPGA device. However, because of the hefty licensing fees involved, the per-unit breakeven point is only reachable with very large production volumes.

If you plan to use FPGAs, much as with microcontrollers you will find that the manufacturer-recommended evaluation boards and commercial development tools can be very expensive. Trenz electronic (www.trenz-electronic.de) is one possible source of lower-cost FPGA boards. However, you might not even need an evaluation board—FPGAs are, after all, *field*-programmable, and the interior functionality is controlled by the firmware you upload to them, so you can be fairly confident about dropping an FPGA directly onto a first-run prototype PCB and debugging your design in-circuit. If you've never used FPGAs before, however, I would advise getting a small evaluation board with which to experiment. Connect the I/O lines to pushbuttons, LEDs, or perhaps an RS232C level-matching IC like the Maxim MAX232A and play with the device to see what you can achieve with it.

Since I'm talking about field-programmable logic, I should also mention Opencores (www.opencores.org), an invaluable resource of free, open source intellectual property ready to be compiled into your FPGA. If you need a core of some sort—a UART, for example, or a DRAM controller—then before starting to write your own, you should visit Opencores to see if there is already a free core available for you to adapt. Opencores is something like the Linux of hardware; at the time of writing, there are free cores for SDRAM controllers, UARTs, cryptographic hardware, microcontrollers, a VGA/LCD controller and many others.

7.4 Whose Development Hardware to Use—Chicken or Egg?

The textbook development cycle recommended by chip vendors is as follows:

1. Choose a microcontroller from the vendor's selection matrix.
2. Buy the vendor's evaluation board for this part.
3. Buy one of the commercial compilers and possibly a hardware debugging module recommended for the evaluation board.
4. License one of the operating systems recommended for the evaluation board.
5. Develop your application *in vitro* on the evaluation board.
6. Develop your hardware.
7. Port the operating system and your known-good application to the real hardware.

One of the driving ideas behind this methodology is that the software team doesn't have to wait for the hardware team to finish designing and debugging the circuit. Unfortunately, as with most textbook descriptions, the cycle described above ignores some important realities, not the least of which is that in many small shops, the job of both the software and hardware "teams" will be performed by a single person.

The evaluation board and software tools recommended by the chip manufacturer are usually expensive, for reasons touched upon in the introduction to this book. Additionally, if you intend to use complex off-chip functionality, it can be extremely difficult to attach this to an evaluation board. For instance, if you intend to implement a PCMCIA socket in your appliance and the microcontroller evaluation board doesn't include one as an option, it could be hard to hand-build a PCMCIA interface board and harder still to graft it onto the evaluation board. The majority of 32-bit parts are quite closely targeted at specific applications; evaluation boards tend to have all the hardware required to demonstrate the maximum possible bells-and-whistles configuration of the CPU's intended application, and this can get in the way of adding your own peripherals to the evaluation board. For example, I was once evaluating a chip targeted at the PDA market. The appliance I intended to build wasn't a PDA, so I didn't need most of the hardware on the evaluation board—audio I/O, Ethernet, color LCD, touch screen, USB interface, etc. Not only did I have to pay for all these peripherals (this particular evaluation board is US$1500, and the microcontroller itself only costs about US$12), but I had to cut several dozen traces, remove a 160-pin surface-mounted chip, and add literally a couple of hundred patch wires in order to be able to bolt on my own peripherals.

Finally, and following on rather neatly from that anecdote, you should remember that the time required to understand the memory map and any special quirks of the evaluation board, and to get its specific combination of hardware running, is time that you are "stealing" from the task of getting your own circuit debugged. This is an acceptable price when you have a large team working simultaneously on the hardware and firmware of the final product, but in a smaller or even one-person environment working on a tight time budget, it is often more efficient to design your own circuit and start working directly on your own hardware.

There are three major ways around these problems, in roughly increasing order of difficulty:

- Locate a third-party demonstration platform for the part of interest.

- Locate a consumer appliance based on the chip that interests you and reverse-engineer it enough to load your own firmware and patch on your own hardware.

- Design your own PCB and have it etched and populated either locally or (if this is a commercial project) by your factory; develop your firmware on this board while debugging the hardware at the same time.

The first option is rarely available but is usually well supported by the board manufacturer. I should point out that in some cases it can be difficult to use these development boards unless you also possess a hardware debugging module such as a JTAG pod. Most difficulties center around how to upload initial bootstrap code to the board. Some microcontrollers, such as the Cirrus Logic CL-EP7212 and 7312 parts, contain a tiny on-chip bootstrap ROM that allows you to upload code to RAM over a serial port. You can implement your own Flash loader quite easily using this method and thereby load your own code onto any board that has a serial port. Some evaluation board vendors will supply the board preloaded with a ROM monitor such as Angel or gdb stubs, and you can communicate with this monitor over a serial link. In a few instances, the board will feature socketed EPROM or Flash memory devices, which you can simply remove and reprogram with your own code. Unfortunately, in a handful of cases, the board is shipped with blank, soldered-down Flash memory and there is no way of getting new code into it short of buying a JTAG pod or some other specialized hardware device. Third-party "demo platforms" tend to be devices that were originally designed for some specific purpose, then later sold to hobbyists with no housing but more detailed technical documentation.

Easy field reprogrammability with minimal external equipment may not have been a design criterion of the original appliance.

Repurposing consumer appliances can vary in complexity from extremely simple to downright impossible, depending on the microcontroller you're interested in and its target market. It can be exceedingly difficult to locate a consumer appliance based on the specific chipset it contains, and you will often need to do quite a lot of reverse-engineering to determine memory maps and so forth. It also isn't necessarily cheap to cannibalize a brand-new appliance, though it's almost always cheaper than buying an expensive evaluation board. The repurposing approach does have advantages for projects that meet certain prerequisites; in particular, it works best when you have a fairly good idea of the hardware capabilities you need (at a macroscopic level, e.g., "Must have Ethernet," "Must have TV output") but you don't much care what specific parts are used in your hardware platform. As a result, this method is particularly attractive for hobbyist and student projects that are very price-sensitive and don't need to worry about ongoing component availability. People in this category can revel in the rich variety of items available on today's surplus market.

The third development option, prototyping directly on your own circuit and debugging the hardware and firmware simultaneously, is the option I personally use most often. Although this method is common for low-speed 8-bit circuits, it is fairly rare in the development of 32-bit systems. However, I find it necessary to work this way because most of the projects I work on involve bringing together several fairly complex devices that aren't found together on any pre-existing evaluation platform. This method does have the advantage that you can tweak the hardware design to simplify firmware development right up until the last PCB revision before manufacture. Unfortunately, it also has the disadvantage that any bottleneck in the hardware development timeline is also a bottleneck in the software development timeline, which unavoidably pushes your delivery date further out.

I should warn you that prototyping like this is similar to bungee jumping: just one catastrophic failure, and you won't get a second chance. If you make a really fatal, unpatchable error in your PCB, in the worst-case scenario you will have to throw it away (and more than likely the parts on it too; hand-reworked surface-mount devices have high failure rates) and halt firmware development until the next batch of boards arrives. This can make the process expensive, but with careful fault analysis and rigorous checking of your work before submitting a PCB layout for manufacture ("measure twice, cut once"!), you can keep the expense to a minimum.

To summarize the above choices succinctly:

- *If your code can be developed on a readily available, affordable development board* (either third-party or direct from the chip manufacturer), you should use this development board as your prototype hardware platform.

- *If you are building a one-off piece* (e.g., a student project or technology demonstration), if you are *certain* you will never need to build more such units, and if you don't need to build around any specific component, your easiest route may be to repurpose a piece of consumer equipment with appropriate hardware features.

- *If you are designing around a specific component or combination of components and either the available evaluation boards are too expensive or it isn't feasible to add the peripherals you need to them,* your best option is to design your own circuit, make a couple of prototype PCBs, and debug the application directly on your own hardware.

If none of the above options seems to be right for your application, I suggest that you develop and demonstrate your software on an embedded PC type platform and use this demonstration to secure sufficient funding to pursue one of the options above.

7.5 Recommended Laboratory Equipment

One question that arises frequently at this point is "What other equipment do I need to buy to equip my laboratory?" There seems to be a fairly widespread belief that developing high-end embedded systems requires a great deal of expensive specialized hardware: storage oscilloscopes, logic analyzers, in-circuit emulators, and so on. Although this equipment can sometimes be useful, the truth is that expensive state-of-the-art equipment is only absolutely necessary for a few special applications. For example, when developing cellular phones, to test your device without causing annoyance to local cellular carriers and the public, you need to be able to emulate a cellular network. To debug circuits that have extremely high-speed buses or delicate RF or analog sections, you might also need some extra equipment, but for a large number of embedded designs, your needs are unlikely to exceed the following major appliances:

- *A reasonably feature-rich multimeter.*

- *A good analog oscilloscope.* Steer clear of generic, no-brand, entry-level scopes intended for the hobbyist market (even if you *are* a hobbyist). You'll find much better value in a refurbished piece of name-brand equipment. A quick search of the Internet will show you a large number of dealers who specialize in sales and rental of refurbished test equipment. (You can also buy secondhand equipment from auction sites like eBay, but secondhand test equipment from a private seller frequently needs recalibration, especially after being shipped a long distance. It is often worth the additional cost to buy a certified, properly packed unit from a reputable vendor of refurbished equipment.) Brand-name units (Tektronix and Hewlett-Packard are the two most popular) that were state-of-the-art three to five years ago are now very affordable and more than adequate for most tasks. Your exact needs will obviously depend on what you're developing, but I would recommend a minimum 150 MHz bandwidth two-channel scope and 10x probes. Look for scopes with many triggering options—these options give you different ways of focusing on the specific section of the waveform you're interested in, and the more flexibility you have there, the better.

- *A laboratory power supply.* It should have at least two independently adjustable DC current-limited outputs (30 V is the maximum you're likely to need), with inbuilt current and voltage indicators.

- *A bench-mounted illuminated magnifier.* This item is mandatory when working with surface-mounted parts, and it's useful even when working on larger packages.

- *A temperature-controlled soldering iron.* Always keep a few spare tips on hand, also—especially if you work with surface-mount packages, you will want to keep at least one tip filed to a very fine point. This point will erode quickly and you'll need to keep filing it down as necessary.

If you're working on something that will be powered from household wall current and that you intend to distribute to other people, it's also a wise idea to have a variac on hand so that you can test how your device will behave in mains brownout conditions, but this isn't essential.

Note that I haven't mentioned a digital oscilloscope. If you do want to buy one, by all means do so, but I suggest you make it a secondary purchase after acquiring a good analog unit. The main reason for this is simply cost; the same money will buy a much more capable analog than digital scope. Digital oscilloscopes are a time-saving luxury rather than an essential for many applications. I have a reasonably powerful digital scope on my workbench, and I rarely power it up. In fact, I most commonly use it when I run out of channels on my analog scope and I need to look at a large number of signals simultaneously.

I also recommend, in general, against the false economy of oscilloscope add-ons for PCs. The quality of the analog-digital converter side of these software/hardware packages is critical to the usefulness of the device. Expensive, high-speed data acquisition cards are outside the cost range of interest to the average reader of this book; cheap 8-bit digitizer devices with no internal buffering (typical of low-end PC oscilloscopes, especially those sold in kit form) are not money well spent, in my view. This type of hardware might be useful if you know you will be spending a lot of time looking at and storing signals at audio frequencies (up to a few tens of kilohertz); you can use the device as a poor man's logic analyzer. As a primary signal inspection tool, I feel this hardware lacks flexibility and, at worst, may be very misleading and counter-productive because it hides information that might be vital for debugging purposes.

7.6 Development Toolchains

A large majority of 8-bit and smaller embedded systems in the real world use proprietary (if any) operating systems, often written using a monolithic assembler/linker package.

("Proprietary" in this context means "developed specifically for one product or family of products," rather than the more general English meaning of "exclusively owned.") A great deal of literature for the embedded field deals with specifics about close-tolerance timing (cycle optimization of code) and single-byte memory-saving techniques. Professional debugging toolchains for these parts often center around using a hardware in-circuit emulator for the microcontroller to simulate the processor *in vivo*, capturing and analyzing its behavior in real time by means of an attached PC.

Design processes and priorities are usually very different when we're targeting 32-bit parts. To begin with, these parts are so fast that hardware emulators are unfeasibly expensive and almost all debugging is performed on the real microcontroller. (Sometimes the microcontroller itself is used as a kind of in-circuit emulator using the JTAG interface. However, this serial interface is too slow for full real-time debugging.)

Also, particularly in the case of a demonstration or hobbyist project, the designer would probably like to avoid handcrafting all the code necessary to bring up a complex system, which implies that some kind of ready-made operating system will be used where possible. RAM and ROM are usually plentiful, making it unnecessary for users to spend a great deal of time squeezing a few extra bytes' efficiency out of their code. Algorithms are also much more complicated and have more points of interaction with each other and the external environment, requiring a significantly different style of design rigor.

As for cycle-exact performance issues, pipeline and cache features on these more advanced processors make hand-optimizing assembly language programs *extremely* difficult; in fact, instruction timing on a cached, pipelined CPU core under varying system load can be so complex that these systems sometimes actually appear to be nondeterministic. Optimization for speed is generally best left to a high-level language compiler on 32-bit platforms. Only if observed performance is inadequate and *actual profiler results* point to a specific area of the code is it generally worth the effort of hand-optimizing in assembly language.

Given these differences, which tools do we choose for our exciting new 32-bit project? With a few rather rare exceptions, the choice of embedded operating system will mandate the choice of a particular toolchain. Despite the proliferation of fairly well-defined binary file standards such as ELF, COFF, and PE, differences in such compiler- and linker-specific behavior as symbolic debugging information, special directives for memory allocation, and C++ name mangling semantics usually make it very difficult to move operating systems from their intended compiler to an alien compiler. This problem is even worse with operating systems that are shipped partly or wholly precompiled, without source code. Although it is possible, in some cases, to force specific combinations of products to work together (e.g., object files compiled with the ARM Developer Suite can be massaged to link with code generated by gcc), this is rarely a wise expenditure of time.

Keep in mind that this interrelationship works in reverse too—in other words, if you don't want to spend the money on a costly commercial toolchain, this is probably going to limit your choice of operating systems. This section focuses on platforms that are supported by free compilers. For all practical purposes, this means platforms supported by the GNU tools; gcc et al. There exist a few free, manufacturer-supplied proprietary compilers, but these vary widely in quality and are generally nonstandardized. Unless your chip or operating system vendor is going to supply you with a huge variety of free, useful intellectual property in the form of libraries that can only be linked with the proprietary compiler and for which you can't obtain open source equivalents, I strongly advise that you stay on the far better-traveled path of GNU tools. It's hard to imagine any algorithm from cryptographic applications to video decoding for which GNU or other open source intellectual property isn't already available. Freely available source probably won't be optimized for your hardware platform and will require some tweaking for best results, but even so the benefit of having the source code is very significant.

I should pause here to point out that if you are using the Intel x86 family for your platform, there are at least two other viable free compiler options for you. Borland has released the command-line version of Borland C++ 5.5 as a free download, and the Watcom C++ compiler (now owned by Sybase®) is in the process of being released as an open source product named OpenWatcom (www.openwatcom.org). OpenWatcom is not available for general download at the time of this writing, but when it does finally make it to the outside world, it should be a very exciting product. Watcom C++ supports numerous Intel targets—Win32, Win16, OS/2®, Novell® NLMs, and both 16-bit and 32bit DOS. With a little external massaging, it can be used to develop almost any x86 code for embedded platforms, especially when combined with a free operating system like FreeDOS (www.freedos.org). In the heyday of DOS, Watcom C++ was also famous for generating highly speed-optimized object code for DOS-based games, which may be an interesting advantage for your application.

The GNU suite is a software-only toolchain, meaning that we need to establish our own link to the target hardware for code uploading and debugging. Most of the 32-bit parts we mention in this book, and certainly virtually all ARM-cored parts (including ASICs), include on-chip JTAG hardware debugging support. For those who haven't used it, this is a simple serial interface that allows external hardware to halt the processor core and inspect and manipulate its state. (The JTAG interface can, of course, be used to directly manipulate other on-chip hardware. However, doing so would require device-specific knowledge of the on-chip peripherals. By taking control of the core, we can generate read and write cycles to access other system hardware without proprietary knowledge of each different microcontroller.) Through this mechanism, it is possible to generate read/write cycles that appear to originate from the core and thereby operate and examine on-chip peripherals and external hardware. To use this interface, you need a JTAG pod; these range in complexity from fully autonomous standalone units that connect to your computer via Ethernet to simple devices that level-match and buffer your PC's parallel port signals onto the target's JTAG port pins. The only readily available

JTAG pod I have found that lies within a small budget is the Macraigor Wiggler, illustrated in Figure 7.1.

Figure 7.1: Macraigor JTAG Wiggler.

The Wiggler belongs to the category of simple parallel port devices; however, it is an extremely powerful tool. With it, you can halt the processor and inspect its state, as well as being able to read and write hardware registers and other memory locations. This capability will save you a lot of time when you're working out how to bring up a new system; instead of having to recompile, upload, and test your bootstrap code iteratively, you can simply connect the debugger and tinker with the hardware registers directly until the peripherals are behaving the way you want them to. Moreover, because the JTAG interface is entirely hardware-based (on the microcontroller end), you can use it to breathe the "kiss of life" into a board with blank Flash memory.

There are a few hardware projects that duplicate the Wiggler's functionality (it's a very simple device); however, the really tricky part is not the hardware but learning the scan chain codes for the chips you intend to debug. This information is usually closely guarded by the chip manufacturers, and you really need to be a large corporate entity to have access to it. For this reason, I recommend sticking with a hardware vendor like Macraigor that has good relationships with the chip vendors, to ensure ongoing support for new parts.

7.7 Free Embedded Operating Systems

Having introduced the common choices for free development tools above, let's briefly explore some of the operating system choices available to us. Fortunately, the open source movement has generated a plethora of free or nearly free operating systems, probably the best-known of which is Linux. One great advantage of Linux is that not only has it been ported to a great many architectures, but the install process for many reference platforms is relatively well documented. Being able to download a working, precompiled kernel and fairly precise installation instructions will save you an enormous amount of frustration at the start of a new project.

In the last year or two, Linux has also attracted quite a lot of attention from the embedded world, and as a result we are starting to see some embedded-specific features emerging in the mainstream Linux code. For example, current kernel versions directly support ROM-based file systems (including compressed file systems) as well as several forms of Flash technology, including NAND Flash (SmartMedia et al) and M-Systems DiskOnChip.

For some applications, it may be valuable to note that "pure" Linux has three important limitations:

1. It requires a hardware memory-management unit (MMU) in the target processor.

2. It is not, strictly speaking, a real-time operating system.

3. It is licensed under the GNU General Public License, which may have privacy implications for your own code.

The reason I qualified the second point above is because off-the-shelf Linux can often be thought of as "real time, for small values of real time." In other words, stock Linux may be quite real-time enough for your needs, especially if you are willing to massage the kernel a little. Developers who are accustomed to working with actual real-time operating systems will doubtless cringe at my cavalier treatment of this issue, but for many noncritical applications, simply using a fast enough processor and removing unnecessary background tasks will be sufficient to ensure that your application gets enough processor time to *appear* to be working in real time. The difference between this and a true RTOS is that the RTOS will have APIs to *guarantee* that, for example, a level 0 interrupt will be serviced within 2 ms of the hardware receiving the interrupt request, or that a given process will always get at least 25 ms out of every 100 ms of processor time. Whether or not you can get away with a non-real-time operating system depends on your application; principally, if physical or financial safety depends on your appliance being truly real time, then you *must* either use a true RTOS or modify your existing OS so that you can guarantee that any critical code will be allowed to run when it needs to.

If you need a truly real-time version of Linux, there are a few options open to you, but probably the best-known is a commercial distribution called Hard Hat Linux from Monta Vista Software (www.hardhatlinux.com). Monta Vista also makes a specialized version of Hard Hat Linux for the telecommunications industry. Another option, and one that you can download freely, is RTLinux (available at www.fsmlabs.com). Despite what you might be told, the "real-time" versions of Linux are not really fundamentally different operating systems; they essentially consist of a small real-time subsystem melded to a normal Linux system. If your real-time needs are modest, you might be able simply to add your own minor patches to the Linux kernel to run your own critical tasks when necessary, rather than inheriting any idiosyncrasies of someone else's "two-pronged-kernel" real-time Linux design.

If you need a version of Linux that will run on microcontrollers lacking a memory-management unit, there is also a version to accommodate you: ucLinux (www.uclinux.org). ucLinux is a public project with a strong leaning toward projects that involve repurposing existing appliances such as Palm PDAs. The ucLinux website also features links to some interesting, moderately priced hobbyist 32-bit development boards based on processors such as the Motorola Dragonball series.

Without a doubt, Linux is the operating system *de rigeur* in the hobbyist arena. Partly because of the percolation of hobbyists into the commercial world and partly simply due to the operating system's own merits, there is large and growing commercial use of and support for Linux-based embedded solutions. For some examples of this support, you should visit www.linuxdevices.com, which is probably the most comprehensive portal site for news of the embedded Linux world. There are a surprising number of product announcements from major vendors aiming at the consumer electronics market. Linux's position as the server operating system of choice on the Internet seems to have helped to make it the top contender to run the next generation of networked home entertainment and other appliances. (It's also well worth visiting linuxdevices.com when you're searching for a ready-built hardware platform for some embedded application or even just for prototyping purposes. The site contains numerous interesting articles and product lists for various embedded computing platforms that can run Linux, and of course there is no reason that you couldn't load your own operating system onto one of those boards. Some reviewers of this text have pointed out to me that there's almost nothing at this portal site that you can't find by some reasonably diligent Web searching, but after all, the primary purpose of a portal is to collect audience-targeted information into one convenient location so you don't have to do the searching legwork yourself.)

Another popular free UNIX variant is NetBSD (www.netbsd.org). This operating system has one major advantage over Linux: It is unconditionally free. (Although the NetBSD operating system kernel is covered by a virtually unrestricted free license, individual components of a distribution may be covered by different licenses such as GPL.) Like Linux, NetBSD has been ported to a huge variety of platforms and supports a wide range of miscellaneous hardware. The main disadvantage to NetBSD is that it has not attracted very much attention from hardware OEMs, at least compared with Linux. The Linux community is sufficiently large and vocal that hardware vendors generally provide at least token support, whereas NetBSD is a poor cousin, relatively speaking. There is a fair amount of code interchange (within licensing limits) between NetBSD and Linux, and a large number of Linux projects can be rebuilt on a

NetBSD base, but overall if you are looking for sheer breadth of ready-made hardware drivers and availability of peer support, Linux is probably a better choice. However, if it is important to you to keep every line of code you write secret, you should look more closely at NetBSD. Although with due care and attention to licensing details you can build a Linux system that doesn't require much (if any) disclosure, you may find it easier to get NetBSD past a reluctant management team who has been frightened by or is otherwise doubtful about the legal status of open source projects. You can simply tell your managers that NetBSD is unambiguously free, there is no disclosure of source code required, and that will (hopefully) be the end of those managerial objections.

Linux and NetBSD are both very "heavy" operating systems; they require a relatively large amount of RAM and nonvolatile storage space (ROM, Flash memory, or another device such as a hard disk). This can be mitigated to a certain degree by very carefully pruning the kernel and deleting unnecessary binaries and libraries, and by using special slimmed-down system libraries, but neither product was originally designed for embedded systems. Both products are also very flexible general-purpose operating systems, and of course this flexibility comes at a price.

A slightly lesser-known free operating system but one with growing popularity is eCos from RedHat (sources.redhat.com/ecos). The great advantage of eCos is that it was purpose-built from the ground up as an embedded operating system, unlike Linux and NetBSD. Although it is monolithic in the sense that it compiles into a single library that you link with your own program (as opposed to being a heterogeneous collection of executables, configuration files and libraries that need to be stored in some kind of file system), eCos is a very well-designed modular operating system. The presence or absence of drivers for various hardware, and all configuration options, are controlled easily with conditional compile macros. RedHat even includes the unaccustomed luxury of a graphical configuration editor that lets you set all the build options with checkboxes, drop-down lists and so forth, and build the operating system library with a single keystroke.

eCos can also be compiled for operation from RAM (extremely useful for debugging; you leave the ROM monitor in control of the board and simply upload new versions of your application as you debug it), ROM (useful when you go to burn the firmware into your device!) or a combination of RAM/ROM startup, where the code is initially located in ROM but relocates itself to RAM for performance reasons. The operating system is supported by a highly flexible bootloader called RedBoot; this bootloader is a very interesting product in its own right, since it offers a simple command-line loader accessible over serial or Ethernet (where supported), Flash rewriting commands, and other useful functionality.

At the time of writing, there are basically two publicly available versions of eCos and its support tools: an ancient "official release" and the current CVS version. (CVS is a version-control tool commonly used in the free software world.) If you intend to play with eCos, download the current CVS version. Instructions for doing this can be found at the eCos website, sources.redhat.com/ecos. The "official release" version is ancient; the CVS version, though it is something of a moving target (since it is not a frozen version, it changes frequently), supports many more hardware platforms and has many more features than the old release. If you're using Windows, however, I do suggest you download and install the official eCos release and then update it with the latest CVS version. By doing this, all the necessary default configuration information, registry values, and so on can be initialized by the automated installer. Be sure to read all the download pages carefully, however—all the old utilities supplied with the release version of eCos must be updated manually with newer versions if you are using the CVS version of the operating system source code.

Another operating system which isn't truly free but is *effectively* free is the Palm OS. The reason I describe it as "effectively" free is that the only way you're likely to be using this operating system in a shoestring-budget project is if you're implementing your project as an application running on a dedicated Palm device (or third-party compatible; Sony Clié, Visor, IBM WorkPad, etc). Since the operating system comes bundled with the hardware platform and free development tools and documentation are available, shipping applications based on this OS is basically free. In fact, quite a few niche market products work precisely this way; you pay for an off-the-shelf Palm device preloaded with custom application software and possibly some special external hardware such as a GPS receiver, barcode reader, or digital camera. An obvious advantage of implementing your project in this way is that as new and more powerful hardware platforms become available, you can upgrade to them quite painlessly; Palm will handle all the work of porting their operating system to the new hardware and you can reap the benefits. (A similar situation applies to Windows CE. At least at the time of writing, you can download free Windows CE compilers at Microsoft's website.) This technique, however, barely falls under the heading of embedded systems development, and so I will not discuss it further here.

Of course, depending on what functionality you require, it might not be necessary to port and bring up an entire operating system simply to acquire some ready-rolled functionality.

Some vendors provide modular packages for specific functions (these are usually supplied as precompiled libraries, so make sure that they can be linked with your toolchain of choice). For example, US Software (now owned by Lantronix) sells standalone modules for TCP/IP networking (USNet®) and DOS/Windows-compatible VFAT file systems (USFiles®), in addition to several embedded operating systems.

It might also be feasible for you to "mine" small fragments of code out of an existing operating system and create your own libraries. If you are thinking that this latter path is the best route for your own project, remember that as the size of the code piece you're extracting increases, so do the number of structural assumptions you're inheriting. For example, if you want to borrow a file system driver out of an operating system, you will have to either modify it heavily to fit your own code or duplicate the file descriptor semantics at the top end and the low-level disk-access device driver semantics at the bottom end, not to mention task synchronization primitives and so on. Effectively, you may find yourself emulating or rewriting large segments of the operating system from which you borrowed your "single" piece of code.

Remember also that even if you start out building a prototype around a ready-made OS, it is entirely possible to "wean" your code off that OS at a future date—though it will be much easier if you start out by designing your code with this intention in mind. When you start a new project from scratch, it is very helpful to have some piece of code around that you can trust to work properly, even if only for reference purposes. This is especially valuable if that piece of code can teach you the correct method and order of initializing the components of a complex system. For example, I once worked on a project based around an ill-documented Super-VGA controller IC. The chip vendor actually supplied free reference source code to bring up the SVGA chip, but it wasn't complete and didn't work. Fortunately, they also provided a working RTOS preloaded on the evaluation board. I obtained the necessary magic register values to get my own code working by booting up the vendor's proprietary RTOS, letting it initialize the display control registers, and then dumping the entire chip state (including, as it transpired, many undocumented registers!) to a serial port for inspection.

Because of the possibility of issues like this, you might want to use a ready-made operating system (on your real hardware) to get your application up and running quickly, and gradually replace parts of that operating system with code of your own until eventually you have duplicated all the desired external functionality in your own application. This is an exceptionally valuable method of doing things when the only operating system that explicitly supports your reference platform has expensive royalty fees but is free for in-house research use. You can simultaneously cut your per-unit costs and your development time by starting your program out as an application on top of the expensive OS. Once you've determined what services you actually need out of the operating system, you can go through your code replacing all the operating system calls with your own hand-written versions of the same functionality. Once you're done, you have a shippable proprietary application that doesn't use any of the expensive third-party code. Obviously, this is a lot more work than simply writing your program around a ready-made operating system, but on the other hand it does save you a lot of debugging work in the initial bring-up stage, and it avoids potentially large operating system license fees.

Note that this technique is subtly different from the technique of prototyping your application code on some arbitrary hardware platform, with the intention of porting it to real hardware once the algorithms have been verified on the demonstration hardware. Using the method above, we are developing on our real hardware (or at least the reference platform we are using to develop the real hardware). At any point, we could bundle together the current code base, load it onto a piece of real hardware, and call it a shippable product (at least from a functional perspective); the only delay is caused by the need to remove expensive licensed code. By contrast, the *in vitro* code prototyping system doesn't result in a shippable product until the very end of the prototyping and porting process.

In the simplest case, you might not need to use an operating system or third-party libraries at all; you can roll your own entirely proprietary code.

7.8 GNU and You: How Using "Free" Software Affects Your Product

In the modern era, almost any nontrivial embedded project of the type we are discussing will require an enormous volume of essentially boilerplate code; TCP/IP networking, data compression, file systems (particularly MS-DOS-compatible FAT file systems; "Where can I get code to read a FAT-formatted hard disk?" is a frequently asked question in embedded newsgroups), audio/video codecs, and GUI libraries are common examples. Of necessity, therefore, implementing such a project from the ground up involves reinventing many wheels. At the very least, this is an inefficient use of your expert time. At the worst, it can mean a project that never gets off the ground because you don't have the manpower needed to get the pedestrian code finished so you can move on to building the value-added magic that makes your product something special and saleable.

In the past, these unpleasant facts could be worked around only by purchasing expensive commercial RTOS packages. However, in recent years, many free alternatives have become available and viable, and the use of open source "free" software in commercial ventures has been greatly legitimized.

> **Note:** "Open source software" is a politically loaded term with multiple more or less widely accepted meanings. In this context, I am using the phrase to mean "royalty-free software for which the complete source code is readily available without payment of fees."

Despite this legitimacy, there is still a state of confusion in the minds of many embedded engineers and entrepreneurs alike as to just what it means to use open source software—what rights and benefits it confers and what obligations it entails. This situation is not ameliorated by the fact that most of the outspoken experts in this field are vigorously pursuing commercial or political agendas and in many cases intentionally obscuring the facts. To fully understand the implications of using some of this free software, it is therefore necessary to be armed with at least small amount of background information about this political situation. Please note that this is intentionally only a brief description, and of course it constitutes neither formal legal advice nor a complete analysis of the social and legal issues surrounding any particular software license.

The reason you need to read this chapter is that when you're implementing a complex project, sooner or later you will be forced to choose between a proprietary operating system or a free product covered by some kind of "open source" license. Chances are good that you will be facing one or more salespersons and free software advocates, each of whom will not necessarily present you with complete information to make your decision. Depending on your organization's structure and history, you may also be combating misconceptions in management about the implications of using "free" software in your product. Free software, used properly, can be part of any totally reliable, legally sound, high-performance product; this approach to software development can no longer be considered trailblazing, and it remains only to select which type of free software you should be using.

The most popular free software license (in terms of lines of code freely available on the Internet, at any rate) is unquestionably the GNU General Public License, commonly abbreviated GPL. Most Linux software, for instance, including the Linux kernel itself, is released under this license, and most free software controversy in the public press centers around GPL. The rationale behind GPL, simply stated, is to force all derivative works of open source products to remain open source. (The actual rationale goes somewhat deeper than this; it is based on the idea that all software should be free, in the philosophical sense of the word; a true free software purist abhors the concept of closed-source applications.) The two aspects of the GPL that will affect you most are:

1. You can experiment with GPLed software as much as you want in private. You only "accept" the license and therefore become bound by its provisions once you "distribute" products derived from GPL code.

2. If you distribute a product that is derived from or closely linked to GPL code, your code must also be released under the GPL. This means that you must release source code (or disclose a means of obtaining the source code) to anyone who requests it. There is an important exception to this rule for the Linux kernel: You do not need to GPL a piece of software whose only link to the Linux kernel is that it calls kernel services using documented interfaces. The original intent of this rule was to allow people to develop Linux device drivers for products whose hardware documentation is covered by nondisclosure agreements (an intent largely nullified by later philosophical changes in the license), but it also provides a very useful way of allowing profit-making use of the large amount of engineering in the Linux kernel.

For in-house prototypes and private experimental research of all kinds, the first rule above is a largely unrestricted free ride. You can take an existing mostly GPL project (like a Linux distribution) and use it as the foundation for your prototype without restrictions. Once you're satisfied that your code and/or hardware are working nicely, you can decide exactly how to bring the product as a whole to market and remain license-compliant. However, you should plan now for what you intend to do when you commercialize your product. Otherwise, you'll demonstrate a fantastic but legally unsaleable prototype at a trade show, people will come to you ready to write orders, and you'll have a huge auditing and rewriting job before you can cash their checks. Your options are as follows, in ascending order of person-hours typically required:

- Release your entire product under GPL. This option can make a lot of sense, particularly when your product is largely special hardware that just happens to require control software (as opposed to general-purpose hardware running special software, where all the value lies in the bundled software). If you take this route, you can also ride a certain amount of bonus publicity from the free software movement, which will be only too happy to promote your product as an example of embedded engineering done right. This extra goodwill can be very useful in some markets. However, sometimes there can be other issues—typically, nondisclosure agreements required by other product vendors you work with, patents and various other trade secret problems—that preclude this option, even if you are personally willing to try it.

- Establish a clear separation between GPL and non-GPL code in your product, and open source only the GPL components of your software bundle. This technique is exceedingly useful when your product is based around Linux, because the Linux kernel exception to rule 2 mentioned above gives you a convenient place to draw the "GPL vs. non-GPL" line in your software bundle. The Linux-based TiVo digital video recorder appliance and Sharp's range of Linux-based PDAs (such as the Zaurus SL-5600) are excellent contemporary examples of this technique. All you are required to release are the special device drivers and other kernel modifications you may have written to get Linux up and running on your hardware; your application code remains secret.

- Determine exactly what GPLed functionality you're using, write your own implementation of all that functionality (or buy someone else's proprietary implementation), and remove all GPLed code from your software bundle prior to release. This is obviously the brute-force approach. I've listed this option last because it is usually the most labor-intensive, but this isn't necessarily true for all applications. If you're very careful to maintain an abstraction layer between your application and external libraries and operating system calls, or your application is of such a nature that it doesn't require many external services, this option might be the best for you. However, the applications that are easy to "de-GPL" in this way are precisely those applications that probably wouldn't have required importing a whole operating system in the first place.

Besides the special rules for the Linux kernel, there are some other varieties of GPL. One of the most useful is the "LGPL," which originally stood for Library GPL but is now referred to as the Lesser GPL. The LGPL is very similar to the Linux kernel license, except that it refers to a single library rather than the kernel itself. Libraries that are licensed under the LGPL can be used by your program without triggering a requirement to GPL your own code, as long as you only use documented calling mechanisms.

One of the most common licensing cases that people ask about in embedded discussion forums is exactly how they can build Linux into their system without having to release all their source code. The answer to this is that the code you write will fall into three categories, with different licensing implications for each:

1. *Kernel modifications.* This includes patches you have made to the public sources as well as additional loadable kernel modules you may have written. Source code for these must be disclosed.

2. *Modifications to LGPL libraries.* You will need to disclose all your source code for these.

3. *Your own application.* As long as you only use the kernel's documented interfaces, and documented interfaces to any LGPL libraries you use, your application code can remain secret. You must not use any libraries or other modules that carry a full GPL license, or you trigger a full GPL disclosure requirement on your own code. You must also avoid any undocumented interfaces to LGPL libraries or the kernel itself.

In practice, a large majority of embedded Linux projects will use exactly one library—glibc, or a cut-down variant of it such as uclibc—without ever needing to modify it, so the caveats in cases 2 and 3 above are never encountered.

At the opposite end of the spectrum from GPL you will find the NetBSD license. This is a refreshingly simple license that allows you to download the free source code, experiment with it, use it and release derivative products, with or without source code disclosure as you see fit. The only real limitation is that your product and its advertising materials must acknowledge the original author, typically with a phrase such as "This product includes software developed by X." (Some variants of the NetBSD license have dropped this last requirement.) There are also some common-sense requirements that are in no way onerous: You agree not to use the original author's name to promote your derivative product, and you agree that the code you received has no warranty. The NetBSD license is literally something for nothing; you get the source code for free, you can distribute your binaries and charge money for them if you wish, and there is no requirement for source code disclosure (though of course it is encouraged to release as much as you can). A splendid example of NetBSD in widespread commercial use is Apple's latest generation of Macintosh operating systems.

There are innumerable other open source licenses, many of which are associated with just one specific product. For example, the RedHat eCos operating system described in the previous chapter is released under the "RHEPL" license similar in philosophy to the NetBSD license. All these miscellaneous licenses lie in a spectrum roughly bounded by GPL at one end and NetBSD at the other (in terms of source code disclosure requirements vs. recognition of proprietary trade secret rights), with special conditions in some cases. However, you will find that the majority of the interesting open source projects in the world are GPL-licensed.

One incidental pitfall does bear mentioning: There is a surprisingly large amount of open source material which implements patented algorithms. For whatever reason—a love of academic freedom of speech, a desire to avoid expensive legal action, or simple lassitude—the owners of these patents often don't see fit to enforce them for free products. Even if you comply with the license agreement for the freeware product, that does *not* imply that you have somehow inherited a right to use the patented intellectual property in your project. For instance, there are freeware DVD playback programs readily available on the Internet. (For the benefit of those who know about such things, let's leave the thorny issue of DeCSS and the evil MPAA out of the equation and consider only an unambiguously legal freeware MPEG-2 player capable of playing unencrypted, legal DVD content such as you might produce if you use a consumer DVD recorder to convert your home movies from VHS to DVD format.) Notwithstanding the free nature of the code license, if you use one of these players as the core of your own consumer electronics DVD player project, you'll find the DVD consortium knocking on your door very quickly indeed looking for monies related to use of the DVD video trademark. You'll also be facing litigation on a raft of patent issues surrounding the MPEG-2 decoder.

Several jurisdictions are currently evaluating possible changes to the way software patents are granted, with the abolition of such patents (or at least limiting them to a few years) being one of the options under consideration. Until such an enlightened, forward-thinking step occurs, however, you need to be willing to research possible patent protection of the algorithms used in your project, regardless of what the licensing conditions might be on any particular implementation that you have referenced in your code.

Whatever third-party intellectual property you wind up using—even if you don't include third-party code in your final product release—it is absolutely essential to maintain an audit document for your software. This need not be a particularly onerous task; at its simplest, it can be a document listing each item you have included in your product (operating system kernel, third-party libraries, example source code, clip art, fonts and so on) along with a copy of the license agreement that accompanied each of these items when you obtained them. This latter is particularly important because some licenses evolve over time (GPL is an example)—the license you obtain today may not be the same license that you would obtain by downloading the software tomorrow. With this document in hand, you have a documented legal defense against any accusations of license violations.

One last note: With the plethora of useful open source code floating around the Internet, free for the downloading, there might be a temptation simply to download and use whatever you please and assume that nobody will ever know because nobody will ever see your source code. Even ignoring the moral issues, this is suicidal folly. Anything from a disgruntled (or simply talkative) staff member to an interested hacker or a competitor reverse-engineering your product will destroy your company; discovery is inevitable, particularly if your project turns out to be a success. At the time of writing, several major American corporations are writhing in the throes of government investigations into accounting fraud; if your major product contains plagiarized code, discovery will lead to similar consequences. Worse—and this also applies to privately held companies, because it's not just a stock price issue—you may be unable to ship any more units without an expensive major rewrite of your operating system. Don't take this kind of risk. If you use free code, honor the license.

CHAPTER 8
The Essence of Microcontroller Networking: RS-232

Fred Eady
Creed Huddleston

8.1 Introduction

Now let's explore the RS-232 protocol. Knowing how to manipulate data with RS-232 will help you master more complex communications protocols. You'll also find RS-232 techniques to be invaluable in the development phase of your projects.

The information you see in the terminal emulator window in Figure 8.1 was generated by some very simple firmware and a not-so-complicated, off-the-shelf, two-buck microcontroller. I used a tiny 8-bit microcontroller that does not contain a built-in hardware USART to transfer the ASCII characters you see in Figure 8.1 from one of its I/O pins to an RS-232 converter IC. A serial cable connected between the microcontroller/RS-232 converter IC circuitry and my personal computer's serial port allowed the ASCII characters to flow from the little microcontroller's firmware out of the microcontroller's I/O pin, through the RS-232 converter IC,

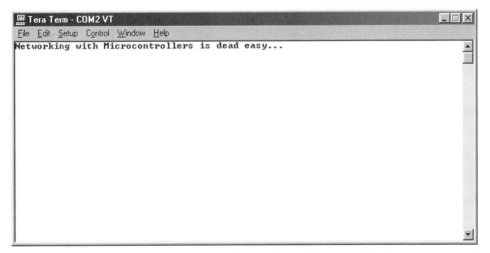

Figure 8.1: Effecting RS-232 communications with a microcontroller is a snap. You will find that knowing how to implement simple RS-232 with a microcontroller can assist you in building and debugging more complex microcontroller projects.

across the serial cable to the personal computer's USART/RS-232 circuitry, finally ending up in the terminal emulator window you see in Figure 8.1.

What I've just described is one of the simplest forms of microcontroller networking. It is commonly known as *serial* or *RS-232 communications.* As you can see in Figure 8.2, RS-232 was designed to tie Data Terminal Equipment (DTE) and Data Communications Equipment (DCE) devices together electronically to effect bidirectional data communications between the devices.

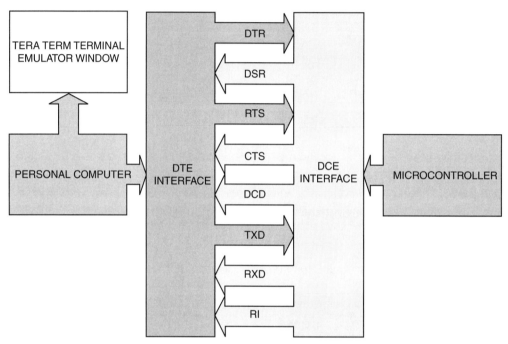

Figure 8.2: The DTE and DCE interfaces usually consist of some sort of voltage-conversion circuitry to translate RS-232 voltage levels to voltage levels that are compatible with the computing equipment on each end of the communications link. The simplest form of an RS-232 link uses only the TXD and RXD signals with a common ground.

An example of a DTE device is the serial port on your personal computer. Under normal conditions, the DTE interface on your personal computer asserts Data Terminal Ready (DTR) and Request to Send (RTS). DTR and RTS are called *modem control signals.* A typical DCE device interface responds to the assertion of DTR by activating a signal called Data Set Ready (DSR). The DTE RTS signal is answered by Clear to Send (CTS) from the DCE device. A standard external modem that you would connect to your personal computer serial port is a perfect example of a DCE device.

8.2 Some History

In May 1960 it was evident that a standard was needed to identify the electrical interface between computers and modems. It was decided to establish a standard voltage with standard signal parameters and a standard nomenclature to identify the conductors in the cable that connected computers and data sets. Even today, you will sometimes hear the term *data set* applied to modems and DCE equipment.

To compete as well as exist in the current communications environment, telecommunications vendors needed common ground to assure that each vendor's equipment set could talk to any other vendor's telecommunications equipment set. In other words, the industry needed a working standard. Without a standard, the whole teleprocessing industry could come to a grinding, nonstandardized halt.

To help establish some harmony, a committee named the Electronic Industries Association (EIA) was formed. The EIA drafted a standard known as EIA RS-232(X). Though it was a great idea, the original specification was broad in meaning and didn't guarantee compatibility. The new RS-232 specification also had a competitor outside the United States, known as the CCITT, or Consultative Committee on International Telegraphy and Telephony, recommendation V.24.

The RS-232 proposal defined a logical and physical interface between DTE equipment and DCE equipment. The computer's DTE serial port presents both a physical and a logical interface to a modem or data set's DCE port and consists of several conductors for controlling, transmitting, and receiving data. Timing and clocking signals are also intermixed within the RS-232 interface. The logical and physical attributes of the RS-232 proposal eventually became a set of standards known today as the EIA RS-232 interface.

Once the signals reach the DCE device, a second interface provides a physical path to the communication channel (RF link, telephone line, fiber optic link, satellite link, and so forth). For most of you, that second interface is a standard two-conductor analog telephone line, which is terminated inside your modem.

The EIA standard originally identified seven interface conductors and no specific connector. Signal voltages were defined as at least 3 V but not greater than 20 V with respect to ground.

In October 1963, RS-232 became RS-232-A and was modified to include a 25-pin connector with a maximum cable length of 50 feet. This revision established fixed relationships between a circuit and specific pin numbers on the 25-pin connector. Also, an alphabetic coding system for each type of interface circuit was presented. The first character of the coding system designated A for ground, B for data, C for control, and D for clocking. Table 8.1 lays out the pinout and various names for each RS-232 signal.

Table 8.1: Specifications list for RS-232 interface.

Pin	Line Label	Line Name	Signal Direction	Level
1	AA	Positive ground	N.A.	A,B C
2	BA	Transmitted data	To DCE	A B,C
3	BB	Received data	To DTE	A,B,C
4	CA	Request to send	To DCE	A B,C
5	CB	Clear to send	To DTE	A B,C
6	CC	Data set ready	To DTE	A B,C
7	AB	Signal ground	N.A.	A B C
8	CF	Received line signal detector (RS-232); data carrier detect (RS-232A/B)	To DTE	A,B,C
11	N.A.	Select standby	To DCE	C
12	SCF	Secondary receive line signal detector	To DTE	C
13	SCB	Secondary clear to send	To DTE	C
14	SBA	Secondary transmitted data	To DCE	C
14	N.A.	New sync	To DCE	A,B,C
15	DB	Transmitter signal element timing	To DTE	A B C
16	SBB	Secondary received data	To DTE	C
17	DD	Receiver signal element timing	To DTE	A,B,C
18	N.A.	Test	To DCE	C
19	SCA	Secondary request to send	To DCE	C
20	CD	Data terminal read	To DCE	A,B,C
21	CG	Signal quality detector	To DTE	C
22	CE	Indicate ring/calling	To DTE	A,B,C

There are a couple of confusion points. Note the total lack of logic when associating DB-25 pins with DB-9 pins. In addition, this table is based on the DTE side of the circuit. To get things to work, you must switch the TD and RD pins on the DCE side of the circuit. When you do, the switch that puts the DTE TD pin's data into the DCE RD pin and the DCE's TD pin's data into the DTE RD pin. If you're using the modem signals, you have to tie them together properly between the DTE and DCE as well.

The original seven basic circuits and the signal-level definition of -3 V for mark and $+3$ V for space were retained intact, adding 10 additional optional circuit definitions. The maximum permissible open-circuit voltage was changed to 25 V, and a current maximum between any two conductors, including ground, was set at 0.5 ampere. Conductors that permit auto-answer capability were introduced in this revision.

October 1965 brought about RS232-B, which defined terminating impedances that permitted circuit designers to build hardware with greater reliability. Open-circuit signal levels remained unchanged at −3 to −25 V as mark and +3 to +25 V as space, but revision B added an important voltage specification: By specifying that signal ground on pin 7 be tied to frame ground on pin 1 in the DCE equipment, a definite signal reference is established between DTE and DCE devices.

The Interface Between Data Terminal Equipment and Data Communication Equipment Employing Serial Binary Data Interchange specification was released in August 1969. It further clarified conductor definitions and stated that properly terminated RS-232 circuits shall not exceed ±15 V.

RS-232-C came along later and defined the interface between DTE and DCE. In the early days, a piece of DTE hardware was usually a dumb terminal. The Digital Equipment Corporation (DEC in those days; Hewlett-Packard/Compaq these days) VT100 was and is the most well-known dumb terminal and is still emulated today.

As you would imagine, a standard DTE device should be capable of emitting and receiving a serial data stream. As you have already seen, that includes microcontrollers and personal computers in the "could be a DTE" category. Although DCE equipment can also transmit and receive a serial data stream, the primary purpose of DCE equipment is to receive the DTE-generated bit stream over an RS-232 interface and convert it to a form that's suitable for transmission over a telecommunication medium. In the case of a personal computer modem, that telecommunications medium is most likely a voice-grade telephone line.

Ever noticed that every serial port interface on your personal computer is male and every modem serial port interface you've ever seen is female? There's a reason for that. The RS-232-C standard states that physical DTE port connectors will be male and physical DCE port connectors will be female.

Older personal computers and modems used a 25-pin connector. Today's 9-pin serial connectors aren't really standards, although they have become so by proxy. The 9-pin interface first appeared commercially on AT-class PCs in the early 1980s.

8.3 RS-232 Standard Operating Procedure

Today the majority of commercially available equipment is based on the RS-232-C or RS-232-D standard. (The CCITT V.24 and V.28 standards are also common and widely used.) There are 25 circuits defined in the RS-232 standard. The good news is that most of the 25 RS-232 circuits don't have to be used to effect an asynchronous communications session between a DTE and DCE device. Things could be different for synchronous communications sessions that employ complex communications protocols, and that's why the timing and clocking signals are defined in the RS-232 standard. There's a good reason that a 9-pin connector is on your personal computer instead of the standard appointed 25-pin connector. You only need nine RS-232 signal lines to communicate asynchronously using a standard asynchronous modem. Let's look at them from a "commented" standards point of view.

- *Pin 1 (Protective Ground Circuit, AA)*. This conductor is bonded to the equipment frame and can be connected to external grounds if other regulations or applications require it.

  Comment: Normally this is either left open or connected to the signal ground. This signal is not found in the DTE 9-pin serial connector.

- *Pin 2 (Transmitted Data Circuit BA, TD)*. This is the data signal generated by the DTE. The serial bit stream from this pin is the data that's ultimately processed by a DCE device.

  Comment: This is pin 3 on the DTE 9-pin serial connector. This is one of the three minimum signals required to effect an RS-232 asynchronous communications session.

- *Pin 3 (Received Data Circuit BB, RD)*. Signals on this circuit are generated by the DCE. The serial bit stream originates at a remote DTE device and is a product of the receive circuitry of the local DCE device. This is usually digital data that's produced by an intelligent DCE or modem demodulator circuitry.

  Comment: This is pin 2 on the DTE 9-pin serial connector. This is another of the three minimum signals required to effect an RS-232 asynchronous communications session.

- *Pin 4 (Request to Send Circuit CA, RTS)*. This signal prepares the DCE device for a transmit operation. The RTS ON condition puts the DCE in transmit mode, while the OFF condition places the DCE in receive mode. The DCE should respond to an RTS ON by turning ON Clear to Send (CTS). Once RTS is turned OFF, it shouldn't be turned ON again until CTS has been turned OFF. This signal is used in conjunction with DTR, DSR, and DCD. RTS is used extensively in flow control.

  Comment: This is pin 7 on the DTE 9-pin serial connector. In simple three-wire implementations, this signal is left disconnected. Sometimes you will see this signal tied to the CTS signal to satisfy a need for RTS and CTS to be active signals in the communications session. You will also see RTS feed CTS in a null modem arrangement.

- *Pin 5 (Clear to Send Circuit CB, CTS)*. This signal acknowledges the DTE when RTS has been sensed by the DCE device and usually signals the DTE that the DCE is ready to accept data to be transmitted. Data is transmitted across the communications medium only when this signal is active. This signal is used in conjunction with DTR, DSR, and DCD. CTS is used in conjunction with RTS for flow control.

Comment: This is pin 8 on the DTE 9-pin serial connector. In simple three-wire implementations, this signal is left disconnected. Otherwise, you'll see it tied to RTS in null modem arrangements or where CTS has to be an active participant in the communications session.

- *Pin 6 (Data Set Ready Circuit CC, DSR)*. DSR indicates to the DTE device that the DCE equipment is connected to a valid communication medium and, in some cases, indicates that the line is in the OFF HOOK condition. OFF HOOK is an indication that the DCE is either in dialing mode or in session with another remote DCE. When this signal is OFF, the DTE should be instructed to ignore all other DCE signals. If this signal is turned off before DTR, the DTE is to assume an aborted communication session.

  Comment: This is pin 6 on the DTE 9-pin serial connector. DSR is sometimes used in a flow control arrangement with DTR. Some modems assert DSR when power to the modem is applied, regardless of the condition of the communications medium.

- *Pin 7 (Signal Common Circuit, AB)*. This conductor establishes the common-ground reference for all interchange circuits, except Circuit AA, protective ground. The RS-232-B specification permits this circuit to be optionally connected to protective ground within the DCE device as necessary.

  Comment: This is pin 5 on the DTE 9-pin serial connector and is the only ground connection. This is the third wire of the minimal three-wire configuration. Thus, an RS-232 asynchronous communications session can be effected with only three signals: TX (Transmit Data), RX (Receive Data), and signal ground.

- *Pin 8 (Data Carrier Detect Circuit CF, DCD)*. This pin is also known as Received Line Signal Detect (RSLD) or Carrier Detect (CD). This signal is active when a suitable carrier is established between the local and remote DCE devices. When this signal is OFF, RD should be clamped to the mark state (binary 1).

  Comment: This is pin 1 on the DTE 9-pin serial connector. Normally in use only if a modem is in the communications signal path. You will also see this signal tied active in a null modem arrangement.

- *Pin 20 (Data Terminal Ready Circuit CD, DTR)*. DTR signals are used to control switching of the DCE to the communication medium. DTR ON indicates to the DCE that connections in progress shall remain in progress, and if no sessions are in progress, new connections can be made. DTR is normally turned off to initiate ON HOOK (hang-up) conditions. The normal DCE response to activating DTR is to activate DSR.

  Comment: This is pin 4 on the DTE 9-pin serial connector. Unless you specify differently or run a program that controls DTR, usually it is present on the personal computer serial port as long as the personal computer is powered on. Occasionally you will see this signal used in flow control.

- *Pin 22 (Ring Indicator Circuit CE, RI)*. The ON condition of this signal indicates that a ring signal is being received from the communication medium (telephone line). It's normally up to the control program to act on the presence of this signal.

  Comment: This is pin 9 on the DTE 9-pin serial connector. This signal follows the incoming ring to an extent. Normally this signal is used by DCE auto-answer algorithms.

That is all that's needed RS-232 signal-wise to establish a session between a DTE and a DCE device. Now that you have a feeling for what each RS-232 signal does, let's review how they react to each other with respect to the transfer of data between a DTE and DCE device:

1. Local DTE (personal computer, microcontroller, etc.) is powered up and DTR is asserted.

2. Local DCE (modem, data set, microcontroller, etc.) is powered up and senses the DTR from the local DTE.

3. Local DCE asserts DSR. If the DCE device is a modem, it goes off-hook (picks up the line). If a dial-up session is to be established, the DTE sends a dial instruction and phone number to the modem.

4. If the line is good and the other end (remote DCE) is ready or answers the dial-up from the local DCE, a carrier is generated/detected and the local and remote DCE devices assert DCD. The session is established.

5. The transmitting DTE raises RTS.

6. The transmitting DCE responds with CTS.

7. The control program transmits or receives data.

In our historical review, the DTE or personal computer and DCE or modem took care of converting the RS-232 signal levels to appropriate personal computer circuitry levels. To perform RS-232 asynchronous communications with microcontrollers, we must employ a voltage translation scheme of our own. Fortunately, there are many ways to do this and all of them are easy to implement.

8.4 RS-232 Voltage Conversion Considerations

RS-232 converter ICs like those made by Maxim and Sipex convert the negative RS-232 voltages to positive logic voltage levels that microcontroller circuits can understand. The positive

RS-232 voltages are converted to a microcontroller's logical 0 (zero) voltage level. If the microcontroller circuitry is powered by +5 VDC, then an RS-232 1 or mark is converted to a Transistor-Transistor Logic (TTL) *high* or 1 and an RS-232 0 or space is translated into a TTL *low* or 0. With the advent of 3 V logic, special RS-232 converter ICs that can operate at the 3 V power supply levels have been introduced. The bottom line is that the RS-232 marks and spaces must be converted to voltage levels the microcontroller can understand before any communications and data transfer can be realized between devices.

In reality, the full-positive and negative voltage swing called out by the RS-232 standard doesn't have to be employed to effect RS-232 communications links. With the right cable an RS-232 voltage of -3 V is sufficient to generate a 1 or mark while $+3$ V will produce a 0 or space. The area between -3 V and $+3$ V (shown in Figure 8.3) is a transition zone and is where most of the nasty line noise can and should be found. By defining this ± 3 V threshold, the signal-to-noise ratio of the RS-232 physical link is improved. If a high-quality serial cable is used and the distance between stations is relatively short, RS-232 voltages that resemble microcontroller logic voltages can be used to transfer information between a DTE and DCE device. In addition, using a high-quality cable could extend the 50-foot maximum cable length specified by the RS-232 specification. Reducing the speed of the data transmission can also extend the maximum cable length between a wired set of DTE and DCE devices as well.

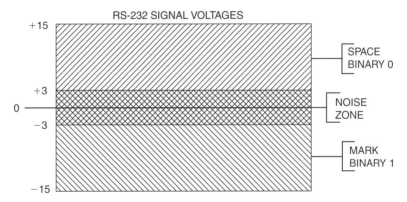

Figure 8.3: Cheap RS-232 implementations dare to use the 0 VDC to +5 VDC region for marks and spaces with 0 VDC being a mark and anything over +3 VDC representing a space. The "NOISE ZONE" I've marked is actually called the transition zone.

The good news is that you don't have to know the nitty-gritty details of the RS-232 specification to use RS-232 as a means of communicating with a microcontroller. In fact, I've already given you more RS-232 history and theory than you really need to know to make a microcontroller talk asynchronously. Our goal is the practical application of RS-232 as it pertains to microcontrollers. So, let's look at some RS-232 hardware and the firmware behind it.

8.5 Implementing RS-232 with a Microcontroller

Now that you've completed "RS-232 History 101," let's look at implementing RS-232 on a microcontroller. We'll use the Microchip® PIC12F675 as our RS-232 engine and we'll power our RS-232 engine with code written with the Custom Computer Services C Compiler.

You can build the circuits in this chapter from scratch. I've chosen to use the Microchip PICkit™ 1 as my "breadboard" because it contains circuitry to program the PIC12F675 and an experimenter area that is perfectly suited for additional RS-232 circuitry.

8.5.1 Basic RS-232 Hardware

Let's begin by looking at a simple microcontroller implementation. In its most basic form, an operational microcontroller-based circuit consists of the microcontroller, a simple power supply, and a clock source. For this project, we'll use the most basic of microcontrollers, an 8-pin Microchip PIC12F675.

The PIC12F675 has an internal clock source but does not contain a USART. That means we will have to implement the functionality of a hardware USART in the PIC12F675's firmware. To do that, we need to know just a bit more about RS-232 signaling. Let's begin by designating the desired RS-232 signaling speed, or *baud rate*. A common baud rate is 9600 bps (bits per second), and most everything RS-232 can operate at this speed. So, 9600 bps it is.

At 9600 bps, our data packet bit width is the reciprocal of the baud rate, which is 104 μS (104 microseconds). The idea is to try to see if the incoming RS-232 bit is a 1 or 0 by having the PIC12F675 microcontroller USART program check the incoming bit in the dead center of the 104 μS bit width. Since our baud rate is 9600 bps and our bit width for 9600 bps is 104 μS, that means we must have the microcontroller check the incoming bit stream every 104 μS.

There are still other things to consider. For instance, how does the microcontroller know when to start and stop the 104 μS bit check intervals? For the answer, let's draw again from the RS-232 specification. We assigned a speed of 9600 bps for our data stream. However, we must also specify the number of data bits that will be transmitted and received in a data packet and the number of stop bits that will indicate the end of the data packet. We do have a choice as to the number of data bits we can stash into a data packet. The data packet bit length choices are 5 bits, 7 bits, 8 bits, and 9 bits. Since the PIC12F675 is an 8-bit device, let's designate a data packet as 8 bits in length. Designating an 8-bit data packet allows the transfer of all readable ASCII characters plus control codes and hexadecimal or Binary Coded Decimal (BCD) data. We could have chosen 7 bits for ASCII transmission as well, but 8-bit data packets are more common, and choosing a 7-bit packet inhibits sending a byte of miscellaneous information in a single-data packet.

The PIC12F675's built-in oscillator operates at 4 MHz, which equates to an instruction execution time of 1 μS. That means that the PIC12F675 can theoretically execute 104 instructions during a stop-bit width, which is the same as the data-bit width of 104 μS. That time could be used to do some other processing if necessary—104 μS is a long time in microcontrollerland, so for us a single-stop bit will be sufficient.

There's another RS-232 component that can also be defined, called *parity*. To keep things easy, we will not assign a parity bit. Parity bits are used to check the integrity of the data packet by inserting an extra bit to make the number of data packet marks even or odd, depending on how the user has set up the communications equipment.

Now we have an asynchronous data stream consisting of a start bit, 8 data bits, no parity bit, and 1 stop bit. The word *asynchronous* here means that the data packet can begin at any time without regard to any predetermined timings. If receiving, the presence of a start bit signals the PIC12F675 that a data packet is starting. So far, so good; we haven't done or defined anything out of the RS-232 ordinary.

Let's walk through the voltages that are generated when an RS-232 data packet is sent containing the ASCII representation of the number 2. A 2 is represented in ASCII by hexadecimal 0x32 or binary 00110010. An idle RS-232 signal is defined as having the voltage on the transmit pin maintain a marking condition for a time that exceeds one data packet bit width. For 9600 bps, the steady marking condition must be greater than 104 μS in length. As you already know, a mark is a negative voltage between -3 and -25 V and represents a 1 in RS-232 lingo.

To signal the start of a data packet, the transmitting device will drive the RS-232 transmit pin positive into the space voltage region of $+3$ to $+25$ V. This transition from a steady marking state that is greater than or equal to one data packet bit width to a spacing state is called a *start bit*. Since we are running at 9600 bps, our start bit width is 104 μS, which is equal to our data packet bit width for a baud rate of 9600 bps.

Now here's where things get a bit tricky. Remember that the idea is to sample the incoming bits as closely to their center as possible to determine whether the bit is a 1 or a 0. Under ideal conditions, the start bit is recognized immediately by the receiving microcontroller. If the 104 μS interval begins at the same instant that the start bit is sensed, the microcontroller will sample at the end of the start bit time, which is 104 μS. The first data bit in the incoming data packet will be lost and so will the rest of the data bits because the microcontroller will be sampling the bits on their leading edges instead of in their centers.

A valid marking condition must exist before a start bit is initiated. So, with that we have a very good idea as to when a start bit should occur. We also know from the RS-232 specification that every valid RS-232 data packet starts with a start bit and ends with at least one stop bit. So, to sync up with the incoming data bits within the incoming RS-232 data packet, the receiving microcontroller is instructed to wait 1.5 RS-232 data packet bit width times after sensing a valid start bit. This allows the receiving microcontroller to begin the bit sampling in the center of the first incoming data bit. From there all the microcontroller has to do is sample every 104 μS seven more times to get the full 8 bits contained within the incoming RS-232 data packet.

A stop condition is indicated by the transmitting device when the RS-232 voltage being transmitted is returned to the marking state for at least one data bit width time, which is 104 μS for 9600 bps, after the correct number of RS-232 data packet bits are generated. This stop condition, or marking state, is actually the stop bit. Everything just described down to the microsecond is summed up in Figure 8.4.

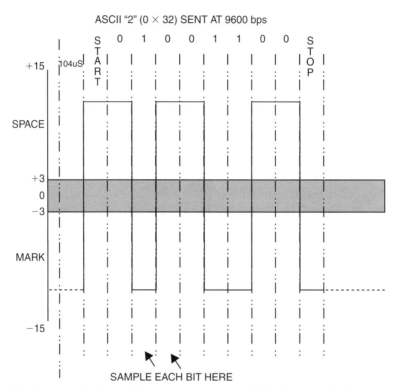

Figure 8.4: This is a graph of a 9600 bps asynchronous RS-232 transmission versus time. The time between each vertical double-dotted line represents 104 μS. Since we are only sampling for each bit one time, the idea is to try to sample as close to the center time of each bit as possible.

Let's run through it again. The transmitting microcontroller is holding it's RS-232 transmit pin in a marking condition. We know that this marking condition must be at least 104 μS in length to satisfy our bit timing for a 9600 bps baud rate. In fact, the marking condition can exist for hours, days, or forever as the receiving microcontroller is continually looking for a valid start condition.

The transmitting device drives its transmit pin to a space condition for one data bit time (104 μS for 9600 bps) to indicate the start of an RS-232 data packet. The receiving microcontroller senses the start bit on its receive pin and waits for 156 μS (1.5 × 104 μS). At the 156 μS interval, the receiving microcontroller samples what should be the center of the least significant bit of the incoming RS-232 data packet, bit 0. The microcontroller samples the second bit of the incoming RS-232 data packet 104 microseconds later. The receiving microcontroller samples every 104 μS until the most significant bit of the RS-232 data packet is sampled (bit 7 since we are sending 8-bit data packets).

The receiving microcontroller has 8 bits of data and expects to see its receive line go to a marking condition indicating a stop condition or stop bit. Note that the receiving microcontroller and the sending microcontroller sync up on every RS-232 data packet using the start bit. From there, every bit inside the RS-232 data packet is expected to be sent and arrive on time according to the baud rate. Later, you'll see that microcontrollers with internal USARTs will perform all the start bit and receive/transmit timing tasks automatically for you. For now, let's do it caveman style.

8.5.2 Building a Simple Microcontroller RS-232 Transceiver

To convert the RS-232 theory I've presented into real-world events, let's assemble some hardware and implement a simple three-wire RS-232 session between our PIC12F675 microcontroller and a personal computer.

A personal computer is most always configured as a DTE device. Recalling what we already know about the RS-232 specification, that implies that the personal computer's serial port uses a male 9-pin or male 25-pin connector. From here on out, unless I say otherwise, we'll use the 9-pin connector and pinout for both DTE and DCE devices. So, with that, pin 3 is the DTE transmit pin and pin 2 is the DTE receive pin. For the record, on a 25-pin male serial connector, pin 2 is the DTE transmit pin and pin 3 is the DTE receive pin. The third wire in our three-wire RS-232 connection is the common ground connection. For a 9-pin male serial connector, the ground pin is pin 5 for both DTE and DCE devices and is designated signal ground in the RS-232 specification. From your history lesson, you know that the 25pin DTE serial connector's signal ground is found on pin 7.

Applying logic (and your knowledge of the RS-232 specification) to the gender of the personal computer's serial connector would lead one to believe that since a DTE device is represented by a male connector then a DCE device would most likely support a matching female connector. Once again, logic prevails, as that is the real-world case. Again, using commonsense logic, one would be led to conclude that since the personal computer is a DTE device, our PIC12F675 would be the center of attention in a DCE device. If that is also true, which it is, that means I can literally plug the personal computer's male DTE serial interface directly into the PIC12F675's female DCE interface and pass data between the personal computer and the PIC12F675. What makes this possible is the DCE serial connector pinout versus the DTE connector pinout. Basically, the DCE device's transmit pin is connected directly to the DTE device's receive pin and the DTE device's transmit pin is wired directly to the DCE device's receive pin with signal ground being common between the DTE and DCE interfaces. Don't confuse this with a "null modem" arrangement; a null modem circuit is intended to attach a DTE device directly to another DTE device by tying complementary modem signals to each other. Therefore, that makes pin 3 on the DCE side the receive pin and pin 2 the DCE transmit pin. Using the standard DTE and DCE pinouts on my connectors means that I can now communicate PIC to personal computer without the need for any special "crossed over" cables. In fact, all I need is three wires.

8.5.2.1 RS-232 Interface Hardware

Because true RS-232 signals are not TTL compatible, the incoming RS-232 voltage levels must be converted to voltage levels compatible with the circuitry behind the serial connector. On the other side of that, the outgoing TTL voltage levels must be shifted to RS-232 signal levels for transmission between the DTE and DCE devices. The easiest way to effect the RS-232 voltage translation process and stay within the RS-232 specification's guidelines is to use a special RS-232 converter IC. One such IC is the industry standard Maxim MAX232CPE.

In the past, if you really wanted to adhere to the RS-232 specification you designed in a ± 12 V or ± 15 V power supply to drive the MC1488 (now called the DS1488) quad line driver. The negative supply voltage coupled with the MC1488 made the marks possible, while the positive 12 V provided the voltage level necessary to produce a space. On the receiving side, an MC1489 (these days it's called a DS1489) picked up the marks and spaces, converted them to TTL levels and fed them to the device's UART.

The DS1488 and DS1489 are still in production and are great choices for low-cost RS-232 interfaces if the power supply voltages are already in the design anyway. However, to really keep it simple and within specification, using a MAX232CPE or similar IC at each end of the RS-232 link is the way to go. The MAX232CPE requires a single +5 VDC and with the help of four common 1 μF capacitors, the MAX232CPE internally generates the voltages necessary to effect marks and spaces on the transmit pin using an internal charge pump. Not only does the MAX232CPE perform the TTL-to-RS-232 conversion duties, it is the "other side" also converting the incoming RS-232 signals into TTL voltages. The MAX232CPE charge pump is capable of producing ± 10 VDC when no significant load is present.

You can build the PIC12F675-based RS-232 transceiver from scratch or you can take a value-added and easier way out by using the Microchip PICkit™ 1 FLASH Starter Kit. Before we move on, let's stop and talk a little about the PICkit 1 (see Schematic 8.1 and Figure 8.5).

The PICkit 1 FLASH Starter Kit is designed to allow easy and inexpensive evaluation of Microchip's new 14-pin Flash-based PICs and some of the legacy 8-pin flash parts like our PIC12F675. The PICkit 1 FLASH Starter Kit programming hardware is centered on the PIC16C745, which contains a USB engine in addition to the normal stuff you would find in a PIC microcontroller.

The Essence of Microcontroller Networking: RS-232

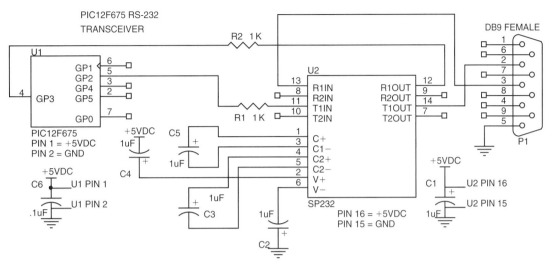

Schematic 8.1: This is the "formal" way to do it. Capacitors C2-C5 help the Sipex SP232ACP's internal charge pump provide the RS-232 voltages that adhere to the RS-232 specification. The PICkit 1 uses this formal approach.

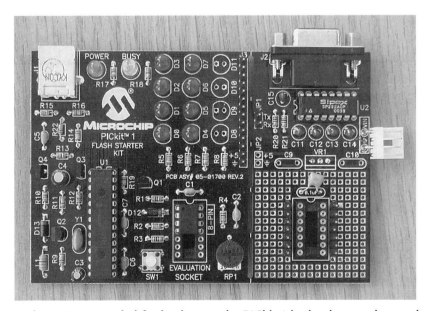

Figure 8.5: Intended for beginners, the PICkit 1 is simple to understand and operate. An 8-pin PIC12F675 is mounted in the evaluation socket. All the USB circuitry is to the far left of U1, a PIC16C745.

Along with the hardware and firmware contained in a USB microcontroller, the magic of USB is performed within the Windows operating system. Special programs and drivers running under Microsoft Windows form an alliance between the microcontroller's I/O ports, the microcontroller's USB interface, and the application that is running under the Microsoft Windows operating system. In effect, all the work is done up front and all the pent-up USB programming in the microcontroller and on the personal computer is unleashed when the user plugs a USB device into a personal computer's USB port.

A really neat feature of the PICkit 1 FLASH Starter Kit is that after you have initially downloaded a hex file, you can compile the file again and as long as you tell the compiler to always replace the old hex file after a compile, the PICkit 1 will automatically bring in the newly compiled hex file for programming when you click on the Write Device command button. The PICkit 1 FLASH Starter Kit programming interface does this by checking the timestamp of the loaded hex file and loading in the latest time-stamped hex file of the same name.

The target PIC's power is controlled (on or off) by clicking on the Device Power button in the Board Controls box. I used this feature extensively to turn off the PIC12F675 after programming it so I could move it over to the snap-off board socket to run the spin of code I had just compiled and programmed.

The PICkit 1 FLASH Starter Kit hardware communicates with the PICkit 1 FLASH Starter Kit programming interface (Figure 8.6) that runs under Microsoft Windows. The PICkit 1 programming interface allows the user/programmer (that's us) to view PIC Program Memory and EEDATA Memory in hexadecimal format. The Program Memory and EEDATA Memory windows contain the contents of a standard Intel hex file the user/programmer loads into the programming interface that has been generated by either a compiler like PicBasic™ Pro Compiler or Custom Computer Services C Compiler or an assembler like PicBasic Pro's PM or Microchip's MPASM™.

The idea is to generate an Intel hex file, load it into the PICkit 1 FLASH Starter Kit programming interface, and "burn" or program the binary code into the physical PIC device in the PICkit 1's evaluation socket. A compiled program file (Intel hex file generated by the compiler or assembler) is downloaded into the PICkit 1 FLASH Starter Kit programming interface using the Import HEX menu item. When the file download is complete, the data contained within the downloaded hex file will appear in the Program Memory and EEDATA windows. At this point, the user/programmer can click on the Write Device button and burn the downloaded code into the target PIC. If all goes well, a green banner will be displayed at the bottom of the PICkit 1 FLASH Starter Kit programming interface window. A red banner signifies that something went wrong in the program cycle.

Providing that the target PIC has not been code protected, the user/programmer can read the contents of the target PIC and save the data as a hex file using the Export Hex menu item.

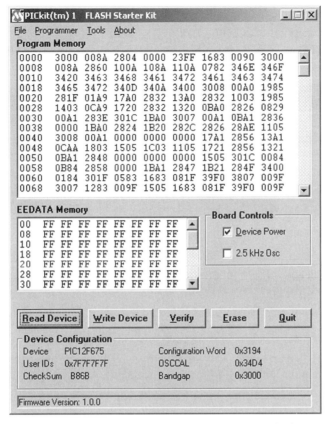

Figure 8.6: Once you load a hex file for programming, each time you issue a Write Device command, the PICkit 1 program finds and reloads the latest version of the hex file you originally specified before programming the PIC.

Two other command buttons allow the user/programmer to verify existing code in a PIC mounted in the PICkit 1 program socket with the contents of a hex file and to erase the target PIC part.

The PICkit 1 FLASH Starter Kit shown in Figure 8.6 is a preassembled PIC development board with an unpopulated snap-off experimenter board. The PICkit 1 FLASH Starter Kit is unique in that it doubles as a PIC programmer, but not just any old PIC programmer. A special Visual Basic program that runs on a host personal computer controls the PICkit 1 FLASH Starter Kit. The personal computer is attached to the PICkit 1 FLASH Starter Kit via USB. The bonus is that all the source code for both the Visual Basic personal computer program and the USB interface is included, in addition to the PIC tutorial and project source code. So, if you're curious about how PIC programmers work and have an interest in how USB works, the PICkit 1 FLASH Starter Kit is a must-have device.

I left the snap-off experimenter board attached to the PICkit 1 FLASH Starter Kit and rigged a standard personal computer's diskette drive power connector to get +5 VDC and ground to the snap-off board. These days, personal computer power supplies are cheap, and using a personal computer power supply gave me a power switch and keyed power receptacle for the experimenter board side of the PICkit 1 FLASH Starter Kit while eliminating the need to solder in a 7805 +5 VDC regulator and its supporting circuitry.

I also substituted a pin-for-pin compatible Sipex SP232ACP for the MAX232CPE, since I don't have a through-hole MAX232CPE in my parts inventory. I completed the assembly of my PICkit 1 FLASH Starter Kit experimenter board by installing the TX (transmit) and RX (receive) header pins and the 14 header pins around the PIC socket. Installing the headers will allow easy connections between the Sipex SP232ACP and the PIC12F675.

Even though the pins of the 14-pin socket on the programmer side of the PICkit 1 are connected directly to LEDs, you can still use the pins to run our RS-232 transceiver project. Just solder in the J3 header and use a jumper wire to connect the programmer side TX and RX pins to the snap-off board's TX and RX pins. This allows you to program and execute the programs without having to move the PIC12F675 from the programming socket to the snap-off test socket.

Although the PICkit 1 is nice to have, if you already have a PIC programmer that will burn the PIC12F675 you can build up the "formal" circuit shown in Schematic 8.1 or you can get down and dirty with the "dirty" RS-232 implementation shown in Schematic 8.2.

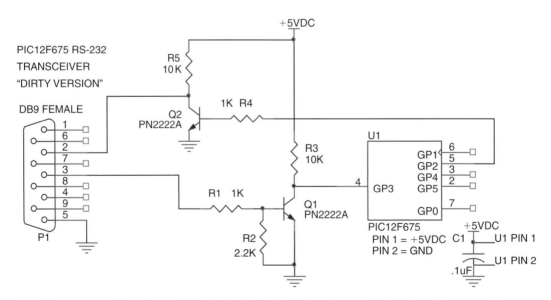

Schematic 8.2: If you don't have a MAX232 or Sipex SP232ACP on hand or if you want to save some bucks and have some fun at the same time, lash up this "dirty" RS-232 transceiver.

In the "dirty" version, Q1, Q2, and the five resistors perform the RS-232 voltage conversion. Any positive voltage coming in on P1's pin 3 that is capable of turning on Q1 will be considered "RS-232 OK" and will pass as a binary 0, or space, to the PIC12F675's GP3 receive pin. If the incoming RS-232 voltages are up to specification and the RS-232 cable is of good quality, this receiver circuit formed by Q1, R1, and R2 will work very well in most instances. The same is true for the transmit circuit, which is driven using Q2, R4, and R5. If the RS-232 cable is not too long and is of a high quality, Q2 will send a "dirty" mark (0 VDC instead of −3 VDC or better) when it is turned on by the PIC12F675's transmit pin, GP2. A clean space will be transmitted when Q2 is off. If your project can tolerate possible RS-232 bit errors, the "dirty" RS-232 circuitry shown in Schematic 8.2 is a cheap and easy way to implement an RS-232 link.

8.5.2.2 Writing Some Simple RS-232 Firmware

No matter which direction you took, "dirty" PICkit 1 or homebrew "formal," I'm sure you'll agree that the RS-232 hardware was easy to obtain and assemble. The RS-232 code for our minimal RS-232 system is just as easy to write.

A variety of C compilers on the market target the Microchip PIC® family of microcontrollers. I've chosen to use the Custom Computer Services C Compiler for Microchip PIC microcontrollers to write the code for the PICkit 1 FLASH Starter Kit RS-232 circuit I've assembled. The inexpensive Custom Computer Services C Compiler is easy to use and has features that take the pain out of writing code for PICs. I've written a couple of programs that simply send the ASCII character *A* to a HyperTerminal™ or Tera Term Pro™ terminal emulator program.

For those of you who don't "do" C, I've selected the PicBasic Pro Compiler from microEngineering Labs to represent the RS-232 firmware on the BASIC side of the house. Like Custom Computer Services C Compiler, the PicBasic Pro Compiler is dedicated to producing clean and tight code for Microchip PIC microcontrollers.

Before I describe the code, let's make sure you have your terminal emulator set up correctly. HyperTerminal is included as an accessory communications program with the Microsoft Windows operating system. It's fairly easy to prepare HyperTerminal to receive our RS-232 data. Once you open HyperTerminal, the first thing you want to do is name your session. In Figure 8.7, I named my HyperTerminal session "Simple PIC RS-232."

After you name your session, another window like the one in Figure 8.8 will appear, asking which COM port you want to use. That all depends on what's available on your machine. In my case, I had both COM ports 1 and 2 open and chose COM 1.

Figure 8.7: Doing this allows you to save the HyperTerminal session with a name for later use.

Figure 8.8: Select an open COM port on your personal computer here.

The final step in setting up your HyperTerminal session is the definition of the communications parameters. We defined those earlier as 9600 bps, no parity bit, 8 data bits, and 1 stop bit. Set up your serial port as it is shown in Figure 8.9.

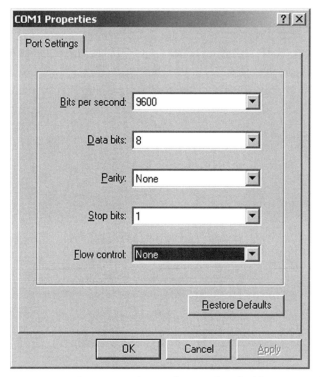

Figure 8.9: No modem control or software signals (flow control) are needed in a simple three-wire RS-232 connection.

Flow control hasn't been covered yet, and for this project we'll assume there isn't any. Flow control comes in a multitude of flavors. Normally, flow control is implemented using the modem control signals CTS and RTS. Flow control can also be initiated using software commands like those used to implement XON/XOFF flow control. One could also use a logic signal from a standard I/O pin to effect an unofficial flow control. Flow control excepted, the goal is to end up with a blank terminal emulator window and a blinking cursor in the upper left corner of the terminal emulator window.

Unless you purchase some upgraded HyperTerminal software, you won't be able to do much more than open a HyperTerminal emulator session and send or receive data with the version that is bundled with Windows. Another terminal emulator called Tera Term Pro provides a bit more functionality and flexibility than HyperTerminal and it costs nothing but your time to download it from the Internet. Tera Term Pro setup is similar to that of HyperTerminal, and as you will see in the pull-down menus, there are some things Tera Term Pro can do that the stock HyperTerminal can't. Tera Term Pro's most useful feature is the scripting language that is built into it. Tera Term Pro's script commands provide a means of automating the process of transferring and receiving files. We won't need any Tera Term Pro scripting for our simple RS-232 project.

Editing the TERATERM.INI file, which resides inside the Tera Term Pro directory, can be used to set up all Tera Term Pro's communications parameters. Here, I'll show you how to get a basic Tera Term Pro emulation session to work on your personal computer manually. The first thing you want to do is tell Tera Term Pro that you will be using a serial interface. As you can see in Figure 8.10, Tera Term Pro is capable of doing many other things on differing interfaces.

Figure 8.10: Use the Serial side of Tera Term and enter a COM port number that's open on your personal computer.

Under the Setup pull-down menu you will find an entry for Serial Port. Selecting the Serial Port menu item will bring up a window like the one depicted in Figure 8.11 and allow you to manually set the communications parameters, which are identical to the communications parameters we set in HyperTerminal (9600 bps, 8 data bits, no parity, 1 stop bit).

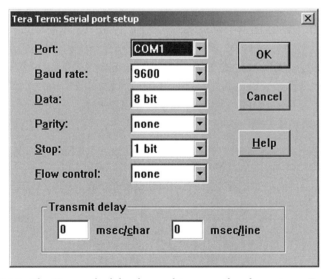

Figure 8.11: The Transmit delay is used to pace the characters. For instance, changing the msec/char field to a 1 would send a character wait 1 ms and then send another character and so on.

Again, just as with HyperTerminal, you should end up with a blank terminal emulation window with a flashing cursor in the upper left corner. To complete the personal computer and terminal emulator setup, all that's left to do is to attach a pin-for-pin (pin 1 to pin 1, pin 2 to pin 2, etc.) 9-pin male-to-female cable between the personal computer's serial (COM port you selected in the setup) port and the PICkit 1 FLASH Starter Kit's 9-pin serial connector on the PICkit 1's snap-off experimenter board. Now, let's pick apart the RS-232 C code.

I'm not going to assume you know every nuance of C, so this time I'll take us through line by line. The `#include` lines at the top of the listing tell the compiler about the physical attributes of the PIC12F675. The "physical attributes" of a microcontroller device may include the number of I/O pins or the types of special purpose modules that reside inside the microcontroller, such as analog-to-digital converters or timers. The include files also define associations. For instance, for operations that need to express a TRUE or FALSE condition, it's much easier to remember TRUE for 1 and FALSE for 0. Using real words also makes the code easier to read and follow. Another example of what include files do involves equating I/O port names. Instead of having to remember that PORTA is actually address 0x005, the `#include` allows you to simply type in "PORTA" when you are performing tasks against address 0x005.

The C include files are readable and you can examine them as you would any other text file. Perusing a microcontroller's data sheet and include files are a good way to learn about what the microcontroller can really do for you. The Custom Computer Services C Compiler comes with an include file for each PIC microcontroller it supports. If there are physical attributes you need to access and they aren't already included in the stock include file, there's nothing to stop you from putting together your own include file. I used the PIC12F675 data sheet to build the f675.h include file, which includes definitions and associations from the PIC12F675 data sheet that were not included in the canned PIC12F675 include file.

```
#include <12F675.h>
#include <f675.h>
```

The data sheet is the most important tool in working with any microcontroller device. Checking the PIC12F675 data sheet tells us that the PIC12F675 is equipped with an on-chip oscillator that does not require an external crystal or resonator. Another look at the PIC12F675 data sheet tells us the internal clock speed of the internal oscillator is a nominal 4 MHz. Another plus in using the Custom Computer Services C Compiler is that once the clock speed is defined to the compiler, things like delays and baud rates are automatically calculated and applied inside the compiler routines that rely on the microcontroller's clock speed. So, the line `#use delay (clock = 4000000)` sets the PIC12F675 clock rate at 4 MHz and tells the compiler to use 4 MHz for its delay and baud rate calculations.

Bits inside fuse words are used to turn on or turn off certain special-purpose modules, functions, or features that the PIC12F675 offers to the programmer. Again, checking the

PIC12F675 data sheet, we know that the PIC12F675 can be instructed to use the internal oscillator or depend on an external crystal arrangement. The `INTRC_I/O` fuse instruction sets a fuse bit that activates the PIC12F675's internal 4 MHz oscillator. In addition to selecting the clock type, the `INTRC_I/O` bit deactivates the clock signal from being accessible via a PIC12F675 I/O line.

```
#fuses INTRC_IO,NOWDT,NOMCLR,NOPROTECT,NOCPD,NOBROWNOUT
```

The next fuse instruction, NOWDT, deactivates the PIC12F675 watchdog function. Watchdog timers are commonly used to monitor the microcontroller's execution of instructions. If the microcontroller "hangs" or "loops" and the watchdog timer doesn't get reset, the microcontroller is forced to reset itself and restart the application that is programmed into it. For simple programs like this one, the watchdog timer function is not necessary.

As you've probably already figured out, the "NO" in front of the rest of the fuse instructions turns off a particular PIC12F675 function. NOMCLR saves an I/O line on the PIC12F675 by not requiring the MCLR reset pin to be offered to the programmer externally. Instead, the MCLR pin function is performed internal to the PIC12F675.

Activating code protection makes it impossible to read the PIC12F675's program memory with a PIC programmer. Since I haven't written any code that would stop an alien attack, NOPROTECT and NOCPD allow the code loaded into the PIC12F675 program memory to be accessed by the standard methods.

I'm also not anticipating my personal computer power's supply voltages to dip or "brown out" under load, so there is no need for brownout protection, and NOBROWNOUT is pretty obvious as to how I feel about that.

While we're on the fuse bit subject, the Custom Computer Services C Compiler has a really nice pull-down View menu feature that describes and lists the valid fuses for the microcontroller you're writing code for. In that same pull-down View menu, the compiler also gives you access to the microcontroller datasheets, which are stored in a directory as standard PDF files. The scope of this book isn't really about teaching you C or tutoring you on how to use the Custom Computer Services C Compiler. However, as we continue on this networking hop, I'll point out goodies inside the compiler packages that will help you write the best code with the least effort. If you're not a C person, who knows—you might pick up enough C to become proficient with the language.

The Custom Computer Services C Compiler does many things behind the scenes to assist you but sometimes it comes at the expense of extra code that is generated by the compiler. If you're a control freak like I am, I want to be in command as much as possible. So, the `#use fast_io(A)` code line tells the Custom Computer Services C Compiler to allow me and not the compiler to determine the direction (input or output) of each PIC I/O line.

Our simple RS-232 C program actually consists of three subprograms: TX_program_1, TX_program_2, and TX_program_3. Each program does the same thing—transmits the ASCII character 0x41, or A. By simply placing each subprogram between a set of `#ifdef` and `#endif` preprocessor statements, I can compile one of the subprograms at a time by "defining" which program is active during the compilation time. The subprogram to compile is chosen by "commenting out" the other subprograms I don't want to be compiled. For instance, to select TX_program_1 in Code Snippet 8.1, I comment out `#define TX_program_2` and `#define TX_program_3`. When I run the Custom Computer Services C Compiler, all that will be included in the final output file will be the common code plus all the code between `#ifdef TX_program_1` and its corresponding `#endif` preprocessor statement. I've used the Custom Computer Services C Compiler to write more complex programs, and you'll get a taste of that as we progress.

```
//*********************************************************
// COMMENT OUT THE PROGRAMS YOU DON'T WANT TO RUN
//*********************************************************
#define TX_program_1      //this program will be compiled
//#define TX_program_2    //this program will not be compiled
//#define TX_program_3    //this program will not be compiled
```

Code Snippet 8.1: When you begin to write larger C microcontroller programs, you'll use the // to comment out parts of code instead of deleting them.

All C programs have a *main* function like the one shown in Code Snippet 8.2. The main microcontroller application program actually flows inside the main function braces. In our RS-232 code, any code that is not fenced in by `#ifdef TX_program_x` and a related `#endif` is always compiled and can react with the selected TX_program_x code segment.

```
//*********************************************************
// MAIN PROGRAM STARTS HERE
//*********************************************************
// This code fragment will always be compiled
void main() {
  setup_adc_ports(0);
  setup_adc(ADC_OFF);
  setup_timer_1(T1_DISABLED);
  setup_comparator(NC_NC_NC_NC);
  setup_vref(FALSE);
                         //PORTA pin 2 = TX line
  SET_TRIS_A(0b00001000); //PORTA pin 3 = RX line
```

Code Snippet 8.2: The Custom Computer Services C Compiler program wizard generated all the setup statements.

In addition to the on-chip analog-to-digital converter, the PIC12F675 also contains an analog comparator, a voltage reference, and some timers. Since we won't be using any services provided by these modules, the `setup_xxxxx` lines of code are there to turn off TIMER_1, the analog-to-digital converter, the comparator, and the voltage reference. Executing the "setup" lines will also free up any I/O pins that the service modules may have wanted to use.

All the subprograms have a few things in common; each subprogram transmits the letter *A* and each subprogram uses the same PIC12F675 I/O pins for transmitting and receiving. That means that I can set the I/O direction of the PIC12F675's receive and transmit pins in the common code. The `SET_TRIS_A(0b00001000)` code line completes the manual I/O direction task and feeds my control freak animal because I, not the compiler, set the PIC12F675's I/O pin direction.

8.5.2.3 A Bit of RS-232 Transmit Code

Earlier I talked about how each of the data bits inside a data packet must be 104 μS in duration to be recognized as a 9600 bps bit stream. The first program, TX_program_1, is a crude 9600 bps algorithm that uses delays and bit voltage levels to transmit the ASCII character *A*. To make things a bit easier to read in the TX_program_1 main code, I've defined the TX (transmit) pin, PIN_A2, and the RX (receive) pin, PIN_A3, in the PIN DEFINITIONS area before the main program code, as shown in Code Snippet 8.3.

```
//*******************************************************
// TX_PROGRAM_1 PIN DEFINITIONS
//*******************************************************
#ifdef TX_program_1
#define TX PIN_A2
#define RX PIN_A3
#endif
```

Code Snippet 8.3: It's best to keep the C code human readable.

TX_program_1 begins by placing the TX line in a marking state for 1 ms. The `output_high(TX)` instructs the PIC12F675 to present a TTL high (binary 1) to the Sipex SP232ACP's TTL input. The Sipex SP232ACP inverts that to present a RS-232 mark on pin 2 (DCE transmit pin) of the communications cable. The `while(1)` statement says that while the tested condition is 1 or while the tested condition is TRUE, the code between the braces ({}) will run. Since 1 never changes value and 1 represents TRUE, the code will run in this loop inside the braces forever. This is one way of creating a continuous loop. I could have also used `for(;;)` to accomplish the same thing. I've included both statements in the source code and Code Snippet 8.4 for you to try.

```c
//*********************************************************
// TRANSMIT PROGRAM 1
//*********************************************************
#ifdef TX_program_1

  output_high(TX);                          //mark for more than 104uS
  delay_ms(1);

  while(1)
  //for(;;)
  {
     output_low(TX);                        //send 0    START BIT
     delay_us(104);
     output_high(TX);                       //send 1    LSB of 'A'
     delay_us(104);
     output_low(TX);                        //send 0
     delay_us(104);
     output_low(TX);                        //send 0
     delay_us(104);
     output_low(TX);                        //send 0
     delay_us(104);
     output_low(TX);                        //send 0
     delay_us(104);
     output_low(TX);                        //send 0
     delay_us(104);

     output_high(TX);                       //send 1
     delay_us(104);
     output_low(TX);                        //send 0    MSB of 'A'
     delay_us(104);
     output_high(TX);                       //send 1    STOP BIT
     delay_us(104);
     delay_ms(1000);                        //pace the transmission
  }
#endif
```

Code Snippet 8.4: I like to use while(1).

The first `output_low(TX)` is a start bit. The TTL low (binary 0) from the PIC12F675 I/O pin is inverted by the Sipex SP232ACP and comes out as a space on the RS-232 side. Note that the ASCII *A* is transmitted to the personal computer's least significant bit first. Eight bits and eight `output_XXX/delay_us(104)` sequences later, Tera Term Pro displays the *A* it received in the terminal emulator window I've captured in Figure 8.12.

I've put a pacing statement at the end of the loop. This will allow you to see the characters as they appear in 1-second (1000 milliseconds = 1 second) intervals in the Tera Term Pro emulator window. You can comment this statement out to see the *A*s zip by.

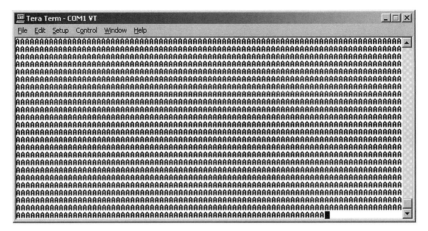

Figure 8.12: Notice I "paced" the transmission in Code Snippet 8.4. After the first A was sent, each A thereafter was sent one per second (delay_ms(1000)).

Let's comment out #define TX_program_1 and #define TX_program_3 to select TX_program_2. Note the #use rs232 statement. This is Custom Computer Services' way of having the compiler set the baud rate and assign the RS-232 I/O pins for you without having to consult the datasheet to make the adjustments manually on a bit-by-bit basis. Remember, the baud rate here is calculated based on the microcontroller's clock speed which is defined at the beginning of the program using the *#use delay(clock=4000000)* statement. Since the PIC12F675 has no internal USART, we can choose almost any pair of I/O pins to be TX and RX. Note that I said that "almost" any pair of PIC12F675 I/O pins could be chosen. The PIC12F675 has an input only pin (GP3) and since this is an input only pin, it can't be used as an output and thus can't be used as a transmit pin. You're probably also wondering where I'm getting these PORTA definitions when the PIC12F675 datasheet states that GPIO is used to define the PIC12F675 I/O port names. That's a Custom Computer Services C compiler thing. It uses PORTA designations instead of GPIO names. GPIO and PORTA are both located in their respective data memory maps at location 0x05. So, it's only a name difference. The whole of TX_program_2 is shown in Code Snippet 8.5.

```
//*******************************************************
// TRANSMIT PROGRAM 2
//*******************************************************
#ifdef TX_program_2

#use rs232(BAUD=9600, XMIT=PIN_A2, RCV=PIN_A3)

  while(1)
  {
   printf("A");
   //printf("Your first name here");
```

```
    delay_ms(1000);
  }
#endif
```

Code Snippet 8.5: Wow! Consider doing this in PIC assembler. Are you beginning to like C?

TX_program_1 consists of 23 lines of C statements (25 if you include the defines for the TX and RX lines). TX_program_2 is comprised of only three C statements and does the exact same thing as TX_program_1. What gives? The trick is the plenty powerful `printf` statement. I'm not going to explain the coding in detail, but you can see for yourself that using `printf` has more advantages than drawbacks. Replace the *A* with your first name and compile and run the program again. Cool, huh? That's what the C compiler `printf` services buys you. Of course, in the embedded world nothing is free. So, to gain the ease of use of the `printf` function, you pay in the increased amount of code the function generates and the additional amount of program memory that is consumed. To get an idea of how much extra code is generated, the Custom Computer Services C Compiler allows you to view the assembler code it generates. Compile TX_program_2 yourself and take a look at the list file to get an idea of what I'm talking about. Even though more code is generated, it's only generated once and placed in memory for use by other calls to the `printf` code. So in the long run, for the price of a little additional code, you get increased functionality with a minimum of coding effort.

The TX_program_3 in Code Snippet 8.6 is a simplified version of TX_program_1. However, it is very similar to TX_program_2 as it is short and sweet and it sends a single *A* to the Tera Term Pro emulator window. Compile and run TX_program_3 to see the *A*s sequence through. Then comment out the *putc* (put character) line and try to compile and run with the *You can't put but 1 character here* line. The compiler will choke and tell you that you can't do this. Why? Because *putc* is an abbreviation for *put character*. That means a single character and not a string of characters.

```
//*******************************************************
// TRANSMIT PROGRAM 3
//*******************************************************
#ifdef TX_program_3
#use rs232(BAUD=9600, XMIT=PIN_A2, RCV=PIN_A3)
  while(1)
  {
    putc('A');
    //putc('You can't put but 1 character here');
    delay_ms(1000);
  }
#endif
```

Code Snippet 8.6: Use putc when you want to conserve program memory and have small canned messages or single characters to send.

As you can see from the example code, using C for RS-232 work removes most of the housekeeping hassles associated with setting up RS-232 hardware and lets you concentrate on getting your data transferred from point A to point B. If I haven't convinced you that C is the easier road to RS-232 happiness and if you just really have to write some assembler to transmit a byte, Code Snippet 8.7 is a working example of C-less assembler RS-232 transmit routine:

```
;*********** RS-232 TRANSMIT SUBROUTINE
;
SENDIT
    MOVWF       XMTREG          ;LOAD BYTE TO TRANSMIT
XMTR
    MOVLW       8               ;LOAD NUMBER OF BITS TO SEND
    MOVWF       COUNT
    BCF         RS232,TX        ;WRITE 0 TO SERIAL PORT
    CALL        DELAY1          ;WAIT 1 BIT PERIOD
XNEXT
    BCF         STATUS,C        ;CLEAR CARRY
    RRF         XMTREG,F        ;ROTATE TRANSMIT REGISTER RIGHT THRU CARRY
    BTFSC       STATUS,C        ;CHECK CARRY STATUS AFTER THE ROTATE
    BSF         RS232,TX        ;IF CARRY IS SET, WRITE A 1 TO SERIAL PORT
    BTFSS       STATUS,C        ;CHECK CARRY STATUS AFTER THE ROTATE
    BCF         RS232,TX        ;IF CARRY IS CLEAR, WRITE A 0 TO SERIAL PORT
    CALL        DELAY1
    DECFSZ      COUNT,F         ;DECREMENT THE COUNT REGISTER
    GOTO        XNEXT           ;NOT DONE, GO GET NEXT BIT AND SEND IT
    BSF         RS232,TX        ;Send Stop Bit
    CALL        DELAY1          ;WAIT ONE BIT PERIOD
    RETLW       0               ;DONE, RETURN TO CALLER
DELAY1
    MOVLW       BAUD            ;104uS for 9600 BAUD
STARTUP
    MOVWF       DLYCNT
REDO1
    NOP
    NOP
    NOP
    DECFSZ      DLYCNT,F
    GOTO        REDO1
    RETLW       0
```

Code Snippet 8.7: This homegrown code was all I had when I started writing microcontroller RS-232 communications functions.

To make the assembler transmit routine work, all you have to do is calculate the bit delay time (number of cycles to expend) versus the clock frequency your project is using and plug your results into the BAUD variable. Remember, if you choose to do this as a C program, the C compiler and its related RS-232 libraries perform the automagic RS-232 setup work.

Now that you have an idea of the hows and whys of sending data with a minimal microcontroller like the PIC12F675, let's figure out how to make that PIC12F675 receive RS-232 data.

8.5.2.4 Some RS-232 Receive Code

One would believe that we could take what we know about data packet timing and write a few lines of C code akin to TX_program_1 to receive some characters from our Tera Term Pro session. That cannot easily be done, however; even though your RS-232 receive C code will consist of mostly C statements, you'll probably still end up writing the time-critical routines in assembler. We actually got lucky in TX_program_1 because our delay loop overhead was small enough to not disrupt our data packet bit timing. Why reinvent the wheel by writing RS-232 receive code from scratch? Let the C compiler and RS-232 libraries do the work. Code Snippet 8.8 is an example of writing a receive routine in Microchip assembler for our PIC12F675.

```
;*********** RS-232 RECEIVE SUBROUTINE
;
GETBYTE
    CLRF        RCVREG
    BTFSC       RS232,RD        ;LOOK FOR A START BIT
    GOTO        GETBYTE

    CALL        STARTBIT        ;go do start bit delay
RCVR
    MOVLW       8               ;load W with 8
    MOVWF       COUNT           ;load w to count
R_NEXT
    BCF         STATUS,C        ;clear the carry bit
    BTFSC       RS232,RD        ;look for data bit
    BSF         STATUS,C        ;if 0..skip this instruction
    RRF         RCVREG,F        ;ROTATE BIT FROM CARRY INTO RECREG
    CALL        DELAY1          ;go wait 104 uS
    DECFSZ      COUNT,F         ;decrement COUNT..skip if 0
    GOTO        R_NEXT          ;skip this instruction if COUNT=0

    RETLW       0

STARTBIT
    MOVLW       STARTDLY        ;DELAY FOR 156uS
    GOTO        STARTUP
DELAY1
    MOVLW       BAUD            ;104uS for 9600 BAUD
STARTUP
    MOVWF       DLYCNT
```

```
REDO1
    NOP
    NOP
    NOP
    DECFSZ      DLYCNT,F
    GOTO        REDO1
    RETLW       0
```

Code Snippet 8.8: Timing is very critical in this code, and the faster the baud rate, the more critical the timing becomes.

Again, to make this code return a character you have to calculate the value of the BAUD variable, which depends on the microcontroller's clock frequency and the amount of loop overhead in the code. In short, you have to count instruction cycles and translate them to elapsed time to set the BAUD value correctly. This is how I used to do it before the introduction of C for PIC microcontrollers. I can tell you that if you don't have a way to view the register values in the debugging process, you will be forced to use time-consuming, trial-and-error coding techniques.

What if you wanted to transmit a random character and not just the character A? I ask this question because if we are to continue with our building of simple RS-232 routines, we must be able to view the results of our receive algorithms. Assuming we would want to test the assembler RS-232 receive code you were just introduced to, how would we transmit the received character to our Tera Term Pro emulator session?

What if we chose to use TX_program_1 to echo the character received by our RS-232 receive assembler program? TX_program_1 would need some heavy-duty modifications to scan the received character's bits and translate them to `output_low` or `output_high` states used in the TX_program_1 algorithm. The overhead of the code needed for the TX_program_1 modification would most likely interfere with the RS-232 data packet bit timing and cause the RS-232 transmit character code to fail or operate erratically. In that case, incorporating the assembler transmit routine would be a better choice than modifying the TX_program_1 code.

Although there is nothing wrong with either the assembler transmit code or the assembler receive code, a couple of simple C statements can eliminate a truckload of RS-232 coding grief. Those little C statements are `putc` and `getc`. The `getc` instruction performs the same task as our RS-232 assembler receive routine. Code Snippet 8.9 an example of how the `getc` function is written in a C program.

```
#use rs232(BAUD=9600, XMIT=PIN_A2, RCV=PIN_A3)
int8 character_in;
//Receive a character
character_in = getc();
```

Code Snippet 8.9: This simple concept will take you far when you're writing your own microcontroller RS-232 communications programs.

The variable `character_in` is a byte, which is defined by the `int8` (8-bit integer) data type descriptor. The `getc` function returns a character, which in this code snippet's case is placed in the `character_in` memory location.

Let's write a C program called RX_program_1 that receives a keyboarded character from our Tera Term Pro session and echoes it back to the same Tera Term Pro session. Don't blink or you'll miss it. The whole program consumes three lines of actual code in Code Snippet 8.10.

```
//*****************************************************
// RECEIVE PROGRAM 1
//*****************************************************
#ifdef RX_program_1
#use rs232(BAUD=9600, XMIT=PIN_A2, RCV=PIN_A3)
   while(1){
   putc(getc());
   }
#endif
```

Code Snippet 8.10: *The getc function is called first, and as soon as a character is received, the putc part of the statement pushes the character out of the microcontroller's serial port.*

Ah—the beauty of C! In RX_program_1, the `getc` function is executed first and returns an 8-bit character. The `putc` function sends the results of the `getc` function, which is the keyboarded and received ASCII character, out to the Tera Term Pro session. The `while(1)` statement assures that this get and put operation will continue until power is removed from the PIC12F675.

You are trained to and can now write and execute a basic RS-232 routine in either C or assembler using the smallest of microcontrollers. It's also evident (I hope) that C is the easier choice. Notice I used the word "easier" and not the word "better," because there could be situations where C is "too big" for your application. In those cases, assembler can be more efficient and more compact. What if neither programming C nor assembler is comfortable for you? Keep reading. Most of you will be in for a pleasant surprise.

8.6 Writing RS-232 Microcontroller Routines in BASIC

In embedded design, it's difficult for "hardware engineers" to avoid software, or at least firmware. In the preceding chapter, I've attempted to convert those of you that are still writing your microcontroller code in assembler to writing your microcontroller code using the C programming language. However, I learned personal computer assembler first, and then as the personal computer BASIC language evolved I moved to that as my primary personal computer programming language. After getting a grip on just what programming was, I finally ended up using C for most of my personal computer programming needs. Note that I said "most of," not "all of." If the personal computer application fits, I will revert to using some form of the BASIC programming language because BASIC is still a viable and powerful programming tool. This section will prove to you that the BASIC language is just as meaningful and just as powerful in the microcontroller programming world as it still is today in the personal computer programming environment.

At first, this wasn't as "BASIC" as I would have liked. It took some effort to understand the PicBasic Pro system, but once I had a grasp of what was going on, things got better in a hurry.

Like the Custom Computer Services C Compiler, microEngineering Labs' PicBasic Pro has very good intentions about making things easy for the PIC programmer. For instance, in PicBasic Pro the watchdog timer is enabled by default and the PicBasic Pro compiler automatically inserts clear watchdog timer commands into the code at the appropriate locations. Akin to Custom Computer Services C Compiler, some PicBasic Pro instructions actually change the PIC port pin to an input pin or output pin automatically to effect their function. The PicBasic Pro built-in functions SerIn and SerOut are examples of PicBasic Pro instructions that automatically set the I/O pin direction (input for SerIn and output for SerOut) when called.

PicBasic Pro comes with its own special IDE, CodeDesigner Lite, and also melds seamlessly with the latest version of MPLAB IDE. Let's put together a simple PicBasic Pro program using CodeDesigner Lite that receives a character from Tera Term Pro and then echoes that character back to Tera Term Pro.

Just like before, the first thing we must do is provide a means of defining the RS-232 baud rate, parity setting, stop bit setting, and bit inversion setting. In PicBasic Pro, the bit inversion setting is used to emulate an RS-232 converter IC when connecting the serial I/O pins directly to another serial device such as your personal computer's serial port. Remember that the TTL bits in the data packet are inverted and voltage-shifted after passing through the Sipex SP232ACP RS-232 converter on our PICkit 1. A binary 1 becomes a mark or negative voltage and a binary 0 becomes a positive voltage space. If you look at the voltage levels of an RS-232 signal, you'll see that it is possible to fool a serial interface using the PIC's TTL I/O levels. Most serial ports will sense a TTL low (1.8 V or below) as a mark and a TTL high (3 V and above) as a space. If we use a PIC microcontroller to send an *A* without using an RS-232 converter IC, the TTL binary sequence LSB (least significant bit) to MSB (most significant bit) would look like this:

```
START BIT (0) 10000010 (1) STOP BIT
```

If an RS-232 converter IC is used, we know that a binary 1 becomes a mark and a binary 0 translates to a space on the RS-232 side of the converter IC. In this case, we don't have the inversion (and voltage conversion) at the sending serial port because the RS-232 converter IC is not in the circuit. If the receiving side of our serial link uses an RS-232 converter IC, the incoming data will be presented to the receiving device's application inverted. So, after its RS-232 converter at the receiver does its thing, the receiving device would actually see:

```
START BIT (1) 01111101 (0) STOP BIT
```

What a mess. First of all, the start bit will not be recognized and will be seen as the first data bit. Second, the binary pattern is an inverse of the character *A* and because the start bit was detected one bit too late, the data bit are shifted as a result. Finally, the stop bit is wrong and if this bit pattern gets through to the application at all, the stop bit will be incorrectly recognized as a data bit. The receiving application will most likely throw this RS-232 data packet (and any similar to this that follow) in the trash. PicBasic Pro's bit inversion option is used to avoid situations like the one we just discussed that stem from not having an RS-232 converter IC at one end of the link. As you can see, the bit inversion takes the place of an RS232 converter IC and allows direct connection to a "true" serial port. Although you can hook raw microcontroller port I/O pins directly to an RS-232 port, be careful as the RS-232 voltages will damage your microcontroller.

PicBasic Pro's `SerIn` and `SerOut` functions use predefined modes to set up the baud rate. The modes are defined in an include file that comes with the PicBasic Pro Compiler package called `modedefs.bas`. For the `SerIn` and `SerOut` functions, 8N1 (8 data bits, no parity and 1 stop bit) is the default setting for an RS-232 data packet and can't be changed. We'll choose 9600 bps as our baud rate and since we do have a Sipex SP232ACP RS-232 converter IC in our PICkit 1 circuit there is no need to specify bit inversion. Thus, our mode will be specified as T9600, where the *T* stands for True. If inversion were required, our mode specification would change to N9600, with *N* signifying that the TTL data will be inverted.

Another automatic feature of PicBasic Pro is its assumption that the target PIC is running a 4MHz clock. Although other clock speeds can be defined, the 4MHz clock default is a good thing for us since our PIC12F675 is running on its internal 4MHz clock. To maintain as accurate an internal clock as possible, the PIC12F675 uses an oscillator calibration value called OSCCAL. The OSCCAL value is kept in program memory space. The PicBasic Pro "DEFINE OSCCAL_1K" definition automatically moves the OSCCAL value that resides in program memory to the OSCCAL register every time the program is run. If you're not careful, you can erase the OSCCAL value. No worries. The PICkit 1 host program has an option that will rebuild the OSCCAL value for you.

There's another way to generate serial I/O using PicBasic Pro. The `DEBUG` and `DEBUGIN` functions allow most any I/O pin to become a serial transmitter or serial receiver. As you may ascertain from their names, the `DEBUG` functions are primarily intended to help you debug your code by allowing you to insert the `DEBUG` statements at various points inside your code to send variable values to a Tera Term Pro or HyperTerminal session. To use the *DEBUG* functions the I/O port, the I/O pin, the baud rate and the bit inversion mode must be specified. I'll stick with the PIC12F675 RS-232 port, pin, baud and bit inversion values we've used throughout our discussion, which makes GPIO our `DEBUG` and `DEBUGIN` port with GP2 acting as the transmit pin and GP3 doing the receiver duty. Our selected baud rate is 9600 bps and there is no bit inversion. Like the `SerIn` and `SerOut` functions, the number of bits in the RS-232 data packet, the parity and the number of stop bits is set at 8N1 and cannot be altered by the programmer. Code Snippet 8.11 is our PicBasic Pro BASIC code up to this point:

```
INCLUDE "modedefs.bas"
DEFINE OSCCAL_1K 1
DEFINE DEBUG_REG GPIO
DEFINE DEBUGIN_REG GPIO
DEFINE DEBUG_BIT 2
DEFINE DEBUGIN_BIT 3
DEFINE DEBUG_BAUD 9600
DEFINE DEBUG_MODE 0
DEFINE DEBUGIN_MODE 0
DEFINE NO_CLRWDT 1
```

Code Snippet 8.11: Hmmm ... looks kind of like C.

I went ahead and threw in the last `DEFINE` line because it's important if you want to change one of PicBasic Pro's automatic features such as whether the watchdog timer runs or not.

How a PicBasic Pro source file is compiled depends on the assembler and some configuration fuse settings found in the PicBasic Pro PIC12F675 include file. Code Snippet 8.12 is the PicBasic Pro PIC12F675 include file that determines what microcontroller fuses are active versus which assembler is invoked. CodeDesigner Lite uses the native PicBasic Pro assembler, PM. Notice that I have modified the original code turning the watchdog timer off (`wdt_off`) and internalizing the PIC12F675 MCLR function freeing the GP2 pin for I/O use with `mclr_off`. The last `DEFINE` statement, `DEFINE NO_CLRWDT 1`, instructs the PicBasic Pro Compiler not to insert the clear watchdog statements into the final code.

MPLAB IDE allows the use of either the PicBasic Pro's PM or Microchip's MPASM assembler. The code we generate with CodeDesigner Lite will compile without modification under the MPLAB IDE. Notice that I have made the same configuration fuse adjustments in the MPASM header code in Code Snippet 8.12.

```
;****************************************************************
;* 12F675.INC                                                   *
;*                                                              *
;* By          : Leonard Zerman, Jeff Schmoyer                  *
```

```
;*  Notice      : Copyright (c) 2002 microEngineering Labs, Inc.  *
;*                All Rights Reserved                             *
;*  Date        : 09/27/02                                        *
;*  Version *   : 2.43
;*  Notes       :                                                 *
;*****************************************************************
        NOLIST
    ifdef PM_USED
        LIST
        INCLUDE 'M12F675.INC'    ; PM header
        device pic12F675, intrc_osc_noclkout,bod_off, wdt_off,
pwrt_on, mclr_off, protect_off
        XALL
        NOLIST
    Else
        LIST
        LIST p = 12F675, r = DEC, w = -302
        INCLUDE "P12F675.INC"  ; MPASM Header
        __config _INTRC_OSC_NOCLKOUT & _BODEN_OFF & _WDT_OFF & _PWRTE_ON
& _MCLRE_OFF & _CP_OFF
        NOLIST
    EndIF
        LIST
```

Code Snippet 8.12: The trick is to recognize that both the PM and MPASM configuration code is included in this file.

Unlike the Custom Computer Services C Compiler, the PicBasic Pro Compiler does not have preprocessor directives like `#ifdef`. So, I'll sacrifice a byte to simulate the C preprocessor directives and add some PIC12F675 setup information to our PicBasic Pro code. Two bits are assigned to `debug_prog` and `serio_prog` and I've allocated a byte, `chr`, to hold the ASCII character our program will send and receive. Equating a 1 to a bit allows the program represented by that bit to be compiled. Conversely, a 0 assigned to a bit effectively turns that program off to the compiler. Normally, this kind of code would not be something you'd want to include in a professional project. I'm incorporating it here to make the PicBasic Pro Compiler serial communications example easier for you to understand and compile by stuffing two programs into one as we did with the Custom Computer Services C Compiler C source.

Recall that we had to generate some C code to turn off the analog section of the PIC12F675 so we could use those dual-purpose pins for digital I/O. Well, we're using the same microcontroller hardware, a PIC12F675, and we must perform the same analog deactivation process using PicBasic Pro code (adcon0 = 0). Our C code example also included a "TRIS" statement to setup the input or output status of the PIC transmit and receive I/O lines. We don't need to code any BASIC "TRIS" statements here as the DEBUG/DEBUGIN and SerIn/SerOut functions automatically set the selected PIC's I/O pin for input or output depending on the function call. The conveniences provided by the PicBasic Pro Compiler make for a very tidy set of RS-232 echo routines, as shown in Code Snippet 8.13.

```
INCLUDE "modedefs.bas"
DEFINE OSCCAL_1K 1
DEFINE DEBUG_REG GPIO
DEFINE DEBUGIN_REG GPIO
DEFINE DEBUG_BIT 2
DEFINE DEBUGIN_BIT 3
DEFINE DEBUG_BAUD 9600
DEFINE DEBUG_MODE 0
DEFINE DEBUGIN_MODE 0
DEFINE NO_CLRWDT 1

debug_prog VAR BIT
serio_prog VAR BIT
chr VAR BYTE

adcon0 = 0

;PROGRAM TO RUN = 1
debug_prog = 1
serio_prog = 0
loop:
    ;CHARACTER ECHO USING DEBUG FUNCTIONS
    IF debug_prog Then
    DebugIn [chr]
    Debug chr
    EndIF

    ;CHARACTER ECHO USING SER_IO FUNCTIONS
    IF serio_prog Then
    SerIn GPIO.3,T9600,chr
    SerOut GPIO.2,T9600,[chr]
    EndIF

    GoTo loop
End
```

Code Snippet 8.13: A bit wordier than its C counterpart, but that's BASIC no matter where you encounter it. The bottom line is that it works just like the C code.

To me, writing code in BASIC is fun. The neat thing about BASIC is that it's easy to learn no matter what microcontroller or personal computer you're writing an application for. The microEngineering Labs PicBasic Pro package is easy to use and powerful in function. No matter which programming language you choose to write your RS-232 code, you'll need some hardware that is capable of turning your typing into reality. Let's round up some RS-232 components, a PIC microcontroller, and turn on the soldering iron.

8.7 Building Some RS-232 Communications Hardware

Enough theory and coding already—let's solder some stuff. But before we can start connecting components together to form our physical hardware, there are a few more things you need to know about microcontrollers and RS-232.

Up to now we've been sending and receiving under the guidance of the RS-232 standard with a microcontroller that doesn't contain any internal serial communications circuitry. The PicBasic Pro Compiler and the Custom Computer Services C Compiler compensate for this lack of circuitry and allow us to emulate the missing serial hardware using the compiler's firmware. The firmware implementation of a serial port works fine until you have to do other things and look for incoming serial data simultaneously.

8.7.1 A Few More BASIC RS-232 Instructions

When other tasks are being serviced by the microcontroller, it may be possible to miss an incoming RS-232 message. PicBasic Pro handles this situation by allowing the user/programmer to "time out" after checking for incoming serial data. Let's add Code Snippet 8.14, another module to our PicBasic Pro source code to demonstrate how that would work.

```
INCLUDE "modedefs.bas"         ;get mode defs
DEFINE OSCCAL_1K 1
DEFINE DEBUG_REG GPIO
DEFINE DEBUGIN_REG GPIO
DEFINE DEBUG_BIT 2             ;serial out GP2
DEFINE DEBUGIN_BIT 3           ;serial in GP3
DEFINE DEBUG_BAUD 9600         ;9600,N,8,1
DEFINE DEBUG_MODE 0            ;no inversion
DEFINE DEBUGIN_MODE 0          ;no inversion
DEFINE NO_CLRWDT 1             ;don't add clrwdt

debug_prog VAR BIT
serio_prog VAR BIT
serio_wait_prog VAR BIT

chr VAR BYTE                   ;for ASCII character

adcon0 = 0                     ;turn off analog I/O

;PROGRAM TO RUN = 1
debug_prog = 0
```

```
serio_prog = 0
serio_wait_prog = 1
loop:
   ;CHARACTER ECHO USING DEBUG FUNCTIONS
   IF debug_prog Then
   DebugIn [chr]
   Debug chr
   EndIF

   ;CHARACTER ECHO USING SER_IO FUNCTIONS
   IF serio_prog Then
   SerIn  GPIO.3,T9600,chr
   SerOut GPIO.2,T9600,[chr]
   EndIF

   ;CHECK FOR A CHARACTER EVERY SECOND
   IF serio_wait_prog Then
   SerIn  GPIO.3,T9600,1000,no_data,chr
   SerOut GPIO.2,T9600,[chr,13,10]
   GoTo loop
no_data:
   SerOut  GPIO.2,T9600,["no character",13,10]
   EndIF

GoTo loop

End
```

Code Snippet 8.14: The microEngineering Labs PicBasic Pro Compiler has as many built-in tricks up its sleeve as the Custom Computer Services C Compiler does.

As you can see in Code Snippet 8.14, I've added a third module and a corresponding third bit, `serio_wait_prog`. The `serio_wait_prog` code looks for an incoming character for 1000 ms (1 second). If no character is received after waiting for 1 second, the program jumps to the "no_data" label and prints "no character" followed by a carriage return (decimal 13) and a linefeed (decimal 10). If a character is detected within the 1-second window, the received character is sent followed by a carriage return/linefeed (CRLF) sequence.

Things are a bit different on the C side in Code Snippet 8.15, but the results are the same.

```
#include <12F675.h>
#include <f675.h>
#use delay(clock=4000000)
#fuses INTRC_IO,NOWDT,NOMCLR,NOPROTECT,NOCPD,NOBROWNOUT

#use fast_io(A)

int32 timeout;
```

```
//***********************************************
// COMMENT OUT THE PROGRAMS YOU DON'T WANT TO RUN
//***********************************************
//#define TX_program_1
//#define TX_program_2
//#define TX_program_3
//#define RX_program_1
#define SERIO_program

/////////////////////////////////////////
//***********************************************
// SERIO WAIT PROGRAM
//***********************************************
#ifdef SERIO_program
#use rs232(BAUD=9600, XMIT=PIN_A2, RCV=PIN_A3)
   while(1)
   {
   timeout=0;
   while(!kbhit()&&(++timeout<50000))
      delay_us(10);
   if(kbhit())
      printf("%c\r\n",(getc()));
   else
      printf("no character\r\n");
   }
#endif
```

Code Snippet 8.15: This C code is a bit more complicated than the BASIC version.

I added `int32 timeout` to allocate a 32-bit area of PIC memory to hold a timeout value and another `#define` statement to point at our new C source module, `SERIO_program`. Instead of having the PicBasic Pro luxury of inserting a timeout value in the function call, our C source code uses the `kbhit` function to signal the presence of a character. Since we have no "real" serial hardware functionality inside our PIC12F675, the `kbhit` function returns a TRUE after detecting a valid start bit on the PIC12F675's GPIO receive line. So, every 10 μs our C program looks for a start bit and increments the timeout value. After about a second or so, if no character has been detected, our C program sends "no character" followed by a CRLF sequence (\r\n). If a valid start bit is detected and followed by a valid character, the character is retrieved using the `getc` function and sent to the Tera Term Pro terminal emulator using the *printf* function, which also appends the sent character with a CRLF sequence.

If you're working with a microcontroller like the PIC12F675, the serial I/O routines I've just described will work well for you. The only drawback is that you have to continually run the routines in conjunction with your main application to "poll" for an incoming character. There will be times when you won't have enough processing time or processor resources to do the polling. That's when you call in some bigger guns.

8.8 I²C: The Other Serial Protocol

RS-232 is a great point-to-point protocol for communicating between two distinct and sometimes distant pieces of equipment. However, at times you'll need to be able to talk to multiple electronic modules across a communications link that only spans the distance of a single printed circuit board. It would be possible to "network" the board-sharing modules using the RS-232 9-bit addressing protocol, but there are lots of caveats in that approach. Even though you could eliminate the RS-232 voltage conversion circuitry, you would find yourself doing a tremendous amount of USART transmit and receive line housekeeping. For instance, you would have to generate an algorithm to handle collisions between modules attempting to transmit at the same time or collisions that occur in the middle of a message another module is already transmitting. The USART transmit and receive lines are not automatically passive or tri-stated when inactive. Thus, you would also have to write some code to make sure the transmit and receive lines are inactive when they're supposed to be. If you want an RS-232 LAN, it can be done, but there is a better way.

Initially designed for use in commercial audio and video systems, the inter-IC or I²C bus is a Philips Semiconductor creation. Just as its name implies, the I²C bus is a bidirectional two-wire bus that is used to transport data between ICs (integrated circuits). Unlike RS-232, the I²C bus doesn't need any voltage converters or special interface parts. If an IC is I²C-bus compatible, everything needed to operate on the I²C bus is incorporated on-chip within the IC.

If you take another look at our RS-232 schematic (Schematic 8.3), you'll see that there are two bus lines integral to the PIC18F452: a serial data line (SDA) and a serial clock line (SCL). The SDA and SCL bus lines make up the I²C interface, and since these lines are designated as an integral part of the PIC18F452 that makes the PIC18F452 an I²C-compatible device. Being I²C-compatible, the PIC18F452 has provisions for a unique I²C bus address. Using the built-in I²C functionality, the PIC18F452 can act as either the master or slave on an I²C network. If the PIC18F452 is configured as an I²C master, it can act as a master-transmitter or master-receiver. Conversely, if the PIC18F452 is chosen to be a slave on the I²C bus, its internal I²C electronics can act in either slave-receiver or slave-transmitter mode. Remember one of my RS-232 "LAN" caveats and collisions? I²C is a true multimaster bus that includes arbitration safeguards against data collisions, which prevents data corruption on the I²C bus. Like RS-232, I²C is an 8-bit bidirectional serial communications method. That's where the similarity ends. I²C operates at a speed of 100 kbs in standard-mode, 400 kbs in fast-mode and up to 3.4 Mbps in high-speed mode. The only limitation as to how many devices can exist on a single I²C bus is the total capacitance the devices place on the bus.

Advantages of using I²C are numerous, and there are a multitude of various I²C building blocks to choose from. By employing I²C in a design, we can eliminate much of the auxiliary support circuitry such as address decoders and standard logic gates needed for other communications methods.

The Essence of Microcontroller Networking: RS-232

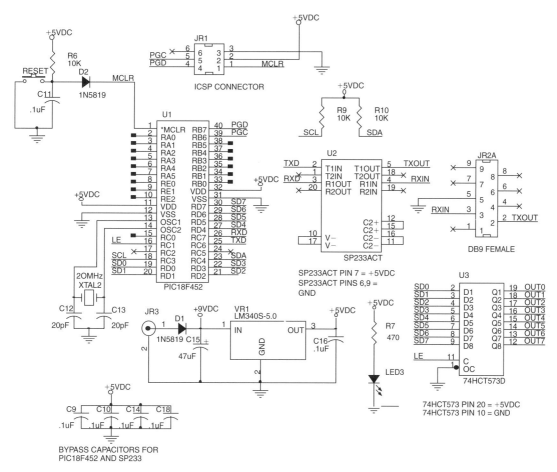

Schematic 8.3: This is a schematic of our partially assembled Easy Ethernet CS8900A. Resistors R9 and R10, plus the PIC18F452's internal I²C engine, are all that's needed to effect an I²C network. Notice we've added a new component.

In this chapter, we're going to use I²C to network our partially assembled Easy Ethernet CS8900A to our partially assembled Easy Ethernet AVR. Since each microcontroller in the network has on-chip I²C resources, we already have a solid basis for an I²C microcontroller network, but before we start slinging solder, let's take a course in I²C.

8.8.1 Why Use I²C?

For folks who make their living designing the neat gadgets we buy at department stores and over the Internet, putting out a product at the least possible cost is paramount. Chances are your television and stereo both contain an I²C bus. Using I²C is cheap because you don't have to do anything special to set up the physical communications link. Two wires or two traces are all that's needed for the physical I²C-bus signal path. Although I²C can operate at a very high speed, most of the time that's not a factor in the design. So, the serial nature of I²C is well suited for low-speed control type applications. I²C also solves a majority of the design problems one would encounter when connecting dissimilar devices on a network. Slow devices must be able to talk to higher-speed devices and vice versa, and everyone on the network must be able to speak the same language. Most important, somebody has to be the network boss, and as in the real world, there could be more than one boss on the bus. I²C has an answer for all these potential problems.

8.8.2 The I²C Bus

As you already know, I²C is built around a two-wire serial bus, SDA (serial data) and SCL (serial clock). Each device on the I²C bus is identified by a unique address. An I²C device can be a microcontroller such as our PIC18F452, a memory device such as a standard I²C EEPROM, or a special-purpose device like an LED display driver. Some I²C devices are capable of transmitting and receiving on the I²C bus; other I²C devices may only be able to receive. In any case, a master-slave environment always exists on the I²C bus. The I²C master device always initiates an I²C bus data transfer and generates the clock signals to make the data transfer happen. The I²C device that responds to the master's calling is considered the slave device.

Microcontrollers are normally defined as masters on a typical I²C bus, with other special-purpose I²C devices acting as their slaves. In our application, even though all the I²C devices are microcontrollers capable of being an I²C master, only one of the microcontrollers will be granted master status.

If more than one master exists on a single I²C-bus, there will be conflict when one of the multiple masters attempts to transmit in unison with another peer master device on the I²C bus. The I²C specification solves this problem with something called *arbitration*. Arbitration is the process of allowing only one master to control the I²C bus at any time. Before I can really explain arbitration to you, there are some basic I²C rules you need to know.

In an ideal world, if the master wanted to communicate with slave, the master would address the slave. The master is now in master-transmitter mode, and the slave is in slave-receiver mode. The master would clock-out data to the slave and terminate the data transfer after all the desired bytes were transmitted (see Schematic 8.4).

On the other side of that, let's say that the master wanted to receive some data from the slave. Again, the master would clock-out an address aimed at the slave. Instead of assuming master-transmitter mode, this time the master would become the master-receiver with the slave acting as slave-transmitter. Data would be clocked-in by the master, which would terminate the transfer after receiving the desired bytes. For every bit of data moved, one clock pulse is generated and the data on the SDA line must be stable during the HIGH period of SCL. The logic level of the data line can only change when the SCL line is LOW.

Note that in either of the aforementioned cases, the master did all of the clocking and controlled the initiation and termination of the I²C session. The I²C master is always responsible for

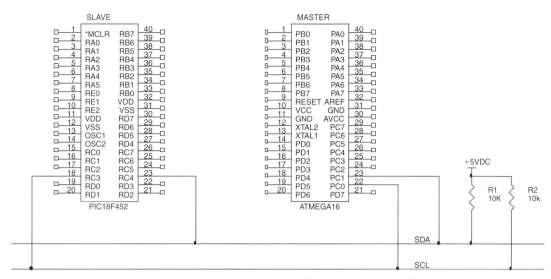

Schematic 8.4: For the sake of simplicity, I've left out all of the standard microcontroller connections to help us focus on the I²C bus. I flipped a coin to choose the master microcontroller.

generating the clock on the I²C bus. For an I²C bus with multiple masters, each master generates its own specific clock. The only things that can alter a master's clock are a slower slave device holding down the clock line or another master I²C device during arbitration.

As you can see in both Schematic 8.4 and Schematic 8.5, the I²C bus SDA and SCL lines are pulled high by a pair of pull-up resistors. To participate on the I²C bus, an I²C device must present an open collector interface to the bidirectional SDA and SCL I²C-bus lines. This type of open collector interface performs a wired-AND function. As long as the 400 pF I²Cbus capacitance limit is not exceeded, any number of I²C devices can coexist on a single I²C bus.

The wired-AND configuration used in I²C could really cause lots of confusion on the bus if it were not for the strict protocol that makes up the logical side of the I²C bus. Remember the start and stop bits you were exposed to in RS-232? Well, I²C has start and stop bits too, but instead of bits they are technically known as I²C START and STOP conditions. The I²C START and STOP logic levels can be seen in Figure 8.13. The SCL line must be in a HIGH state for either a START or a STOP condition to occur. An I²C START condition is defined as a HIGH to LOW transition of the SDA line while the SCL line is HIGH. An I²C STOP condition occurs when the SDA line toggles from LOW to HIGH while the SCL line is HIGH. The I²C master always generates the S and P conditions. Once the I²C master initiates a START condition, the I²C bus is considered to be in a busy state.

I know what you're thinking. I²C has STOP and START bits like RS-232 does, and I²C transfers 8 bits of data in a data packet, just like RS-232 does. That means that an I²C data packet

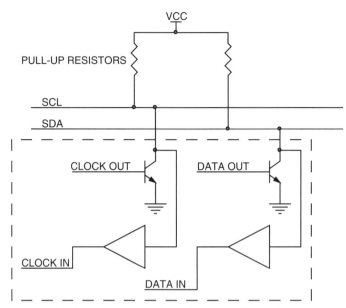

Schematic 8.5: This is a representation of how an I²C device connects to the I²C bus. Note that the type of transistors and associated circuitry would depend on the technology (CMOS, NMOS, bipolar) of the I²C device.

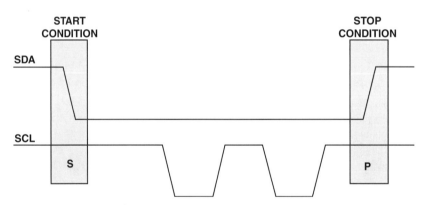

Figure 8.13: I²C itself is an abbreviation, so why not abbreviate START and STOP with an S and a P?

is just like an RS-232 data packet with 8 bits of data sandwiched between a START and STOP bit. Not exactly …

It is true that I²C requires that the data be transferred in bytes. It is also true that I²C starts a transmission with a START condition and ends the transmission with a STOP condition. The difference between an RS-232 transmission and an I²C transmission is that an unrestricted number of data bytes can flow between an I²C START and STOP condition while only a single byte of information can be transferred between the start and stop bits of an RS-232 data packet. Another major difference in I²C and RS-232 is that the data is transferred most significant bit first in an I²C data packet instead of least significant bit first, as it is in RS-232. Regardless of the order in which the bits are transmitted, the real enabler for I²C multibyte transfers is the I²C acknowledge bit. Every byte that flows on the I²C bus must be followed by an acknowledge bit. Since the acknowledge bit is very important for I²C communications, let's get a better understanding of how it works.

8.8.3 I²C ACKS and NAKS

The acknowledge bit (ACK) rides on the master-generated clock pulse train. During an acknowledge, the transmitting device releases the SDA line and uses the wired-AND functionality of the I²C bus to pull the SDA line to a HIGH state. The I²C master generates an acknowledge clock pulse and during the acknowledge bit time (HIGH SCL), the receiving I²C device must pull the SDA line down to a LOW state for the time that SCL is in the acknowledge clock pulse HIGH state. Standard I²C protocol expects the receiving I²C device to acknowledge every byte that is received.

There could be times when the slave can't acknowledge the master. For instance, the slave is busy taking analog readings and "can't come to the I²C phone." In this case, the slave leaves the SDA line in a HIGH state. The I²C master senses this negative acknowledge (NAK) and can choose to either end the transaction with a STOP condition or begin a new transfer by issuing a repeated START condition. The repeated START condition allows the current I²C-bus master to keep control of the I²C-bus to issue another START bit instead of relinquishing the bus and attempting to recapture it to issue another START condition.

What if the slave "answers the I²C phone" in slave-receiver mode but later gets called by a process that doesn't allow the slave to receive any more bytes? When the slave can't continue, it allows the SDA line to go HIGH during the acknowledge bit time, which in turn sends a NAK to the I²C master. At this point, the I²C master can either abort the transfer or attempt a restart.

A NAK condition isn't always a bad thing. When the I²C master is in master-receiver mode, it signals the end of the data transfer from the slave-transmitter by generating a NAK on the last byte it clocked out of the slave-transmitter. The slave-transmitter senses the NAK and releases the SDA line so the I²C master can either generate a STOP condition or a repeated START condition. Logical examples of ACKs and NAKs are depicted in Figure 8.14.

8.8.4 More on Arbitration and Clock Synchronization

Now that you're up to your ankles in I²C theory, let's talk a bit more about arbitration. I²C depends heavily on accurate clocking from each master on the I²C bus, and the wired-AND-based I²C bus connections have a hand in the clock synchronization process as well.

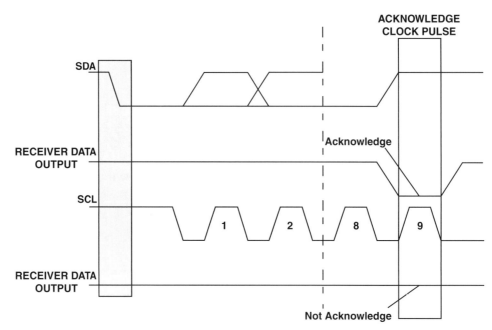

Figure 8.14: The receiver data output is shown twice here to illustrate the difference between an ACK and a NAK.

Data on the I^2C bus is only valid when the SCL line is in the HIGH portion of a clock pulse. Let's use our example I^2C bus with two microcontrollers attached, as shown in Schematic 8.2. If each microcontroller can clock the I^2C bus at a specific speed, that means that the internal master I^2C engine of each microcontroller on the I^2C bus has a means of counting to effect the elapsed times needed to swing the I^2C bus HIGH and LOW at a specific rate.

The AVR being the master of the I^2C bus wants to communicate with the PIC slave. The AVR generates the clock on the SCL line and sends a byte of data. The PIC acknowledges the data and then has to go off to service an external interrupt. If the AVR continues to try to communicate with PIC, the AVR will soon miss the acknowledgement it is expecting from the slave PIC, and the transmission would have to be aborted or restarted by the AVR. This is where the I^2C-bus wired-AND logic comes into play to help avoid such a situation.

Think of the I^2C bus as a simple AND gate. The truth table for a two-input AND gate is shown graphically in Figure 8.15.

Now, in Figure 8.16, let's substitute the AVR's and the PIC's SCL line states for the inputs with the AND gate outputs representing the resultant state of the I^2C bus SCL line.

When the PIC is able to service the AVR's requests immediately, the PIC leaves the SCL line alone by driving its SCL interface HIGH. You can see this in states 1 and 3 of our substituted AND gate example in Figure 8.16. If the PIC needs more time to respond to AVR's requests,

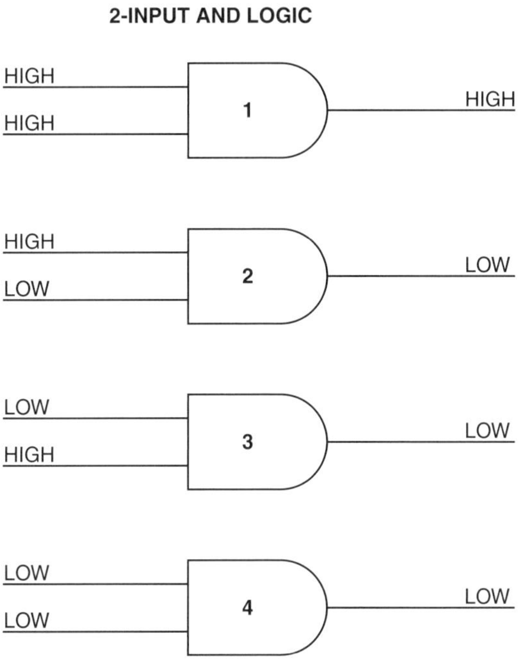

The Essence of Microcontroller Networking: RS-232

2-INPUT AND LOGIC

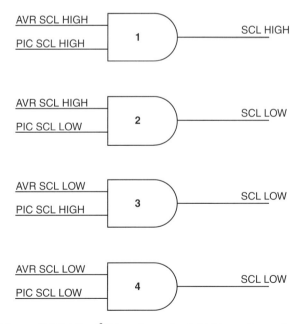

Figure 8.15: This is plain old everyday logic. Any presence of a LOW on either of the inputs results in a LOW on the AND gate output.

Figure 8.16: The I^{2}C bus is a wired-AND configuration.

it can pull the SCL line down to a LOW state. The act of the PIC pulling down the SCL line is called *clock stretching*. As you can see in states 2 and 4, the AVR is unable to change the state of the SCL line when the PIC is holding the SCL line LOW. So, the AVR goes into a HIGH wait state and sits there until the PIC releases the SCL line and allows it to return to a HIGH state. The bottom line is that the SCL line will be held LOW by the I²C device with the longest LOW period. The I²C device with the shortest HIGH period determines how long the SCL line will remain in a HIGH state during clocking. This is how the I²C bus is synchronized.

It is possible for two or more I²C masters to initiate a start condition at the same time. When that occurs, the masters requesting the use of the I²C bus must utilize the I²C arbitration process. I²C arbitration is performed using the SDA line while the SCL line is at a HIGH level. Both the SCL and SDA lines are wired-AND configurations. So, we can apply the same logic to the arbitration process as we did to the I²C bus clock synchronization.

We must assume that both the AVR and the PIC in Schematic 8.2 are masters on the I²C bus. Figure 8.17 shows us that when any master on the I²C bus takes the SDA line LOW, the other masters on the I²C bus are unable to drive the SDA line high. Thus, the I²C-bus arbitration loser is the master that attempts to transmit a HIGH, while another master is transmitting a LOW on the SDA line. The master transmitting a HIGH when the SDA line is LOW senses that the SDA line is not at the same level as it is transmitting and switches off its data output stage. The losing master applies a HIGH to the SDA line and reverts to slave mode if it is configured to perform the slave function. By presenting a HIGH to the SDA line, the losing

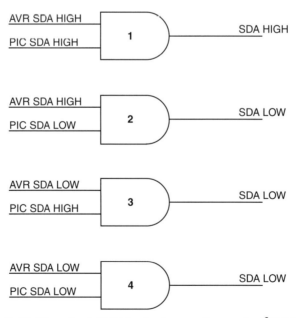

Figure 8.17: The wired-AND logic also applies to the I²C SDA line.

master releases the SDA line to the winning master. Let's say the AVR is the winning master, and the PIC is the losing master. In Figure 8.17, states 1 and 3 define the state of the SDA line while the AVR was in charge of the I²C bus. If the PIC was declared the winner and the AVR the loser, states 1 and 2 would go into effect while the PIC was in control of the I²C bus.

Arbitration can be performed for a number of bits into the transaction. For instance, the masters may all be addressing the same slave in the same manner. In that case, the address bits from each master would be identical. The good news is that the winning master's address and data are the only valid items on the I²C bus and nothing in terms of address and data information is lost in the arbitration process.

Clock synchronization is always going on in the SCL domain while arbitration may be occurring at the SDA level. A slave I²C device can throttle the speed in which it accepts data bytes by dragging the SCL line LOW. In standard mode, any smart device on the I²C bus that can extend the LOW period of the clock can control the speed of other devices on the I²C bus because the device with the longest LOW period determines the top speed of every other master device on the I²C bus.

As long as we follow the rules and use a device with built-in I²C capability, I²C is dead-easy to implement. Before we write some I²C code to go along with our AVR and PIC RS-232 code modules, let's take a look at how data flows across an I²C bus.

8.8.5 I²C Addressing

You already know that a START condition begins the I²C data transfer process. Since multiple devices can coexist on the I²C bus, there must be a way to differentiate them. This is done with I²C addressing. I²C devices can be addressed using a 7-bit or 10-bit format. I²C 10-bit addressing isn't difficult to grasp once you understand 7-bit addressing. So, instead of trying to school you on 10-bit addressing, I'll concentrate on showing you how 7-bit addressing works as we'll only be using 7-bit addressing in our project.

The first byte sent on the I²C bus after the start is usually an address byte. One exception involves sending a "general call" address following the start condition. The "general call" addresses everyone on the I²C bus. Our project doesn't use the "general call." So, let's move on with picking apart the I²C 7-bit address mechanism.

The seven ADDRX bits in the 7-bit address scheme shown in Figure 8.18 are taken from the first seven bits of the address byte that follows the start condition. Remember, in I²C land, the most significant bit is transmitted first. So, bits 7 through 1 of the address byte actually carry the I²C address information. The least significant bit, bit 0, determines whether the I²C operation will be a read or write. A binary zero in bit 0 of the address byte tells the slave that the master will be writing data to the slave device. Conversely, a binary 1 in the least significant bit (LSB) position will allow the master to read information from the slave. Each device on the

MSB							LSB
ADDR6	ADDR5	ADDR4	ADDR3	ADDR2	ADDR1	ADDR0	R/W

Figure 8.18: Think of this as subtracting 1 from the real I^{2}C address to write and adding 1 to the I^{2}C address to read. The ADDRX bits make up the actual slave address.

I^{2}C bus sees the address byte. Only the device that contains the match for the first seven bits of the address byte will ultimately respond to the I^{2}C master's call. If the I^{2}C operation is a write from the master, the slave device enters slave-receiver mode. An I^{2}C-bus read operation will put the addressed slave device into slave-transmitter mode.

Let's write some I^{2}C code.

8.8.6 Some I^{2}C Firmware

Custom Computer Services PIC Compiler easily handles the I^{2}C master chores. Custom Computer Services C for PICs provides built-in code for the standard I^{2}C functions such as `i2c_start`, `i2c_read`, `i2c_write`, and `i2c_stop`. In this section, we're also going to be producing AVR I^{2}C code in parallel with the PIC C code using ICCAVR. The ImageCraft C compiler doesn't have built-in AVR I^{2}C functions, but we can easily write our own. Reading and writing in I^{2}C master mode is straightforward. The real coding work comes in when exercising the slave side of these common I^{2}C functions.

You've already seen Schematic 8.3, which contains the PIC18F452 I^{2}C circuitry. Schematic 8.6 shows the AVR I^{2}C circuitry, which is very similar to the PIC I^{2}C circuitry.

8.8.7 The AVR Master I^{2}C Code

Atmel's term for I^{2}C is Two-Wire Interface (TWI). For I^{2}C master operation, we will deal with only four AVR registers: the Two-Wire Interface Data Register (TWDR), the Two-Wire Interface Control Register (TWCR), the Two-Wire Interface Bit Rate Register (TWBR), and the Two-Wire Interface Status Register (TWSR). The TWBR is set and forget. So, we'll only be exercising the contents of three AVR I^{2}C registers.

You can read data sheets as well as I can, so let's examine the AVR TWI subsystem as we write some code to drive it. To make this easier to digest, we want to write our AVR TWI code to look as much like our PIC I^{2}C code as we can. So, I'll use the Custom Computer Services C Compiler nomenclature for I^{2}C in the TWI AVR ICCAVR C source code.

The first thing we want to do is initialize the AVR's TWI module. The TWCR, which is used rather heavily, is shown in Figure 8.19.

Clearing the TWEN bit of the TWCR disables the AVR's TWI module, and stuffing 0x1E into the TWBR bit puts our I^{2}C bus on the I^{2}C SLOW train. There's a formula for calculating the

The Essence of Microcontroller Networking: RS-232 353

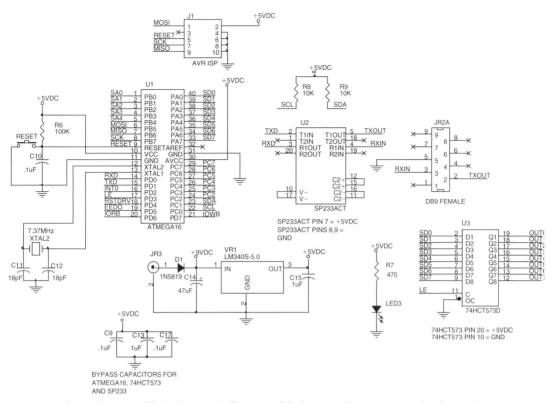

Schematic 8.6: This is the partially assembled Easy Ethernet AVR circuitry with an added 74HCT573D octal transparent latch. For both the PIC and the AVR, the only parts I've added that are really required are the I²C pull-up resistors. In some instances, the AVR doesn't require pull-up resistors because it can pull up the I²C port pins internally. You only need one set of pull-up resistors on the I²C bus.

TWCR

7	6	5	4	3	2	1	0
TWINT	TWEA	TWSTA	TWSTO	TWWC	TWEN	-	TWIE

Figure 8.19: You've already figured out that TW stands for Two-Wire Interface. Bits 7:4 are the busiest bits in this register.

I²C-bus bit rate in the data sheet, but there's an easier application to do the bit rate calculation included with ICCAVR (Figure 8.20).

Once the I²C bit rate is set, we can enable the AVR's TWI module. Our application will be simple enough to preclude the use of interrupts, and our AVR master will not be configured to also act as a slave. Therefore, the TWIE bit will remain clear for now. I've coded the TWI registers in Code Snippet 8.16 to reflect that. The *flags* variable is used to identify certain states of operation in our I²C code.

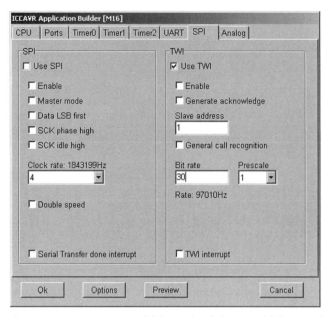

Figure 8.20: Hmmm ... which method do you think I used to get the value for the I²C SLOW bit rate?

```
unsigned char flags;
//****************************************************************
//* INITIALIZE THE TWI
//****************************************************************
void twi_init(void)
{
  flags = 0x00;
  TWCR= 0x00; //disable twi
  TWBR= 0x1E; //set bit rate
  TWSR= 0x00; //set prescale
  TWAR= 0x00; //set slave address
  TWCR= 0x04; //enable twi
}
```

Code Snippet 8.16: Since the AVR will be the master on the I²C bus, we'll leave the slave address at 0x00 for now.

Like its PIC counterpart, our AVR I²C master will need some code to implement the basic elements of I²C that allow it to participate on an I²C bus. Since a START condition is the beginning of every I²C transfer, let's begin by writing the AVR I²C start routine. The Custom Computer Services C Compiler provides a built-in I²C start routine called `i2c_start` (Code Snippet 8.17).

```
#define START_i2c 0x08
//****************************************************************
//*  AVR i2c START
//****************************************************************
void i2c_start(void)
{
  TWCR = (1<<TWINT) | (1<<TWSTA) | (1<<TWEN);
  while (!(TWCR & (1<<TWINT)));
  if ((TWSR & 0xF8) != START_i2c)
    printf("i2c Start Error\r\n");
}
```

Code Snippet 8.17: Note that the 0xF8 masks out the prescale bits in the TWSR. The status codes specified in the AVR data sheet do not include the prescale bit values.

Writing a 1 to the TWINT bit of the TWCR clears the TWINT bit. Everything revolves around the state of the TWINT bit, as when it is set the TWI has finished an operation and is waiting for the application to respond. Normally an interrupt is generated every time the TWINT goes from a low to high state. Since we're not using I²C interrupts, we must poll the TWINT bit after we reset it and look for it to return to a high state.

An I²C START is issued when the TWINT, TWSTA and TWEN bits are set. When the TWINT bit returns to a set state, the I²C START has completed. A successful I²C START condition is signaled by 0x08 in the TWSR. I've added some diagnostic `printf` code to flag an I²C START condition error.

We must also be able to stop the I²C transfer. That is done within the Custom Computer Services C Compiler with a built-in `i2c_stop` function. Guess what we will call our AVR stop function. Our AVR stop code is shown in Code Snippet 8.18.

```
//****************************************************************
//*  AVR i2c STOP
//****************************************************************
void i2c_stop(void)
{
   TWCR = (1<<TWINT)|(1<<TWEN) | (1<<TWSTO);
}
```

Code Snippet 8.18: In slave mode, the STOP condition can be used to recover from an error condition by forcing the slave to release the SCL and SDA lines.

A STOP condition is generated by setting TWINT, TWEN, and TWSTO. The TWSTO bit is automatically cleared once the STOP condition has executed on the I²C bus.

Once a START condition is generated, the next thing that happens in a normal I²C data transfer is the transmission of the slave address and mode bit. The slave address and mode bit are transmitted using an I²C write command. We'll name our AVR code in Code Snippet 8.19 after the Custom Computer Services C Compiler I²C function called `i2c_write`.

```c
#define addrflag          0x01 //00000001
#define clr_modeSLA       flags &= ~addrflag
#define set_modeSLA       flags |= addrflag
#define MODE_SLA          (flags & addrflag)

#define modeMRflag        0x02 //00000010
#define clr_modeMR        flags &= ~modeMRflag
#define set_modeMR        flags |= modeMRflag
#define MODE_MR           (flags & modeMRflag)

#define modeMTflag        0x04 //00000100
#define clr_modeMT        flags &= ~modeMTflag
#define set_modeMT        flags |= modeMTflag
#define MODE_MT           (flags & modeMTflag)
//*****************************************************************
//*   MASTER TRANSMITTER MODE STATUS CODES
//*****************************************************************
#define MT_SLA_ACK  0x18     //Master Transmitter Slave Addr ACK
#define MT_DATA_ACK 0x28     //Master Transmitter Data ACK
//*****************************************************************
//*   MASTER RECEIVER MODE STATUS CODES
//*****************************************************************
#define MR_SLA_ACK  0x40     //Master Receiver Slave Addr ACK

//*****************************************************************
//*   AVR i2c WRITE
//*****************************************************************
void i2c_write(unsigned char datum)
{
   TWDR = datum;
   TWCR = (1<<TWINT)|(1<<TWEN);
   while (!(TWCR & (1<<TWINT)));
   if(MODE_SLA && MODE_MT)
   {
     if ((TWSR & 0xF8) != MT_DATA_ACK)
     printf("i2c Data Transfer Error MT Mode %x\r\n",(TWSR & 0xF8));
     else
     {
       clr_modeSLA;
       clr_modeMT;
     }
   }
   else if (MODE_SLA && MODE_MR)
   {
```

```c
    if ((TWSR & 0xF8) != MR_DATA_ACK)
    printf("i2c Data Transfer Error MR Mode %x\r\n",(TWSR & 0xF8));
    else
    {
      clr_modeSLA;
      clr_modeMR;
    }
  }
  else
  {
    if ((TWSR & 0xF8) == MT_SLA_ACK)
    {
       set_modeMT;
       set_modeSLA;
    }
    else if ((TWSR & 0xF8) == MR_SLA_ACK)
    {
       set_modeMR;
       set_modeSLA;
    }
    else
    {
       printf("i2c Start Error %x\r\n",(TWSR & 0xF8));
       clr_modeSLA;
       clr_modeMR;
       clr_modeMT;
    }
  }
}
```

Code Snippet 8.19: Everything in this snippet flows on status codes.

Before initiating the I^{2}C transmission, the slave address and mode bit are loaded into the TWDR. Toggling the TWINT bit in the TWCR kicks off the slave address and mode bit write process. The TWEN bit is set to ensure that the AVR's I^{2}C interface is activated.

When the slave address and mode bit write has completed without error, status codes of `0x18` (`MT_SLA_ACK`) or `0x40` (`MR_SLA_ACK`) will appear within the TWSR. If the mode bit is set, an I^{2}C slave read operation will be performed and flags will be set to denote this state (`MODE_SLA` and `MODE_MR` for a read operation, `MODE_SLA` and `MODE_MT` for a write operation).

If the mode is set for the AVR to become a Master Transmitter (`MODE_MT`), the next I^{2}C operation will perform the writing of the data. Our application will only send one byte per transmission, and again we will call upon the services of the AVR i2c_write function we just wrote. This time the slave address and mode bit are replaced by the actual data we want to send to the slave. At this point, the AVR is considered a Master Transmitter and the slave is in slave-receiver mode. Our AVR I²C code has set the MODE_SLA and MODE_MT flags indicating that the AVR is in Master Transmitter mode and that the slave has been successfully addressed. A clearing of the TWINT bit sends the data onto the I²C bus. If everything goes as planned, the TWSR will contain 0x28, which says that the slave acknowledged the data transfer. The AVR Master Transmitter then issues a STOP condition to end the I²C session.

8.8.8 The AVR I²C Master-Receiver Mode Code

There will be times with the AVR master must retrieve some information from the PIC slave. That's when we deploy the AVR *i2c_read* function in Code Snippet 8.20.

```
//*****************************************************************
//*   MASTER RECEIVER MODE STATUS CODES
//*****************************************************************
#define     MR_DATA_ACK 0x50    //Master Receiver Data ACK
#define     MR_DATA_NAK 0x58    //Master Receiver Data NAK

#define     ACK_i2c     0x01
#define     NAK_i2c     0x00
//*****************************************************************
//*   AVR i2c READ
//*****************************************************************
unsigned char i2c_read(unsigned char acknak)
{
  if(acknak == ACK_i2c)
  {
    TWCR = 0xC4;
    while (!(TWCR & (1<<TWINT)));
    if ((TWSR & 0xF8) != MR_DATA_ACK)
    printf("i2c Data Transfer Error MR Mode %x\r\n",(TWSR & 0xF8));
  }
  else //acknak == NAK_i2c
  {
    TWCR = 0x84;
    while (!(TWCR & (1<<TWINT)));
    if ((TWSR & 0xF8) != MR_DATA_NAK)
    printf("i2c Data Transfer Error MR Mode %x\r\n",(TWSR & 0xF8));

    clr_modeSLA;
    clr_modeMR;
  }
  return(TWDR);
}
```

Code Snippet 8.20: The important thing to do here is to always send a NAK when reading the last byte from the slave.

Figure 8.21 lays out the bit pattern written to the TWCR after the START condition and slave addressing has successfully completed. The AVR is in master-receiver mode, and the slave is in slave-transmitter mode when the AVR *i2c_read* function is entered.

TWCR

7	6	5	4	3	2	1	0
TWINT	TWEA	TWSTA	TWSTO	TWWC	TWEN	–	TWIE
1	1	0	0	0	1	0	0

Figure 8.21: The TWI Enable Acknowledge (TWEA) bit is a "don't care" bit until we enter master-receiver mode.

Notice that we purposely set the TWEA bit, which we have been ignoring until now. Setting the TWEA bit generates an ACK on the I^{2}C bus when a data byte is received by the AVR master receiver. When things go right, the TWSR will hold the value of the MR_DATA_ACK (0x50) after each byte received by the AVR in Master Receiver mode. Our I^{2}C application is setup to read 4 bytes from the slave device.

The last byte we receive from the slave transmitter must be NAKed. That's where the TWEA bit in Figure 8.22 gets the other 7.5 minutes of its 15 minutes of fame. By writing a 0 (zero) to the TWEA bit, a NAK is generated, which results in termination of the I^{2}C read session between the master receiver and the slave transmitter. The TWSR will contain a 0x58 (MR_DATA_NAK) if all goes well with the NAK operation.

TWCR

7	6	5	4	3	2	1	0
TWINT	TWEA	TWSTA	TWSTO	TWWC	TWEN	–	TWIE
1	1	0	0	0	1	0	0

Figure 8.22: Writing a 0 to the TWEA bit temporarily disconnects the AVR from the I^{2}C bus.

I have a project in mind. Let's combine our AVR RS-232 skills with our newfound AVR I^{2}C skills to transfer data between the partially assembled Easy Ethernet AVR and the Easy Ethernet CS8900A boards. Before we put the whole of the AVR code together, let's write some PIC I^{2}C slave code first.

8.8.9 The PIC I^{2}C Slave-Transmitter Mode Code

To implement I^{2}C on the Microchip PIC, there are only three PIC registers we need to be concerned with: SSPCON, SSPSTAT, and SSPBUF. SSPCON is used to determine whether or not a collision has occurred (WCOL) and to ensure we are not stretching the clock when we shouldn't be (CKP = 1). Clock stretching is legal for an I^{2}C slave device when it can't respond in a timely manner. SSPSTAT gives us the status of the data transfer, whereas SSPBUF is the register that actually transfers the data to and from the I^{2}C bus.

The PIC's Master Synchronous Serial Port (MSSP) does several other things for us, including double buffering our received I^{2}C data using the SSPSR/SSPBUF register combination, providing a holding register for the slave address, and generating I^{2}C interrupts on START and STOP conditions. *Double buffering* is the act of holding or collecting data in an input or output buffer while operating on a totally separate input or output buffer. In short, double buffering allows data to be assembled for transmission while previously accumulated data is being transmitted. Receive double buffering occurs when the microcontroller is working on pulling previously received data from an input buffer while yet another input buffer is taking in new data and holding it until the microcontroller can start processing it.

As simple as the I^{2}C concept is, if you're not careful, you can get your I^{2}C code wrapped around the axel. To make I^{2}C coding more manageable, the I^{2}C transmission and reception process can be broken down into five states. Everything that's normal in I^{2}C begins with a START condition. The START condition must be detected (S = 1) no matter what, and nothing begins until a valid START condition is sensed. Once we have detected a valid START bit, we can use the other bits inside the SSPSTAT register to determine which state the I^{2}C transaction is currently in. We used the TWSR for this in the AVR I^{2}C code. The MSSP issues an interrupt on every byte transfer. This allows us to write I^{2}C code, such as the code presented in Code Snippet 8.21, using the five states to take advantage of the MSSP module's interrupt generation.

```
//****************************************************************
//*    SLAVE RAM DEFINITIONS
//****************************************************************
int1   update_latch;
int8   index,digit;
int8   numbers[] = {0,1,2,3,4,};
//****************************************************************
//*    I2C SLAVE RECEIVE
//****************************************************************
#INT_SSP
   ssp_interrupt ()
{
//#bit SMP =  SSPSTAT.7
//#bit CKE =  SSPSTAT.6
//#bit D_A =  SSPSTAT.5
//#bit P   =  SSPSTAT.4
//#bit S   =  SSPSTAT.3
//#bit R_W =  SSPSTAT.2
//#bit UA  =  SSPSTAT.1
//#bit BF  =  SSPSTAT.0
   int8 dummy;
//--------------------------------------------------
// The I2C code below checks for 5 states:
//--------------------------------------------------
```

```
// State 1: I2C write operation, last byte was an address byte.
//
// SSPSTAT bits: S = 1, D_A = 0, R_W = 0, BF = 1
//
// State 2: I2C write operation, last byte was a data byte.
//
// SSPSTAT bits: S = 1, D_A = 1, R_W = 0, BF = 1
//
// State 3: I2C read operation, last byte was an address byte.
//
// SSPSTAT bits: S = 1, D_A = 0, R_W = 1, BF = 0
//
// State 4: I2C read operation, last byte was a data byte.
//
// SSPSTAT bits: S = 1, D_A = 1, R_W = 1, BF = 0
//
// State 5: Slave I2C logic reset by NACK from master.
//
// SSPSTAT bits: S = 1, D_A = 1, R_W = 0, BF = 0
//
//-------------------------------------------------

//State 1
   if(S && !D_A && !R_W && BF )
      dummy = SSPBUF;
//State 2
   else if(S && D_A && !R_W && BF )
      {
        digit = SSPBUF;
        update_latch = TRUE;
      }
//State 3
   else if(S && !D_A && R_W && !BF )
      {
        index = 0x00;
        while(BF);
        do{
        WCOL = 0;
        SSPBUF = numbers[index];
        }while(WCOL);
        ++index;
        CKP = 1;
      }
//State 4
   else if(S && D_A && R_W && !BF )
      {
        while(BF);
```

```
        do{
        WCOL = 0;
        SSPBUF = numbers[index];
        }while(WCOL);
        if(++index > 0x04)
        index = 0x00;
        CKP = 1;
     }
//State 5
   else if(S && D_A && !R_W && !BF )
   index = 0;
}
```

Code Snippet 8.21: The `update_latch` variable and `numbers[]` array will be used by in our AVR-to-PIC grand I²C ball.

The I²C SLAVE RECEIVE routine is the PIC18F452 I²C interrupt handler code that responds to every interrupt issued by the PIC18F452's microcontroller's MSSP module. I've moved the bit definitions of the SSPSTAT register into the routine's air space for clarity.

Notice that in each of the five defined states that $S = 1$ is common. The bit S is defined as the third bit of the SSPSTAT register. If a valid START condition is detected, this bit will be set.

The slave address byte immediately follows the START bit. Since the slave microcontroller's MSSP will always generate an interrupt if the incoming address byte matches the slave's internally stored address (in SPPADD), the matching address byte just received triggers our first interrupt and its subsequent response. The MSSP module will also automatically issue an acknowledge (ACK) pulse upon detecting an address match.

The D_A bit signals if the last byte received was data or address. In this case, we know that a START bit was generated and was indeed followed by a 7-bit address. Therefore, D_A is cleared to zero, indicating that the last byte received was an address byte.

The R/W bit of the address is cleared for a write operation and set for a read operation. The R_W bit of the SSPSTAT registers reflects the level of the R/W bit in the address byte. Note that if the operation is a write operation, the Buffer Full (BF) bit is always set, indicating that data is in the buffer. The State 1 code runs following the reception of the address byte. The address byte is read and discarded as the slave MSSP module has already digested the address byte's contents. The act of reading SSPBUF also clears the BF bit. If the BF bit is not cleared at this point, the next incoming byte would cause an overflow condition. Let's follow the entire state-by-state chain of events involved with sending some data from the AVR master I²C microcontroller to the PIC slave I²C microcontroller.

Suppose that the AVR master I²C microcontroller needs to send a message via I²C to the PIC I²C slave microcontroller that tells the slave microcontroller to write 0x55 to its onboard 74HCT573 latch. The basic AVR TWI code would consist of what you see in Code Snippet 8.22.

```
i2c_start();
i2c_write(0x18);
i2c_write(0x55);
i2c_stop();
```

Code Snippet 8.22: The Easy Ethernet CS8900A's I²C address is 0x18.

After initiating a START condition, the master microcontroller clocks out the slave microcontroller's I²C address, hexadecimal 18 (0x18). The code `i2c_write(unsigned char datum)` indicates that an I²C write operation has been requested as the R/W bit in the I²C address byte is cleared. At this point in time, every slave microcontroller on the I²C bus is listening on the I²C link looking to match its address against the incoming address byte. Our PIC I²C slave microcontroller compares the incoming address with the address stored in its SSPADD register and detects a match. The slave's BF bit is set, an ACK pulse is generated by the slave microcontroller's MSSP hardware, and an SSP interrupt is generated. The PIC I²C slave microcontroller enters the I²C SSP interrupt routine and using the SSPSTAT bits determines that the I²C transaction is in State 1, which tells us that the last byte received was an address byte. The BF bit is set, which means that the contents of the SSPSR register have been transferred to the SSPBUF register. To avoid an overflow condition, the PIC's SSPBUF register must be read even though we don't have any further use for the address data.

It's the slave microcontroller's duty to translate the incoming I²C data stream.

```
#define le_pin PORTC,1

#define latchdata bit_set(le_pin); \
   delay_us(1); \
   bit_clear(le_pin);
//****************************************************************
//* SLAVE MAIN
//****************************************************************
   do{
       {
   if(update_latch)
   {
       output_d(digit);
           latchdata;
       update_latch = FALSE;
   }
       }
       }while(1);
}
```

Code Snippet 8.23: Now you know what the `update_latch` variable you saw in Code Snippet 8.21 is for.

The data that was sent from the I²C master that is to be output to the slave's 74HCT573 latch was collected into the digit variable in the PIC's I²C interrupt handler routine. In the same stroke, the PIC I²C interrupt handler updated the `update_latch` flag to TRUE.

The code in Code Snippet 8.23 is the main routine that runs continuously inside the Easy Ethernet CS8900A's PIC18F452. The PIC I²C slave's code picks up the state of the `update_latch` variable. If the `update_latch` variable is TRUE, the data within the `digit` variable is output to the 74HCT573 latch by the `latchdata` macro and the `update_latch` variable is cleared to a FALSE condition. Each time a value is received by the slave via the I²C bus, it is transferred to the latch.

If the master microcontroller wants data from the slave microcontroller, State 3 starts things off and the slave microcontroller is coaxed into slave-transmitter mode while the master microcontroller becomes a master-receiver. In Code Snippet 8.24, the master microcontroller initiates a START condition and follows it with a "read" address byte. Since the R/W bit is the least significant bit in the address byte, the write address is simply the base address incremented by 1 (0x19 in our case). Incrementing the address byte has the effect of setting the R_W bit inside the I²C address byte. In this mode the master microcontroller, not the slave microcontroller, generates the I²C ACKs and NAKs on the I²C bus.

```
#define  ACK_i2c    0x01
#define  NAK_i2c    0x00

   i2c_start();
   i2c_write(0x19);
   for(x=0;x<3;++x)
   {
      datum = i2c_read(ACK_i2c);
      printf("datum = 0x%x\r\n",datum);
   }
   datum = i2c_read(NAK_i2c);
   i2c_stop();
   printf("datum = 0x%x\r\n",datum);
```

Code Snippet 8.24: No worries: We read every byte except the last within the `for` loop.

Things on the I²C bus are a bit busier when a master is reading from a slave. We already know that the slave microcontroller has four bytes of information the master can access stored in the `numbers[]` array. Let's use the AVR and the I²C bus to retrieve the four bytes from the slave's `numbers[]` array and print them out to a master Tera Term Pro session.

The slave microcontroller must be ready to send the first byte of data after the ACK following the address byte. The State 3 code attempts to load the SSPBUF with that first byte of data while looking out to make sure the SSPBUF is clear and ready for the byte to be loaded. In our code, the first byte of the array `numbers[]` (0x00) is loaded and sent following the reception of the address byte. The `index` variable is incremented to point to the next element of the `numbers[]` array. Setting the CKP (SCK release control) bit assures that the slave microcontroller is not holding the clock line low and thus "stretching" the clock.

The master microcontroller is coded to collect a total of four bytes. Since the last byte read was not the address byte, we can move on to State 4 in the PIC interrupt handler code. The remainder of the four bytes of data required by the master microcontroller are clocked out of the slave-transmitter microcontroller in State 4. To halt the I^{2}C read operation, the master generates a NAK after the last byte is read. The `NAK_i2c` in the `i2c_read(NAK_i2c)` tells the AVR I^{2}C read function to send the NAK. That brings us to State 5 and the end of the I^{2}C read operation.

8.8.10 The AVR-to-PIC I^{2}C Communications Ball

Let's put everything we've written for RS-232 and I^{2}C for the AVR together with everything we've written for RS-232 and I^{2}C for the PIC and move some data. The source code PIC slave application and the AVR master I^{2}C application is contained within Code Snippets 8.25 and 8.26, respectively.

```
////////////////////////////////////////////////////////////////////
// PIC I2C SLAVE DRIVER
// EASY ETHERNET CS8900A BOARD
// Author: Fred Eady
// Version: 1.0
// Date: 08/25/03
// Description: I2C SLAVE FUNCTION WITH 74HCT573 CODE
////////////////////////////////////////////////////////////////////
#include <18F452.h>
#include <f452.h>
#device ICD=TRUE
#fuses
DEBUG,HS,NOWRT,NOWDT,NOPUT,NOPROTECT,NOBROWNOUT,NOLVP,NOCPD,NOEBTR
#id 0x0812

#use fast_io(A)
#use fast_io(B)
#use fast_io(C)
#use fast_io(D)
#use fast_io(E)

#define  esc   0x1B
```

```
//********************************************************************
//*    I2C SLAVE ADDRESS
//********************************************************************
// LANE ADDRESS IS UPPER NIBBLE
#define i2c_addr 0x18
//********************************************************************
//*    RS232 AND I2C DEFINITIONS
//********************************************************************
#use delay(clock=20000000)
#use i2c(Slave,Slow,sda=PIN_C4,scl=PIN_C3,force_hw,address=i2c_addr)
#use rs232(baud=9600,parity=N,xmit=PIN_C6,rcv=PIN_C7)
//********************************************************************
//*    SLAVE FUNCTION PROTOTYPES
//********************************************************************
void cls(void);
//********************************************************************
//*    SLAVE RAM DEFINITIONS
//********************************************************************
int1   update_latch;
int8   index,digit;
int8   numbers[] = {0,1,2,3,4,};

#define     le_pin      PORTC,1
#define     latchdata   bit_set(le_pin); \
    delay_us(1); \
      bit_clear(le_pin);
//********************************************************************
//* I2C SLAVE RECEIVE
//********************************************************************
#INT_SSP
    ssp_interrupt ()
{
//#bit SMP =  SSPSTAT.7
//#bit CKE =  SSPSTAT.6
//#bit D_A =  SSPSTAT.5
//#bit P   =  SSPSTAT.4
//#bit S   =  SSPSTAT.3
//#bit R_W =  SSPSTAT.2
//#bit UA  =  SSPSTAT.1
//#bit BF  =  SSPSTAT.0

    int8 dummy;
//;--------------------------------------------------
//; The I2C code below checks for 5 states:
//;--------------------------------------------------
//; State 1: I2C write operation, last byte was an address byte.
//;
```

```
//; SSPSTAT bits: S = 1, D_A = 0, R_W = 0, BF = 1
//;
//; State 2: I2C write operation, last byte was a data byte.
//;
//; SSPSTAT bits: S = 1, D_A = 1, R_W = 0, BF = 1
//;
//; State 3: I2C read operation, last byte was an address byte.
//;
//; SSPSTAT bits: S = 1, D_A = 0, R_W = 1, BF = 0
//;
//; State 4: I2C read operation, last byte was a data byte.
//;
//; SSPSTAT bits: S = 1, D_A = 1, R_W = 1, BF = 0
//;
//; State 5: Slave I2C logic reset by NACK from master.
//;
//; SSPSTAT bits: S = 1, D_A = 1, R_W = 0, BF = 0
//;
//;------------------------------------------------
//State 1
   if(S && !D_A && !R_W && BF )
      dummy = SSPBUF;
//State 2
   else if(S && D_A && !R_W && BF )
      {
         digit = SSPBUF;
         update_latch = TRUE;
      }
//State 3
   else if(S && !D_A && R_W && !BF )
      {
         index = 0x00;
         while(BF);
         do{
         WCOL = 0;
         SSPBUF = numbers[index];
         }while(WCOL);
         ++index;
         CKP = 1;
      }
//State 4
   else if(S && D_A && R_W && !BF )
      {
         while(BF);
         do{
         WCOL = 0;
```

```c
         SSPBUF = numbers[index];
         }while(WCOL);
         if(++index > 0x04)
         index = 0x00;
         CKP = 1;
      }
//State 5
   else if(S && D_A && !R_W && !BF )
      index = 0;
}
void main() {

   int8 x;
   SET_TRIS_A(0b11111111);
   SET_TRIS_B(0b11111111);
   SET_TRIS_C(0b11111101);
   SET_TRIS_D(0b00000000);
   ADCON1 = 0x06;     //00000110 all ports set for digital
   ADCON0 = 0;
   update_latch = FALSE;
//*****************************************************************
//*    INITIALIZE COMMON VARIABLES
//*****************************************************************
   SSPSTAT = 0x80;
   SSPCON2 = 0x00;
//*****************************************************************
//*    ENABLE SLAVE INTERRUPTS
//*****************************************************************
   enable_interrupts(INT_SSP);
   enable_interrupts(GLOBAL);
//*****************************************************************
//*    SLAVE MAIN
//*****************************************************************
   do{
      {
         if(update_latch)
         {
         output_d(digit);
      latchdata;
         update_latch = FALSE;
         }
      }
   }while(1);
}
```

Code Snippet 8.25: Don't worry; I've included the code on the CD-ROM so you won't have to burn up your fingers typing code.

You already have a good handle on the inner workings of the PIC I²C slave code in Code Snippet 8.25. However, I've thrown in the kitchen sink in the AVR master code coming up in Code Snippet 8.26. So, I'll break it up and discuss the code parts as they are encountered. Consider the rest of the code in this section as part of Code Snippet 8.26.

```
////////////////////////////////////////////////////////////////////
// AVR I2C MASTER DRIVER
// EASY ETHERNET AVR BOARD
// Author: Fred Eady
// Version: 1.0
// Date: 08/26/03
// Description: RS232 FUNCTIONS AND I2C MASTER FUNCTIONS
////////////////////////////////////////////////////////////////////

#include <iom16v.h>
#include <stdio.h>
#include <macros.h>

#pragma interrupt_handler USART_RX_interrupt:iv_USART_RX
#pragma interrupt_handler USART_TX_interrupt:iv_USART_UDRE
```

Code Snippet 8.26a: There's nothing here you can't talk about intelligently.

It looks like we're going to include some interrupt driven RS-232 on the AVR side. The #pragma statements in Code Snippet 8.26a are a dead giveaway. The confirmation of an RS-232 resurrection is confirmed in Code Snippet 8.26b.

```
//****************************************************************
//*    FUNCTION PROTOTYPES
//****************************************************************
int recvchar( void );
int sendchar( int );
unsigned char CharInQueue(void);
void init_USART(unsigned int baud);

void twi_init(void);
void i2c_start(void);
void i2c_write(unsigned char datum);
unsigned char i2c_read(unsigned char acknak);
void i2c_stop(void);
```

Code Snippet 8.26b: These declarations are a preview of what's to come.

The code in Code Snippet 8.26c should look familiar as well. All of the USART-related code is contained in this snippet.

```c
//*********************************************************************
//*    BAUD RATE NUMBERS FOR UBRR
//*********************************************************************
#define         b9600   47   // 7.3728MHz clock
#define         b19200  23
#define         b38400  11
#define         b57600  7

#define USART_RX_BUFFER_SIZE 16      /* 1,2,4,8,16,32,64,128 or 256 bytes */
#define USART_RX_BUFFER_MASK ( USART_RX_BUFFER_SIZE - 1 )
//#if ( USART_RX_BUFFER_SIZE & USART_RX_BUFFER_MASK )
//#error RX buffer size is not a power of 2
//#endif
#define USART_TX_BUFFER_SIZE 128     /* 1,2,4,8,16,32,64,128 or 256 bytes */
#define USART_TX_BUFFER_MASK ( USART_TX_BUFFER_SIZE - 1 )
//#if ( USART_TX_BUFFER_SIZE & USART_TX_BUFFER_MASK )
//#error TX buffer size is not a power of 2
//#endif
//*********************************************************************
//*    AVR RAM Definitions
//*********************************************************************
unsigned char USART_RxBuf[USART_RX_BUFFER_SIZE],USART_TxBuf[USART_TX_BUFFER_SIZE];
unsigned char USART_TxHead,USART_TxTail,USART_RxHead,USART_RxTail;
unsigned char flags,datum,byteout,cntr;
//*********************************************************************
//* Init USART Function
//*********************************************************************
void init_USART(unsigned int baud)
{
  UCSRB = 0x00; //disable while setting baud rate
  UCSRA = 0x00;
  UCSRC = 0x86;
  UBRRL = baud; //set baud rate lo
  UBRRH = 0x00; //set baud rate hi
  UCSRB = 0x98;
}
//*********************************************************************
//*    USART Receive Interrupt Handler
//*********************************************************************
void USART_RX_interrupt(void)
{
  unsigned char data;
  unsigned char tmphead;
```

```c
  data = UDR;      /* read the received data */
                   /* calculate buffer index */
  tmphead = ( USART_RxHead + 1 ) & USART_RX_BUFFER_MASK;
  USART_RxHead = tmphead;         /* store new index */

  if ( tmphead == USART_RxTail )
  {
    /* ERROR! Receive buffer overflow */
  }
  USART_RxBuf[tmphead] = data; /* store received data in buffer */
}
//*******************************************************************
//* USART Receive Character Function
//*******************************************************************
int recvchar( void )
{
  unsigned char tmptail;
                                  /* wait for incoming data */
  while ( USART_RxHead == USART_RxTail );
                                  /* calculate buffer index */
  tmptail = ( USART_RxTail + 1 ) & USART_RX_BUFFER_MASK;
  USART_RxTail = tmptail;         /* store new index */

  return USART_RxBuf[tmptail];    /* return data */
}
//*******************************************************************
//*   USART Transmit Interrupt Handler
//*******************************************************************
//interrupt [iv_USART_UDRE]
void USART_TX_interrupt(void)
{
  unsigned char tmptail;
                                  /* check if all data is transmitted */
  if ( USART_TxHead != USART_TxTail )
  {
                                  /* calculate buffer index */
  tmptail = ( USART_TxTail + 1 ) & USART_TX_BUFFER_MASK;
  USART_TxTail = tmptail;         /* store new index */

  UDR = USART_TxBuf[tmptail];     /* start transmission */
  }
  else
  {
    UCSRB &= ~(1<<UDRIE); /* disable UDRE interrupt */
  }
}
```

```
//******************************************************************
//*    USART Transmit Character Function
//******************************************************************
int sendchar( int data )
{
  unsigned char tmphead;
                                /* calculate buffer index */
  tmphead = ( USART_TxHead + 1 ) & USART_TX_BUFFER_MASK;
                                /* wait for free space in buffer */
  while ( tmphead == USART_TxTail );
                                /* store data in buffer */
  USART_TxBuf[tmphead] = (unsigned char)data;
  USART_TxHead = tmphead;      /* store new index */

  UCSRB |= (1<<UDRIE);          /* enable UDRE interrupt */

  return data;
}
//******************************************************************
//* USART Character Waiting Function
//******************************************************************
unsigned char CharInQueue(void)
{
  return(USART_RxHead != USART_RxTail);
}
```

Code Snippet 8.26c: We've already examined this code down to the bit level using emulators and in-circuit debuggers.

The code in Code Snippet 8.26d is the full complement of AVR I²C routines we cloned to match the built-in I²C functions provided by the Custom Computer Services C Compiler.

```
#define addrflag         0x01 //00000001
#define clr_modeSLA      flags &= ~addrflag
#define set_modeSLA      flags |= addrflag
#define MODE_SLA         (flags & addrflag)

#define modeMRflag       0x02 //00000010
#define clr_modeMR       flags &= ~modeMRflag
#define set_modeMR       flags |= modeMRflag
#define MODE_MR          flags & modeMRflag)

#define modeMTflag       0x04 //00000100
#define clr_modeMT       flags &= ~modeMTflag
#define set_modeMT       flags |= modeMTflag
#define MODE_MT          (flags & modeMTflag)

#define hexflagbit       0x08 //00001000
#define clr_hex          flags &= ~hexflagbit
```

```
#define set_hex         flags |= hexflagbit
#define hexflag         (flags & hexflagbit)

#define iorwport        PORTD
#define LE_pin          0x08 //PORTD3 00001000
#define set_le_pin      iorwport |= LE_pin
#define clr_le_pin      iorwport &= ~LE_pin

#define latchdata       set_le_pin; \
                        delay_us(1);\
                        clr_le_pin;

#define START_i2c       0x08
#define ACK_i2c         0x01
#define NAK_i2c         0x00
//*****************************************************************
//*   MASTER TRANSMITTER MODE STATUS CODES
//*****************************************************************
#define MT_SLA_ACK      0x18    //Master Transmitter Slave Addr ACK
#define MT_DATA_ACK     0x28    //Master Transmitter Data ACK
//*****************************************************************
//* MASTER RECEIVER MODE STATUS CODES
//*****************************************************************
#define MR_SLA_ACK      0x40    //Master Receiver Slave Addr ACK
#define MR_DATA_ACK     0x50    //Master Receiver Data ACK
#define MR_DATA_NAK     0x58    //Master Receiver Data NAK
//*****************************************************************
//* INITIALIZE THE TWI
//*****************************************************************
void twi_init(void)
{
  flags = 0x00;
  TWCR= 0x00; //disable twi
  TWBR= 0x1E; //set bit rate
  TWSR= 0x00; //set prescale
  TWAR= 0x00; //set slave address
  TWCR= 0x04; //enable twi
}
//*****************************************************************
//*   AVR i2c START
//*****************************************************************
void i2c_start(void)
{
  TWCR = (1<<TWINT) | (1<<TWSTA) | (1<<TWEN);
  while (!(TWCR & (1<<TWINT)));
  if ((TWSR & 0xF8) != START_i2c)
    printf("i2c Start Error\r\n");
}
```

```c
//*******************************************************************
//*    AVR i2c WRITE
//*******************************************************************
void i2c_write(unsigned char datum)
{
  TWDR = datum;
  TWCR = (1<<TWINT)|(1<<TWEN);
  while (!(TWCR & (1<<TWINT)));
  if(MODE_SLA && MODE_MT)
  {
    if ((TWSR & 0xF8) != MT_DATA_ACK)
    printf("i2c Data Transfer Error MT Mode %x\r\n",(TWSR & 0xF8));
    else
    {
      clr_modeSLA;
      clr_modeMT;
    }
  }
  else if (MODE_SLA && MODE_MR)
  {
    if ((TWSR & 0xF8) != MR_DATA_ACK)
    printf("i2c Data Transfer Error MR Mode %x\r\n",(TWSR & 0xF8));
    else
    {
      clr_modeSLA;
      clr_modeMR;
    }
  }
  else
  {
    if ((TWSR & 0xF8) == MT_SLA_ACK)
    {
      set_modeMT;
      set_modeSLA;
    }
    else if ((TWSR & 0xF8) == MR_SLA_ACK)
    {
      set_modeMR;
      set_modeSLA;
    }
    else
    {
      printf("i2c Start Error %x\r\n",(TWSR & 0xF8));
      clr_modeSLA;
      clr_modeMR;
      clr_modeMT;
    }
  }
}
```

```
//*****************************************************************
//* AVR i2c READ
//*****************************************************************
unsigned char i2c_read(unsigned char acknak)
{
  if(acknak == ACK_i2c)
  {
    TWCR = 0xC4;
    while (!(TWCR & (1<<TWINT)));
    if ((TWSR & 0xF8) != MR_DATA_ACK)
    printf("i2c Data Transfer Error MR Mode %x\r\n",(TWSR & 0xF8));
  }
  Else
  {
    TWCR = 0x84;
    while (!(TWCR & (1<<TWINT)));
    if ((TWSR & 0xF8) != MR_DATA_NAK)
      printf("i2c Data Transfer Error MR Mode %x\r\n",(TWSR & 0xF8));

    clr_modeSLA;
    clr_modeMR;
  }
  return(TWDR);
}
//*****************************************************************
//* AVR i2c STOP
//*****************************************************************
void i2c_stop(void)
{
  TWCR = (1<<TWINT)|(1<<TWEN) | (1<<TWSTO);
}
```

Code Snippet 8.26d: Nothing to it so far. You haven't seen anything new unless you "chapter hopped" to this point.

Here's where all our RS-232 and I^2C work comes to fruition. I attached an MPLAB ICD 2 to the Easy Ethernet CS8900A and an AVR JTAG ICE to the Easy Ethernet AVR. The PIC slave code will run under control of MPLAB and the MPLAB ICD 2, and the AVR master code will run on the Easy Ethernet AVR under control of the AVR JTAG ICE and AVR Studio.

I also connected the PIC's I^2C interface (SDA, SCL and ground) to the AVR's TWI. The Easy Ethernet CS8900A has an I^2C "port," whereas the Easy Ethernet AVR's TWI is bundled in with the AVR's PORTC pins. The RS-232 communications will be handled by the AVR I^2C master, and I've attached the Easy Ethernet AVR's serial port to a personal computer Tera Term Pro serial session. All of the in-circuit debuggers are attached to a single personal computer, and Tera Term Pro, MPLAB and AVR Studio are running on that same personal computer. I attached the Microchip MPLAB ICD 2 using USB, and the Atmel AVR JTAG ICE is communicating with AVR Studio using the COM1 serial port. The Easy Ethernet AVR's serial port is attached to the personal computer's COM2 serial port, which is under the control of Tera Term Pro.

OK, here's how it all works!

The slave Easy Ethernet CS8900A is started and is listening on the I²C bus. Once the Easy Ethernet AVR master's USART and TWI are initialized, the Easy Ethernet throws up the "Networking with Microcontrollers is dead easy …" banner in the Tera Term Pro window and waits for a character to be received by the Easy Ethernet AVR's serial port.

If the incoming character is a * (0x2A), the hexflag flag bit is set and the byte counter variable cntr is cleared. The * sets up the Easy Ethernet AVR to take the next two ASCII bytes following the * from the Easy Ethernet AVR's serial port and convert them into a single hexadecimal digit. Once the hexadecimal digit is assembled, the hex digit is sent via I²C to the slave, Easy Ethernet CS8900A, where it is latched out to the Easy Ethernet CS8900A's 74HCT573 latch. The Easy Ethernet AVR sends a message to the Tera Term Pro session informing you what was sent over the I²C bus.

Entering a $ symbol from the Tera Term Pro session puts the Easy Ethernet AVR into master-receiver mode, and the four bytes stored in the slave's number[] array are read into the AVR's memory and displayed in the Tera Term Pro session.

If you don't enter a * or a $ character, everything you type is echoed back to the Tera Term Pro session.

```
//*****************************************************************
//* MAIN MAIN MAIN MAIN MAIN MAIN MAIN MAIN MAIN MAIN MAIN
//*****************************************************************
void C_task main(void)
{
  unsigned char x;
  CLI(); //disable all interrupts
  PORTA = 0xFF;
  DDRA = 0x00;
  PORTB = 0xFF;
  DDRB = 0x00;
  PORTD = 0xFF;
  DDRD = 0x00;

  for(x=0;x<USART_RX_BUFFER_SIZE;++x)
    USART_RxBuf[x] = 'R';
  for(x=0;x<USART_TX_BUFFER_SIZE;++x)
    USART_TxBuf[x] = 'T';
```

```c
    USART_RxTail = 0x00;
    USART_RxHead = 0x00;
    USART_TxTail = 0x00;
    USART_TxHead = 0x00;

    MCUCR = 0x00; //disable sleep modes
    GICR = 0x00; //set interrupt vectors at start of flash
    TIMSK = 0x00; //disable timer interrupt sources
    init_USART(47);
    twi_init();
    SEI(); //re-enable interrupts
    printf("Networking with Microcontrollers is dead easy...\r\n");
while(1){
++cntr;
    while(!(CharInQueue()));
      datum = recvchar();
    if(hexflag)
{
    if(datum >= '0' && datum <= '9')
  datum -= 0x30;
      else if(datum >= 'A' && datum <= 'F')
        datum -= 0x37;
      else if(datum >= 'a' && datum <= 'f')
        datum -= 0x67;
else
{
      cntr = 0x00;
    clr_hex;
}
if(cntr == 1)
  byteout = datum << 4;
    if(cntr == 2)
{
  byteout |= datum & 0x0F;
      i2c_start();
        i2c_write(0x18);
        i2c_write(byteout);
        i2c_stop();
    clr_hex;
        printf("\r\nByte Sent Via i2c = 0x%x\r\n",byteout);
  }
}
if(datum == '*')
{
  set_hex;
cntr=0;
}
```

```
else if(datum == '$')
{
  printf("\r\n");
  i2c_start();
  i2c_write(0x19);
  for(x=0;x<3;++x)
  {
    datum = i2c_read(1);
    printf("datum = 0x%x\r\n",datum);
  }
  datum = i2c_read(0);
  i2c_stop();
  printf("datum = 0x%x\r\n",datum);
}
else
  sendchar(datum);
  }
}
```

Code Snippet 8.26e: This little application shows just how easy it is to move data between multiple devices using RS-232 and I²C.

You've succeeded in building the RS-232 and I²C hardware for both a PIC and an AVR microcontroller. Along the way, you've also written some pretty nifty code to drive that hardware.

8.9 Communication Options

A broad variety of other interfaces allow communication with other devices on the same printed circuit board or with remote systems. These interfaces include:

- The Synchronous Peripheral Interface (SPI)
- The Controller Area Network (CAN) interface

Let's look at the SPI and the CAN interfaces in detail.

8.9.1 The Serial Peripheral Interface Port

The Serial Peripheral Interface, or SPI as it is more commonly called, is a synchronous serial interface that is designed primarily to transfer data between devices that are all located on a single printed circuit board (PCB), although it can be used to communicate between PCBs as well. The interface is fairly simple, consisting of a Serial Data Out (SDO) signal, a Serial Data In (SDI) signal, a Serial Clock signal (SCK), a Chip Select (CS) signal, and a Slave Select (SS) signal. All of these signals are single-ended digital signals, one of the reasons that the

SPI is not well suited to long data links or noisy environments. Because it is so easy to implement and troubleshoot, many devices, both microprocessors and peripheral chips, employ SPI. For example, the dsPICDEM board uses one of its two SPI ports to communicate with the on-board temperature sensor, sending configuration data to the sensor and reading temperature and status values from it.

Over the years, the SPI has evolved to support four basic modes of operation (imaginatively denoted Mode 1, Mode 2, Mode 3, and Mode 4), that operate in basically the same manner but which employ different timing relationships between the SCK clock edge and the SDO and SDI data signals to determine when to transmit data and when data is valid at the receiver. Most devices support only a subset of these modes, so it's important to make sure that both the transmitting and receiving device are able to support at least one common operating mode. Figure 8.23 shows the four possible SPI operating mode combinations. In our examples, we will use Mode 0 since it is one of the more common configurations.

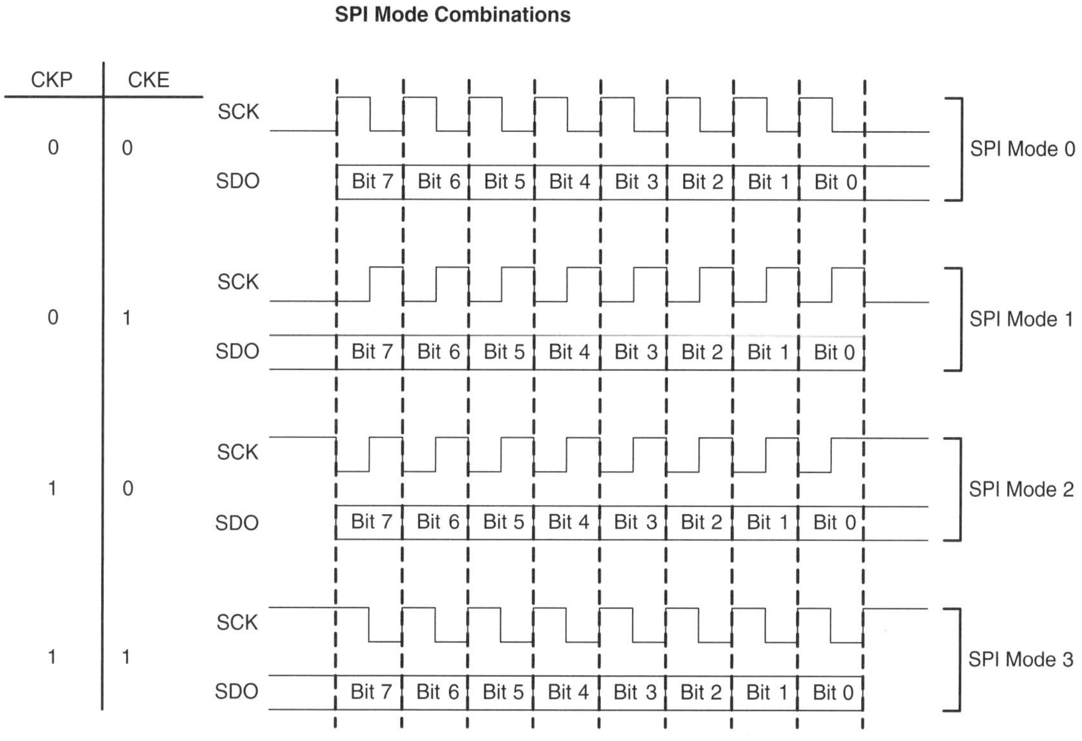

Figure 8.23: The four SPI operating mode combinations.

The Microchip 16-bit Peripheral Library does a good job of implementing a useful framework of functions to control and access SPI ports on its parts. Unlike the interfaces for the UART and the CAN bus that we'll look at shortly, the SPI is usually used to transfer small, often byte-size or word-size chunks of data. Because its transfer rate is so high, this means that we can essentially treat the transfers as in-line operations that are completed in real-time as the code that uses them executes. For instance, if we're reading two bytes of temperature data from a sensor connected via the SPI, often we can afford to issue the request and wait for the response since the data transfer will not significantly slow our operation. This is not the case when transferring large amounts of data through other communication ports (or even through the SPI); generally, we have to implement a buffered, interrupt-driven framework to deal with that situation. Fortunately, the Microchip 16-bit Peripheral Library already has all of the functionality we need.

8.9.2 The Controller Area Network

The Controller Area Network (CAN) is one of the more sophisticated of the serial interfaces. It incorporates a very advanced internal hardware controller that supports moderate speed (up to 1 Mbps) data transfers with built-in hardware error detection, a sophisticated message prioritization scheme, and the ability to set filters that allow only messages of interest to be received, all with very little processor overhead. Widely used in the automotive and industrial-processing world, the CAN architecture offers a robust way to link together multiple nodes on a single network.

With all of these positives, why would anyone *not* use the CAN interface? There are two main reasons: complexity and cost. One big advantage of CAN is that it's highly configurable, but one big disadvantage is that CAN is so highly configurable. Because it's so flexible, a CAN topology can be used in a wide variety of applications using basically the same hardware. (In this case, topology is simply the technical term for the arrangement of nodes in a network.) Unfortunately, that flexibility must be configured fairly precisely or the channel will be either unreliable or completely unusable, and debugging problems with the channel can be both time-consuming and frustrating.

8.9.2.1 Basic CAN Architecture

Developed by Bosch in the early 1980s, the CAN architecture is pretty simple. Although the CAN standard itself is intentionally media-neutral, one of the most common implementations uses a single differential serial bus running at 1 Mbps (1,000,000 bits per second) or less to connect two or more nodes together. Along with the associated ground signal, a reliable interface can consist of only three wires!

> **Note:** Media-neutral simply means that the protocol does not specify the physical medium required to implement the protocol. This was intentionally left out of the specification so that the protocol can operate over a variety of physical media (so long as the media supports the ability to have a dominant and a recessive bit state).

The CAN's communication protocol is a member of the CSMA/CD family, a cryptic acronym that stands for Carrier Sense Multiple Access/Collision Detection. Although the family name is long, the concepts behind it are easy. In a *carrier sense* system, all nodes have to monitor the network for a period of inactivity before they can attempt to send a message. Once this inactive period has elapsed, however, any of the nodes in the network can transmit data, hence the term *multiple access*. As one would expect, there will be times when two or more nodes try to send data at the same time, a condition known as *collision*, so the network has to have some way to perform *collision detection*. Individual members of the CSMA/CD family handle these tasks differently, but all members of a given type (such as CAN) do so in the same manner.

Of these tasks (carrier sensing and collision detection), the more difficult by far is collision detection. The CAN designers came up with an ingenious solution to this problem, one that allows the system designer to prioritize message traffic so that more important messages are always able to gain access to the bus ahead of less important messages (in much the same way that interrupts are prioritized by the dsPIC DSC). Not only does the CAN allow message prioritization, its network *arbitration* scheme is *nondestructive* to the higher-priority message and ensures that the higher-priority message experiences no transmission delay. Since message arbitration is so important, we'll look at that in detail after we first get some more background information under our belt.

> **Note:** In this case, arbitration is the process by which one of two or more nodes that are competing for access to the network is allowed to transmit data. Interrupt arbitration is the process by which the dsPIC DSC's interrupt controller determines which interrupt condition to service. Nondestructive arbitration means that the message that ultimately is transmitted on the bus is left intact. Destructive arbitration would determine which message should be allowed onto the bus, but it would corrupt the message, meaning that the node that is allowed to transmit would have to resend the message from the beginning, which adds to the overall transmission time and reduces the resulting available bandwidth.

Another of the CAN's key features is its built-in error-detection circuitry that flags problems with the bus and that will gradually remove an individual network node from the bus should the node generate too many errors. Although the protocol does not support error correction, its error-detection feature helps avoid the serious problem of a single erroring node bringing down the entire network. Unfortunately, because errors can accumulate quickly when there is a problem, tracking down the source of the problem can be difficult because it may go away once the node stops trying to transmit.

If all of this functionality sounds as though it imposes a severe load on the processor, you can relax; because of its complexity, the vast majority of the CAN interface is contained in two hardware components: a CAN controller state machine that handles all of the arbitration and error detection and a CAN bus driver that drives and monitors the CAN bus physical medium. In most systems, these two hardware components are housed in individual integrated circuit (IC) packages. (ICs are the silicon chips that contain much of the electronic circuitry in a system.) Once the CAN interface circuitry has been configured, it simply presents fully formed data messages and status bits to the receiver and transmits complete data messages to other nodes. Since all error detection and handling are performed in hardware, the processor overhead associated with the CAN interface is minimal.

One last high-level consideration is just how far one can run a CAN bus, and the answer is that the maximum bus length depends on the data rate that the bus must support. Table 8.2 shows the recommended maximum bus lengths for a variety of bit rates.

Table 8.2: Recommended maximum CAN bus lengths.

Bit Rate (Kbps)	Bus Length (m)
1,000	30
500	100
250	250
125	500
62.5	1,000
Source: Microchip Application Note 713, Controller Area Network (CAN) Basics, available on the Microchip website (document DS00713A)	

As the table clearly demonstrates, the maximum bus length drops off rapidly with increasing data rates, but even at 1 Mbps (1,000 Kbps), the maximum bus length is reasonably robust.

8.9.2.2 CAN Data Formats

According to the CAN 2.0 specification (CAN Specification 2.0, Robert Bosch GmbH, 1991), data sent over the CAN bus is in one of four basic data formats, called *frames:*

- The data frame, which transmits data from one node to all other nodes on the bus
- The remote transfer frame, which requests data from another node on the bus
- The error frame, which reports that a communication error has been detected
- The overload frame, which reports that the transmitting node is busy processing a previous message and cannot accept more data at this time

The most commonly used format is the data frame, which comes in two flavors: the standard frame and the extended frame. The two data frame formats, illustrated in

Figures 8.24a and 8.24b, are essentially identical, the only real difference being the shorter arbitration ID of the standard frame. All data frame formats have the following basic elements:

- an arbitration ID field whose size varies with the frame type
- A 6-bit control field
- A data field of 0 to 8 bytes in length
- A 2-byte CRC field
- A 2-bit acknowledge field
- A 1-bit end-of-frame marker

Standard (11-bit ID) CAN Data Frame Format

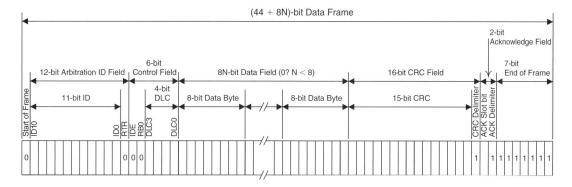

Key Points

1. Each message consists of four main fields and some framing bits:
   A. 12-bit Arbitration ID Field
   B. 6-bit Control Field
   C. 8N-bit Data Field of N data bytes
   D. 16-bit CRC Field

   The user has control of the first three fields only; the CAN controller hardware sets the data in the CRC field and the framing bits.

2. A bit value of 1 is considered to be the recessive state, and a bit value of 0 is considered to be the dominant state.

3. Bus arbitration to determine which node can transmit its message is based on the value of the Arbitration ID field, with the message that has the first dominant bit value in the field having priority and thus being allowed to transmit on the bus. In practice, this means that the lower the value of the field, the higher the message priority.

4. Standard-format data frames have priority over extended-format data frames.

Figure 8.24a: Standard CAN data frame format.

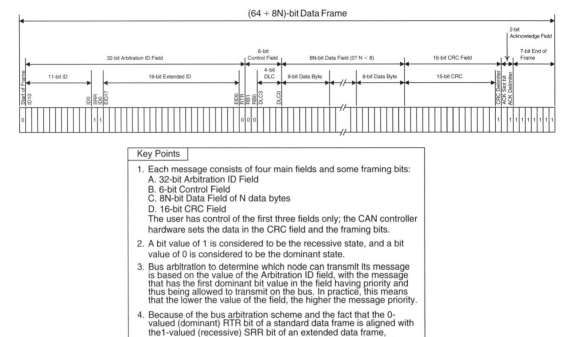

Figure 8.24b: Extended CAN data frame format.

Of these fields, the user has control over the arbitration ID, the control, and the data fields, whereas the CAN controller hardware automatically generates and validates the CRC field, the acknowledge field, and the end-of-frame marker. Let's delve a little deeper into the fields before examining the CAN arbitration technique.

In a standard data frame, the arbitration ID field consists of an 11-bit identifier and a 1-bit Remote Transmission Request (RTR) flag. The extended data frame format is slightly different, but it is designed so that if there is a collision between a standard frame and an extended frame, the standard frame has priority. For an extended frame, the identifier is 29 bits, with the 11 most significant bits being transmitted after the Start of Frame, followed by a 1-bit Substitute Remote Request (SRR) flag, a 1-bit Identifier Extension (IDE) flag, and then the remaining 18 bits of the identifier, with the 1-bit RTR flag completing the field. There are also slight differences in the control field layout for the two data frame formats, although it is 6 bits wide in both cases. In the standard frame, the leading bit of the control field is the IDE flag, which is followed by a single reserve bit denoted as *r0* in the CAN specification. The final 4 bits of the field comprise the Data Length Code (DLC), which specifies the number of data bytes that will follow in the message. Although the DLC is 4 bits wide, it can only assume a value of 0 to 8, since the protocol supports a maximum of 8 data bytes per message.

Because the extended frame includes the IDE flag as part of the arbitration ID field, it has two reserved bits in the control field, r1 and r0. The DLC is the same as in the standard data frame and labors under the same restrictions.

The cyclic redundancy code (CRC) field is not of much interest to us as designers, since it is handled exclusively in hardware and is therefore transparent to the programmer. For the sake of completeness, note that the CRC itself is a 15-bit value, and the CRC field is composed of the CRC value and a 1-bit CRC delimiter bit.

The final field in a CAN message is the 2-bit acknowledgement (ACK) field, which consists of a leading ACK Slot bit that is set to the recessive state (defined in the next paragraph) by the transmitting node and then set to the dominant state by all nodes that receive the message successfully, whether they actually use the message or not. The final bit in the ACK field (and the message) is the ACK delimiter bit, which simply returns the bus to the recessive state to signal that the transmission is complete.

Remote transfer frames are used to request the automatic transmission of data from a node (the data having been already loaded into the CAN module in anticipation of the request), and error frames are generated by a node when it detects an error condition on the bus. Because error frames intentionally violate the timing parameters of the CAN bus, they cause all the nodes that were transmitting data to stop, reset the transmission, and start their transmissions again.

8.9.2.3 Bus Arbitration

As we've already noted, because data transfers are asynchronous, some sort of access arbitration is required to determine which node may transmit if two attempt to send data simultaneously. The CAN designers came up with an ingenious solution to this problem, creating a nondestructive arbitration scheme that uses the value of the arbitration IDs of the colliding messages to decide which node has priority. To understand how this scheme works, we first need to learn two terms that apply to CAN-based systems. Data on the CAN bus is said to be in either a *dominant state* (a logical 0) or a *recessive state* (logical 1). When two bits of different state are transmitted at the same time, the dominant state "wins,"—in other words, that is the resulting state on the bus.

The CAN uses this fact for its transmission access arbitration. Whenever two or more nodes try to transmit a message simultaneously, the dominant bit state is the one that is present on the bus. As each node transmits data onto the bus one bit at a time, it checks to see whether the data on the bus reflects the state of the most recently transmitted bit. If a transmitting node sends a recessive bit but detects that the bus is in the dominant state, the node knows that there is another node that is also transmitting, and the node whose data was recessive knows to get off of the line. The recessive node immediately disables its transmitter and waits until the end of the current transmission before attempting to transmit its own data again.

By handling the arbitration in this manner, the CAN assures both that there is a structured approach to transmission access and that collisions don't result in lost data that forces all nodes to retransmit their messages. Since the dominant state is 0, designers of CAN-based systems select arbitration IDs such that the most important messages have low ID values and thus the highest priorities. For instance, by choosing arbitration ID values of 000H−01FH for alarm conditions and ID values of 020H−7FFH for normal operating messages, the designer ensures that alarm messages always have priority over normal operating messages. The example shown in Figure 8.25, in which an alarm message with an arbitration ID value of 010H is sent at the same time as a normal operating message with an arbitration ID value of 040H, illustrates this. In addition, the scheme allows both standard and extended data frames to reside on the bus, with the standard frame messages having priority over the extended data frames.

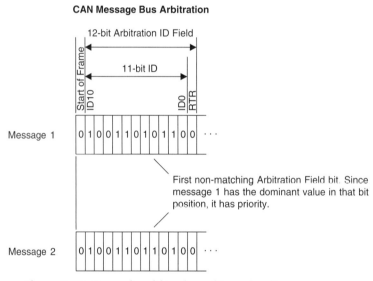

Figure 8.25: Example arbitration of two simultaneous messages.

8.9.3 Acceptance Filters

One optional aspect of the CAN protocol that all CAN controllers implement is *message filtering*, which allows the controller to accept only messages whose arbitration ID fields match a programmable bit-mapped filter value. In this case, when we refer to a filter, we're not talking about a digital filter that processes the digitized signal; rather, we're referring to the process by which only a limited group of messages that meet certain criteria are selected for processing by the CAN controller. Note that even when the controller chooses to ignore the message, it always responds with the ACK Slot bit set appropriately. Filtering is a midlevel technique by which we can reduce the overhead on the processor by limiting the types of messages we choose to handle, whereas the acknowledgment process is a low-level requirement for ensuring the accurate delivery of the network traffic to all nodes.

Filtering the CAN messages consists of two steps, both of which are configurable by the designer but which are executed by the CAN controller hardware. First, we need to set the acceptance filter values (the dsPIC DSC supports up to six different filters), which are logically ANDed with the arbitration ID of each received message on a per-bit basis. The resulting value is then compared to an acceptance mask on a per-bit basis, and if the result of applying the filter matches the acceptance mask, the incoming message is added to the CAN receive buffer (assuming there's room in the buffer).

This can be a point of significant confusion for new (and sometimes more experienced) CAN designers, so an example is appropriate. Let's assume that we want to accept any standard CAN data frame whose arbitration ID field is in the range of 300H to 3FFH. In that case, the acceptance filter is simply F00H and the acceptance mask is also 300H, since ANDing the acceptance filter with the 12-bit arbitration ID field of any received message will make the lower byte of the arbitration ID field a don't care condition (since the entire lower byte will be ANDed with 0), and the filter will pass through the upper nibble. Only arbitration ID fields whose upper nibbles are equal to 3 will match with the acceptance mask and thus be accepted.

Endnote

Microchip Application Note 713, Controller Area Network (CAN) Basics, Microchip website (document DS00713A).

CHAPTER 9
Interfacing to Sensors and Actuators

Kamal Hyder
Bob Perrin

9.1 Introduction

This chapter is concerned with the practicalities of attaching sensors and actuators to digital controllers. The Rabbit RCM3400 prototyping board is used for all the examples in this chapter, but the concepts covered are applicable to most embedded systems.

9.2 Digital Interfacing

There are many books devoted to digital design. Most are concerned with formal methods for logic reduction or techniques used to implement sequential logic. Even with all the available material, device manufacturers are compelled to publish application notes and white papers describing the practical application of their devices.

The working engineer will seldom refer to textbooks discussing canonical equations and logic reduction by Karnaugh map. Engineers are often too busy trying to figure out how to prevent their circuits from being damaged by ESD or being overheated from driving too much current.

We will cover these issues here. We begin with a look at how to bridge the gap between 3.3 V systems (such as most Rabbit-based designs) and 5 V systems.

> **Note:** If you happen to be in need of a truly excellent textbook on combinatorial logic, sequential state machine, and asynchronous state machine design, Richard Tinder's book *Digital Engineering Design: A Modern Approach* (Prentice Hall) will be an excellent addition to your library.

9.2.1 Mixing 3.3 and 5 V Devices

Not so long ago, TTL-based digital systems were designed to operate on 5 V rails. As new CMOS logic technologies have become mature and robust, there has been a natural migration to lower-voltage systems.

Power consumption is proportional to the square of the voltage. In simple DC circuits, we know that:

$$\text{Power} = \frac{\text{voltage}^2}{\text{resistance}}$$

In AC systems, effects of capacitance and operating frequency also enter into the equation. CMOS devices have very small quiescent currents (very high resistance), but the energy stored in their internal parasitic capacitors is governed by:

$$\text{Energy}_{CAP} = \frac{1}{2} \cdot C \cdot V^2$$

As digital states change, these parasitic capacitors must be charged and discharged. The resistive paths through which this charge is moved dissipate power. The more capacitive nodes involved in a system-level state change mean more energy that must be moved and power dissipated. The faster the state changes occur means that more power is dissipated over a given time interval.

This brings us to the equation:

$$\text{Power}_{CONSUMED} \propto k \cdot \frac{1}{R} \cdot C \cdot F \cdot V^2$$

where:

k is a catchall constant

F is the system's frequency of operation

R is derived from quiescent currents

C is determined from dynamic currents

V is the switching voltage

The important bit is that if we drop the voltage by half, we decrease our power about four times.

Energy is related to power by:

$$\text{Energy} = \int_{\text{time}} \text{Power} \cdot dt$$

For the simple case of a static system, we can simply multiply watt ($W = J/s$) by time to get energy.

So by reducing the rail voltage of a system, the power consumed is reduced by an inverse square and so, therefore, is the energy required to operate the system.

In this age of laptop computers, PDAs, and cellphones, energy storage directly translates to weight (and volume). Ultimately, the push for smaller, lighter, portable, energy-efficient devices has pushed the digital world to lower supply rails.

Older 5 V systems are still ubiquitous. Design engineers are often faced with the challenge of interfacing newer 3.3 V technology to legacy 5 V systems.

There are two main issues to consider. When driving 3.3 V inputs with 5 V outputs, the CMOS inputs will be driven above their 3.3 V supply rail. If the 3.3 V device has a high-side ESD diode this will lead to smoke. This will be discussed further in the next section (see Figure 9.5b) when we discuss input protection diodes.

The Rabbit 3000 has 5 V tolerant inputs. A Rabbit powered from 3.3 V rails may have its inputs driven with either 3.3 or 5 V. No damage will occur.

The second issue to consider is how the 3.3 V outputs will drive the inputs of a 5 V device. With a large array of logic families from which to choose the 5 V CMOS device, the issue of noise margin is easily solved.

Since CMOS devices drive their outputs very close to ground, the $V_{IL(MAX)}$ characteristic of the 5 V powered device is seldom a concern. However, $V_{IH(MIN)}$ is often a concern. Table 9.1 shows a comparison of $V_{IH(MIN)}$ for some common logic families.

Table 9.1: Logic families comparison.

	Logic Families						TinyLogic™ Single-Gate Devices		
	HC	HCT	VHC	VHCT	LVT	LVX	HS	HST	UHS
$V_{IH(MIN)}$	0.7·VCC	2.0 V	0.7·VCC	2.0 V	2.0 V	2.0 V	0.7·VCC	2.0 V	0.7·VCC

The "T" families, such as HCT, VHCT, and HST have input stages optimized for interfacing to older TTL devices. This is perfect for interfacing to 3.3 V CMOS systems.

Other families exist with input thresholds fixed at 2.0 V regardless of supply voltage. Fairchild's LVT and LVX families fall into this category.

The potential problem with devices that have a minimum high-input voltage of 0.7*VCC is that with a supply rail of 5 V, the input threshold is only guaranteed valid if it exceeds 3.5 V. This says that if a 3.3 V device is driving a 5 V device with a 0.7*VCC input threshold, the configuration will not work.

For most CMOS families, the 0.7*VCC $V_{IH(MIN)}$ has a bit of a safety margin built in. This means that a lot of the time, a 3.3 V device will drive a 5 V device just fine. This is especially true if the 3.3 V rail is a little hot and the 5 V rail a little low. This is a most unfortunate situation.

Systems designed without proper noise margins may work fine at room temperature on an engineer's workbench. Once these poorly designed systems hit mass production and are shipped to customers, however, invariably problems result. Sometimes the design flaws show up at temperature extremes or when parts from a specific batch of ICs from a particular manufacturer are used.

When interfacing 5 V logic to 3.3 V devices such as the Rabbit 3000, be careful to select a compatible 5 V logic family. The HCT, VHCT, and HST devices are ideally suited for this situation.

9.2.2 Protecting Digital Inputs

Digital devices are susceptible to damage from all manner of electrical stresses. Protecting these devices is both art and science.

The science comes from our ability to model circuits and methodically test our designs. The art comes from the necessity to make sound trade-offs so that our designs are affordable yet well suited to their intended market. Fortunately, an ever-growing array of protective devices is available.

Gas discharge tubes (GDTs), also called spark gap suppressors, are found in telecommunications equipment. These devices are constructed by precisely placing two or three electrodes in a sealed glass chamber filled with specifically selected gasses.

The GDT is placed between a protected line and ground. Under normal line conditions, the GDT looks like an open circuit. When a transient event pushes the voltage between the two electrodes above a spark-over threshold, the gas ionizes and conducts. The GDT diverts the potentially destructive transient energy from the protected electronics to ground.

Once the gas inside the GDT is ionized, it only takes a relatively low voltage, called the *holdover* or *glow voltage*, to keep the device in conduction. This feature precludes most GDTs from AC power protection.

For example, consider a GDT with a 700 V spark-over voltage and a 50 V holdover voltage that is placed on a 110 VAC line. Assume that a transient event causes spark-over to occur. Once the fault clears, the normal AC mains voltage would hold the GDT in conduction. The 110 VAC nominal line voltage is higher than the 50 V holdover voltage. The GDT would cause a short circuit and would self-destruct.

There are GDTs specifically designed for AC operation that will cease to conduct at the AC zero crossing. Most GDTs do not stop conducting quickly enough and are not intended for AC power line protection but rather for telco and other lower-voltage protection.

GDTs are capable of repeatedly shunting thousands of amps for short periods of time. These devices generally cost $1.50–5.00. Figure 9.1 shows several GDTs. These devices are fairly

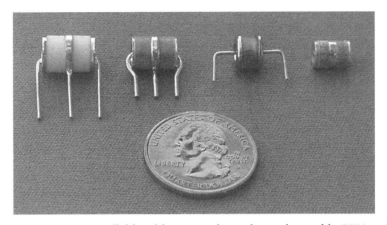

Figure 9.1: GDTs are available with two or three electrodes and in PTH or SMT.

large compared to the components found on a microprocessor's PCB. The right-most GDT in Figure 9.1 is an SMT (Surface Mount Technology) device.

If a system has long external sensor leads, a GDT is good insurance against transient voltage induced by near lightning strikes. GDTs are the slowest of the transient suppression devices we will examine. GDTs are best suited if transient events are expected to last milliseconds or longer.

GDTs operate best when they are used with another form of protection. This cascaded arrangement is called "coordinated protection." The GDT is generally placed nearest the transient event and is considered the primary protective element.

Figure 9.2 shows a GDT combined with a metal oxide varistor (MOV) in a coordinated protection scheme. The designer should select a MOV that has a lower clamping voltage than the

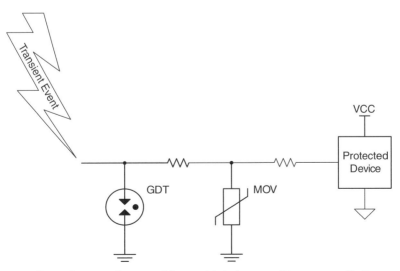

Figure 9.2: Coordinated protection provides multiple layers of incrementally faster protection.

GDT spark over voltage. The resistors limit current and will help to dissipate the transient energy as heat.

MOVs are faster devices than GDTs. They can be purchased in a variety of physical sizes. The larger the MOV, the more energy it can dissipate before suffering permanent damage. Surface mount (SMT) MOVs are small devices and have limited ability to dissipate energy. The pin-through-hole (PTH) devices can be quite large and can dissipate much larger amounts of heat.

MOVs are formed by mashing together tiny bits of zinc oxide until they form a shape to which two electrodes can be soldered. "Secret sauce" ingredients are also added, most consisting of other metal oxides.

Each boundary between zinc oxide particles acts as a little zener diode. The massive combination of random particles statistically acts like one big back-to-back zener diode.

The breakdown voltage for MOVs is less accurate than that for zener diodes. The response time of MOVs is usually slower than zener diodes. A MOV's primary advantage over a zener is the ability to dissipate more power than a zener diode. MOVs are available with breakdown voltages in the 10's to 100's of volts.

Many engineers believe that MOVs, like fuses, are sacrificial devices. A MOV is expected to splatter its guts all over the PCB while valiantly protecting the electronics. This is flawed thinking. MOVs fully recover after a transient event occurs. This assumes that the power dissipated in the MOV was within the MOV's specified safe operating area (SOA).

It is true that the breakdown voltage of a MOV may change a few percent during the first several clamping episodes.

For a MOV to provide long-term protection, the system designer must select a large enough MOV to handle the anticipated currents. Also, the MOV's initial specified breakdown voltage must be high enough that if the breakdown voltage should decrease 10%, it will not drop low enough to fall in the operating voltage range of the protected signal.

Because MOVs have a relatively fast response time and are capable of dissipating large amounts of energy for short periods, they often find application in protecting AC power lines. Figure 9.3 shows a common AC protection scheme.

When a transient over-voltage condition occurs, the MOVs will clamp the high-voltage spike to ground. If the fault is sustained, the sustained high current through the fuses will cause one or both fuses to open. The MOVs protect against momentary transients. In the event of sustained overvoltages, the combination of MOVs and fuses protects both the "protected device" and the MOVs from damage. The only sacrificial protective elements in Figure 9.3 are the fuses.

A device often confused with a MOV is a proper transient voltage suppressor (TVS). The TVS is a semiconductor device that can be modeled as two back-to-back zener diodes. Figure 9.4 shows the schematic symbol for several protective devices.

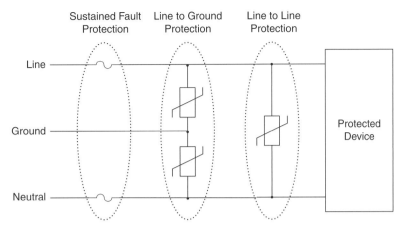

Figure 9.3: Although useful for protecting digital lines, MOVs are often found across AC lines.

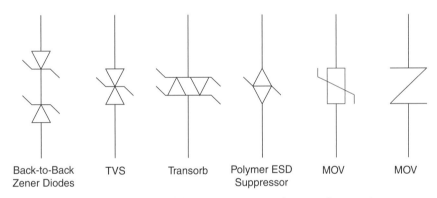

Figure 9.4: The schematic symbol for a TVS, Transorb, or polymer ESD suppressor derives from the schematic for two back-to-back zener diodes.

A TVS is the fastest of the clamping devices. They are also the least capable of carrying large currents for extended periods of time.

Marketing departments are forever trying to differentiate their product from the competition. Years ago, General Semiconductor (now part of Vishay) coined the word "Transorb" to distinguish their TVS from the competition. Figure 9.4 shows the symbol used for a Transorb.

TVS devices are best suited in protecting against electro-static discharge (ESD). These devices are often found as a secondary or even tertiary protective devices in coordinated protection networks.

There are other protective devices. Cooper Bussmann has a device it dubbed Polymer ESD Suppressors with SurgX® Technology. They are small SMT devices and can carry only a few tens of amps.

As technology advances, the number of options open to the system designer for circuit protection increases. One of the simplest tools available to the designer is the diode. These can be used to great effect as an ESD protection device. Many IC's have ESD protection diodes on their I/O pins. A simplified model of on-chip ESD protection is shown in Figure 9.5a.

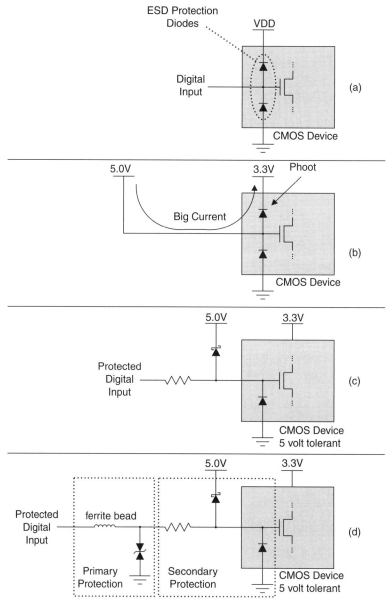

Figure 9.5: On-board ESD protection diodes offer protection but can also complicate the design of multivoltage systems.

The internal protection diodes in Figure 9.5a can be problematic in mixed rail systems. For example, consider the circuit in Figure 9.5b, where a CMOS device is powered from a 3.3 V rail and the device's input is driven from a 5 V rail. The CMOS's VDD side diode will enter conduction under forward bias. Unless a there is a device to limit current, the high-side diode will be damaged. Most likely a significant portion of the device will also be collaterally damaged.

Some device families are "5 V tolerant." A common way to implement this is to remove the high-side internal ESD protection diode. Figure 9.5c shows a 5 V tolerant CMOS device.

To protect the input in a mixed rail system, an external diode will need to be added between the highest rail and the IC's input. Schottky diodes are often used because of their fast switching times. Figure 9.5c shows a configuration suitable for protecting a mixed 3.3 and 5.0 V system.

The resistor in Figure 9.5c limits the current into either diode. Coupled with the parasitic capacitance of the schottky and the CMOS device's input capacitance, the resistor forms an RC low-pass filter. This will slow down high-speed transient events allowing the diodes extra time to enter conduction.

Plain, old-fashioned carbon-composition resistors are the best type of resistors for this application.

Metal film resistors have patterns etched into their film to trim the resistance to the desired value. ESD has a tendency to jump the insulative gaps in the metal film. During a transient event, this reduces the effective resistance of the resistor. Furthermore, if ionized or carbonized paths form, the resistor's value will be altered permanently.

Surface-mount resistors have an added disadvantage over their larger PTH brethren. Under conditions of high current, "hot spots" will form in SMT resistors. This is due to nonuniform current densities in the resistive film. These hot spots can permanently alter the resistor's value. Ohm's law tells us that if we have voltages in the thousands or tens of thousands of volts, we will have high currents during ESD events.

A coordinated network can be constructed to offer protection beyond that of Figure 9.5c. Figure 9.5d shows a two-stage network. The primary protection is a TVS working against the impedance generated by a ferrite bead. The two diodes and resistor provide the secondary protection.

These solutions are fairly expensive in terms of component count, board space, and assembly cost. If protection must be added to a digital input and cost is an overriding factor, a simple RC network coupled with the internal protection diodes can provide a reasonable amount of protection.

The biggest problem with an RC network as a front line of defense against high-speed, high-energy transients is the capacitor's parasitic effective series resistance (ESR) and effective series inductance (ESL). The ESR allows undesired high voltages to develop on the protected node. The ESL reduces the response time of the capacitor. The good news is that the internal protection diodes can usually handle the leading edge of a transient event. This gives the RC network time to act as a filter.

Some integrated circuit manufacturers integrate high-end ESD suppression into their devices. For example, RS-232 and RS-485 transceivers are available with protection that guarantees the device can withstand repeated $\pm 15\,KV$ ESD hits. Analog switch manufacturers now offer devices with similar levels of protection.

The protected RS-232 and RS-485 transceivers are plug-in replacements for older, unprotected transceivers. The protected devices are not significantly more expensive than their unprotected counterparts. In this day and age of CE marks and emphasis on building robust devices, there is little reason not to use an ESD protected transceiver.

The best way to evaluate the level of protection any of these circuit topologies provide is through testing with an ESD generator. Schaffner EMC Inc, has excellent ESD simulators, also called ESD guns. These tools allow an engineer to zap a circuit under test with up to ± 21 kilovolts of simulated ESD.

Testing a few sample circuits on a bench isn't a particularly large sample set. But design is an exercise in trade-offs and risk management. For most products, ESD testing of a handful of model samples is sufficient.

Testing will also unmask hidden problems that will not show up in mathematical models or design equations. For example, consider the sample PCB layout shown in Figure 9.6. ESD will arc between the vias and damage the microprocessor. The best ESD protection can be made useless by sloppy PCB routing. Auto-routers should never be allowed to route protected networks. Unprotected signals should be kept well away from traces that go off board.

During ESD testing, a valuable technique is to darken the lab and while injecting ESD, look at the circuit board and visually check for arcing. Problems such as those shown in Figure 9.6 will become apparent.

9.2.3 Expanding Digital Inputs

One challenge faced by designers is how to add I/O to a processor. Greg Young, while working as a design engineer at Z-World, once said, "Every I/O pin has two struggling to get out." There are numerous techniques for expanding inputs, shown in Figure 9.7. Each has advantages and disadvantages.

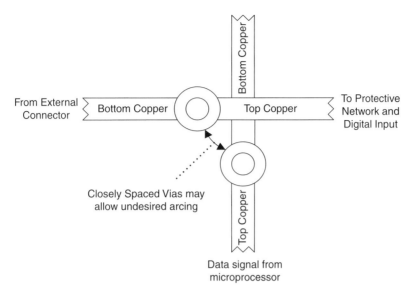

Figure 9.6: Even the best ESD protected input may be defeated by bad PCB layout.

All of the circuits shown in Figure 9.7 refer to HC logic devices. The HC logic family devices operate well from 3.3 or 5 V rails. An abundance of logical functions are available.

An engineer faced with the task of building an interface between a 5 V external system and a 3.3 V processor core will have to consider noise margins. For example, the circuits in Figure 9.7 will work well if the HC devices are powered from 5 V as long as there exists an HCT or HST buffer between the 3.3 V core and the HC device. In most cases this means that HCT should be used for the glue logic.

In other cases, if the engineer can locate HCT or VHCT parts with equivalent logic functions to the HC parts shown, it may be preferable to replace the HC parts with another logic family. From this point on, we will assume that we are not interfacing to 5 V external logic and the HC logic parts are driven from a 3.3 V rail. This will allow us to focus on the issues of capacitive loading, simultaneous sampling, and general interfacing logic techniques.

As we look through these interfacing examples, we should consider that the underlying logical concepts are more important than the particular device implementations shown. The specific devices shown have been quite useful in designs, but there are always newer logic families and alternate devices available.

For example, the 74HC244 shown in Figure 9.7a is an example of an octal buffer. There are other parts available that perform the same function but with different pin outs. For example, the 74HC245 is a bidirectional buffer that is often used in place of the 74HC244. The 74HC245 has a shorter propagation delay than the 74HC244 and may already be a line item

Chapter 9

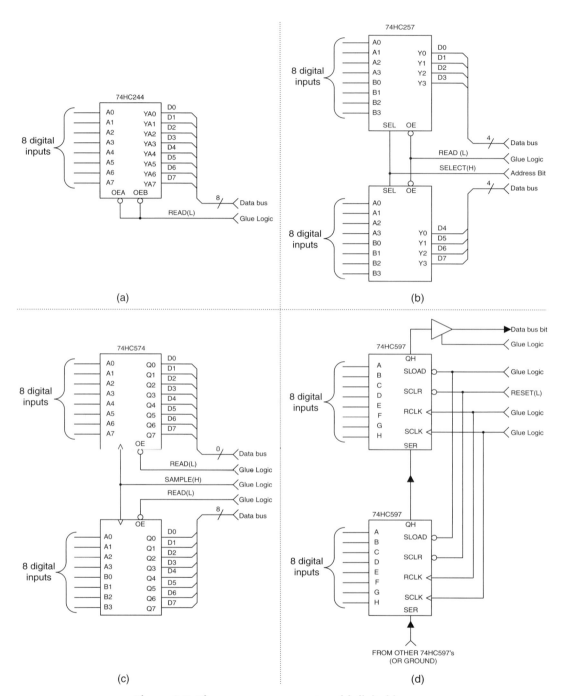

Figure 9.7: There are many ways to add digital inputs.

on a design's bill of material. Of course the designer must decide which way to hard wire a 74HC245's DIR pin to ensure the device operates as a buffer in the correct direction.

Figure 9.7a is one of the most often seen methods for adding digital inputs to a processor's data bus. When a processor wants to read the digital inputs, the 74HC244's output-enable is asserted and data flows through the 74HC244 and onto the data bus. The biggest disadvantage of the 74HC244 is the 20pF worst-case capacitance of a tri-stated output. If more than a couple of these are added to a data bus, the capacitive load on the CPU's data bus may become intolerable.

The Rabbit 3000 offers a helping hand to the designer that wants to plop down a fistful of 74HC244s on the data bus. The Rabbit's Auxiliary Data Bus was added to the processor to minimize the capacitive loading on the high-speed memory bus.

If many 74HC244's are used to expand the I/O in a Rabbit-based design, the designer should consider using the Auxiliary Data Bus.

Figure 9.7b shows a technique that uses multiplexers to implement additional inputs. The 74HC257's SEL signal determines if the A0..3 or B0..3 inputs are presented to the Y0..3 outputs. Each 74HC257 tri-stated output capacitance is 15pF worst-case. Since each output actually corresponds to two inputs, the total data bus loading is 7.5pF per input. This compares favorably to the 20pF per input of the 74HC244 solution.

Sometimes a system is required to simultaneously sample more inputs than the data bus has bits. In this case, a latch can be employed. Figure 9.7c shows how two 74HC574's can be used to simultaneously capture 16-bits of data. The 74HC574's tri-state capacitance is 15pF per pin.

The 74HC574's sister chip, the 74HC374, has the same functionality but a different pinout. Depending on the PCB routing, one or the other IC's will be preferable. The 74HC574 has all of the inputs on one side of the chip, the outputs are on the other side. Most of the time, routing a PCB will be easiest with the 74HC574.

When minimizing capacitive loading on the bus is paramount, the scheme shown in Figure 9.6d should be considered. The serial shift chain only uses a single bit from the data bus. The only capacitive load on the bus is that presented by the tri-state buffer.

Each 74HC597 contains 16 flip-flops comprising two 8-bit registers. One register is the input latch. This is loaded by a rising edge on RCLK. The second register is the shift register. Data is moved from the input latch to the shift chain by asserting SLOAD(L) (active low).

Once the data is in the shift register rising edges on SCLK cause the data to be shifted out through QH. Each time a shift occurs, the "A" bit of the shift register is loaded from the SER input. 74HC597's can, for all practical purposes, be cascaded indefinitely.

SCLR(L) clears all of the registers in the shift-chain. This can be connected to the system's RESET(L) signal if the designer wants the shift-chain to be initialized with all zeros. Many designs will just tie SCLR(L) HIGH (inactive) and save the trouble of routing the trace. The shift chain will be loaded from the input latches on each read.

The concepts demonstrated in Figure 9.7 can be mixed and matched to suit the application. For example, eight shift chains from Figure 9.7d could use a single 74HC244 from Figure 9.7a as the tri-state output buffer.

Now, we can hear all the recent graduates leafing through these very pages and wondering aloud, "Where are the FPGAs and CPLDs? That's what real engineers use to implement digital logic. Right?"

Unless the designer has some reason to use CPLDs or FPGAs beyond expanding digital I/O, they are a bad idea. They are expensive, require programming, are power hungry, and worst of all, are often single-source components (meaning they are available from only one manufacturer).

About the only advantage an FPGA or CPLD has over discrete HC or VHC components is board space. That's only true if the number of I/O pins is fairly high. If only 8 inputs need to be added, the board area-per-input (mm$^2 \cdot$ input^{-1}) is tough to beat for a single 16-pin SSOP. Further marginalizing the FPGA/CPLD density advantage is the fact that most external I/O signals will require ESD protection networks. Since a large board area will be needed, using a high pin-density TSOP doesn't practically buy anything.

Surprisingly, it's hard to beat the simplicity, price, and power consumption of HC and VCH logic parts for expanding digital inputs.

9.2.4 Expanding Digital Outputs

Expanding a system's digital output count is similar to expanding the digital inputs. It boils down to adding flip-flops that retain and present (to other devices) values written by the processor. The same issues of capacitive data bus loading exist for output logic as did for input logic. Additionally, issues of initialization, current drive capability and tri-state ability must be considered.

Figure 9.8 shows four schemes for implementing digital outputs. Each circuit has subtleties that will make it suited to some applications but not to others.

Figure 9.8a shows a simple 8-bit latch (74HC574). This is the same device used in Figure 9.7c to implement digital inputs. One troubling issue is that of initialization. The 74HC574 doesn't have a RESET pin.

The design shown in Figure 9.8a works around the lack of a RESET pin on the 74HC574. Upon system reset, 74HC574's outputs are tri-stated by the external flip-flop. This allows the external resistors to fix the port's state. The advantage of this method is that some of the

Interfacing to Sensors and Actuators 403

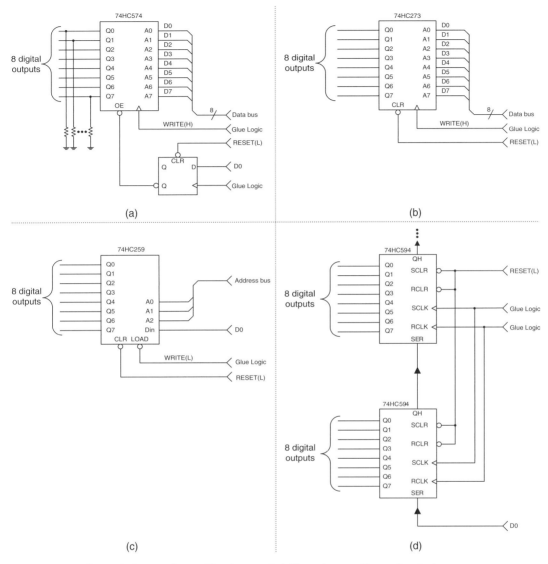

Figure 9.8: A variety of latches and shift registers allows the designer to make trade-offs to meet the application's requirements.

outputs may be tied low and some high. This allows greater flexibility over a latch that uses a RESET (or CLR) pin to fix all the outputs low upon initialization. Of course, the resistors will not be able to source or sink as much current as the 74HC574's outputs.

The added expense of the external flip-flop and resistors may be undesirable. This is especially true if an "all zero" initialization state is required. The circuit in Figure 9.8b shows how a 74HC273 can be used to implement a digital output with an all zero initialization. The CLR(L) pin can simply be wired to the system reset.

Both the circuits in Figures 9.8a and 9.8b suffer from the problem of placing a fairly high capacitive load on the data-bus per digital output. Figure 9.8c shows how to use a 74HC259 bit addressable latch with global CLR(L) to reduce capacitive bus loading. The 74HC259 has one annoying feature: The LOAD(L) signal that writes data into the flip-flops is level sensitive. This places a requirement on the processor bus to hold the data bit (DO in Figure 9.8c) valid after the LOAD(L) is brought high.

Some processors do not expect this hold time requirement. I/O devices are expected to capture their data when the WRITE single is first asserted. The processors have setup and hold time around the leading edge of the WRITE signal assertion. When the WRITE signal is deasserted, some processor datasheets state a zero minimum hold time. The Rabbit has configurable hold times.

Figure 9.8d shows how to use the 74HC594 to implement a series of digital outputs. The 74HC594 does not offer an output tri-state. A similar chip, the 74HC595 swaps the output latch clear (RCLR(L)) for an output enable. Using a 74HC595 and a series of pull-up and pull-down resistors, as in Figure 9.8a, allows individual outputs to be initialized to HIGH or LOW states as the application may require.

There are other schemes for adding digital I/O to a processor. CPLDs and FPGAs are currently "all the rage." For reasons described earlier, they should be used only when absolutely required. Regardless of the technique used to implement the digital output, there is a question of protecting the output from an over-current condition. Most of the CMOS parts presented have a maximum output current rated at 20 mA. If an output is inadvertently shorted to a power rail, damage can occur.

In systems with multiple PCBs, cable harnesses carry digital outputs between boards. If the outputs are used purely for logic functions, then a resistor placed in series with the output is well tolerated. Low input-current requirements ensure that the drop across a series resistor will be minimal. Under a fault condition, such as when a cable harness is incorrectly connected, the series resistor limits the current into or out of the digital output thereby protecting the driver IC.

ESD protection techniques are based on the same principles discussed earlier for digital inputs. GDTs, MOVs, and TVSs are all good options for diverting potentially damaging transient energy away from a digital output pin.

9.3 High-Current Outputs

Digital outputs are fine and dandy, but embedded systems usually need to control actuators with digital ICs. The limited current available from a CMOS output is seldom enough to drive much beyond an LED.

The usual suspects for implementing high-current outputs are bipolar junction transistors (BJTs), Darlington pairs, MOSFETs, electromechanical relays, and solid-state relays (SSRs). We will have a brief look at each of these tools and examine the strengths and pitfalls of each.

9.3.1 BJT-Based Drivers

Bipolar junction transistors (BJTs) are one of the most cost-effective ways to implement a "high-current driver." Discrete transistors are available in PCB mountable packages from the rice-sized SOT-23 to strawberry-sized TO-3.

When a BJT is saturated, $V_{ce(sat)}$ will be finite and nonzero. For small signal transistors switching small collector currents, 100 mV is a good estimate for $V_{ce(sat)}$. As the collector current goes up, so will $V_{ce(sat)}$.

The power dissipated in a transistor due to the collector current will be $V_{ce(sat)} * I_c$. The base current will also contribute $V_{be} * I_b$ watts to the total power dissipated.

Smaller devices can shed less heat than larger packages. Transistors like the MMBT2222A in a SOT-23 can only dissipate about 350 mW @ 25°C. The PN2222A in a pin-through-hole TO-92 is rated for 650 mW @ 25°C.

Transistors such as the Zetex FMMT625 have a combination of low $V_{ce(sat)}$ and high current-transfer ratio (also called beta in saturation, β_{SAT}, or $h_{FE(SAT)}$). This combination minimizes power dissipation.

Figure 9.9a shows the simplest single transistor low-side driver, also called a sinking driver. After power dissipation, the biggest issue to consider in this topology is the base current required to keep the transistor in saturation.

Over temperature, a ten is a conservative value for $h_{FE(SAT)}$. Some higher-end devices such as the FMMT625 have an $h_{FE(SAT)}$ twice that. This is a far cry from the value of 100 that many engineers use.

The reason for this discrepancy is that transistors biased in the active region have a much higher current gain than transistors in saturation. While a β of 100 might be a conservative parameter for amplifier design, it is 10 times too high for most transistors operating in saturation.

The practical implication is that the base drive for a single transistor driver may be a burden on the CMOS output connected to the BJT. For example, if the high-current driver in Figure 9.9a is expected to switch a 500 mA load, a base current of 50 mA is required.

CMOS devices specify a maximum I_{cc} that may be pulled from VCC or sunk into GND. The 74HC574, which is one of the more robust CMOS parts, has an $I_{cc(max)}$ of 70 mA.

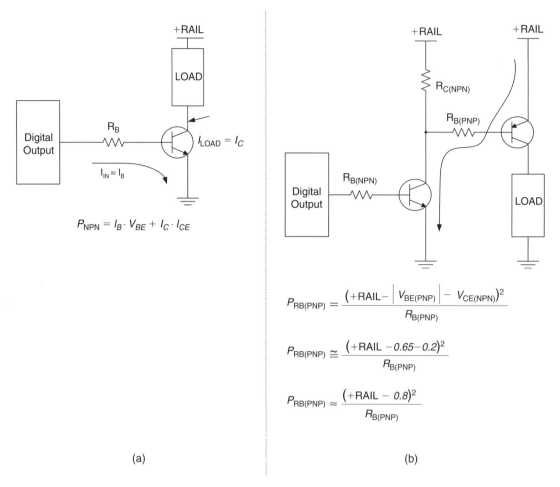

Figure 9.9: Even simple transistor drivers require careful attention to design details.

If all eight outputs are used then only (70 mA / 8 outputs) 8.75 mA per output (on average) is available. That leaves no safety margin for the 74HC574.

Most CMOS devices are less capable than the 74HC574. The 74HC259 has an $I_{cc(max)}$ of only 50 mA. Distributed over eight outputs, only 6.25 mA per output (on average) is available.

Building drivers capable of sourcing or sinking many amperes will usually require multiple stages of current amplification.

Figure 9.9b shows a two-stage driver. Heating in the PNP's base resistor is a key design consideration.

The voltage imposed across the PNP's base resistor is $+\text{RAIL} - |V_{BE(PNP)}| - V_{CE(SAT)NPN}$. This approximates to $(+\text{RAIL} - 0.8)$. For high rail voltages the V^2/R heating in the PNP's base resistor can be very high. For every doubling of voltage imposed across the resistor, the power dissipated quadruples.

For example, if the circuit shown in Figure 9.9b is designed to allow a 50 mA PNP base current and +RAIL is 30 volts, then $R_{B(PNP)}$ will have to be approximately 600 ohms. The power dissipated in $R_{B(PNP)}$ will be:

$$\text{POWER} = I \cdot V = I^2 R = \frac{V^2}{R}$$

$$P_{RB(PNP)} = 0.050 \cdot (30 - 0.8) \approx 0.050^2 \cdot 600 \approx \frac{(30 - 0.8)^2}{600} \approx 1.5 \text{ watts}$$

If eight channels are designed into a device, then (8 * 1.5 watts) 12 watts of heat will have to be removed from the circuit. This may require ventilation and possibly a fan.

The least expensive transistors cost only pennies, but the assembly cost dwarfs the component cost. A good rule of thumb to use for insertion cost is $0.12 per part. This can be used for both PTH and SMT parts. Devices that require a heat sink, like a TO-3 or TO-220, will have additional charges.

Figure 9.9a requires two parts. The resistor and transistor may only cost a couple of pennies, but the insertion cost will be 2 × $0.12 = $0.24. The circuit in Figure 9.9b has five components for a total of $0.12 × 5 = $0.60 in assembly costs. For systems that require many I/O pins, this will quickly become expensive. For applications requiring many high-current drivers, there are ICs available containing eight channels of Darlington drivers.

Figures 9.10a and 9.10b show the ULN2803 and UDN2985 respectively. The ULN2803 is a sinking driver and the UDN2985 is a sourcing driver. Both of these devices may be driven from CMOS outputs.

The ULN2803 can sink up to 500 mA per pin, although the device is limited by the total power the DIP package can dissipate. This means the entire device can sink about 500 mA split up between the eight channels.

The UDN2985 can source around 250 mA. The maximum rail voltage is 30 V. Like the ULN2803, the UDN2985 is limited by the package's heat-shedding ability.

Both devices have integral fly-back suppression diodes. When driving electromechanical relays, the fly-back suppression diodes protect the transistors from the back EMF generated by the relay coil.

One disadvantage of these integrated drivers is the Darlington's inability to pull the output to the rail. In a Darlington, the output transistor is never driven fully into saturation. A Darlington can only pull an output within 1.2–2.5 V of the rail.

Chapter 9

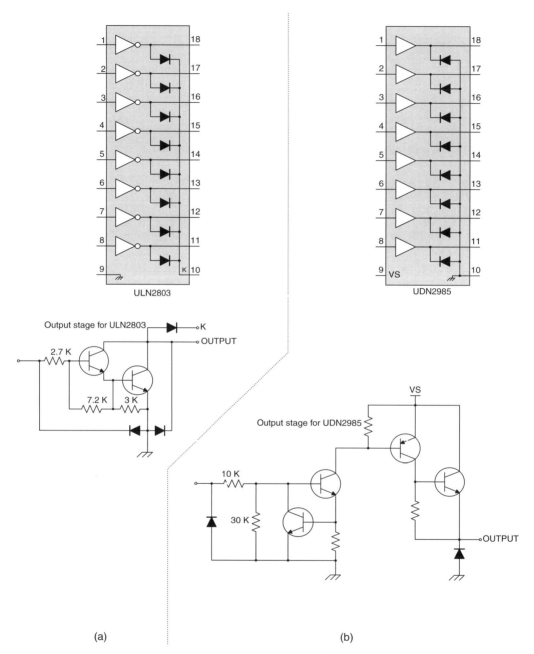

(a) (b)

Figure 9.10: Prepackaged drivers simplify a designer's life.

The UDN2985's output stage is not a proper Darlington configuration but is most accurately referred to as a compound PNP output stage. As with the Darlington configuration, the drive transistor is never saturated and subsequently the output stage will only drive within a volt or two of the rail.

For example, let's say a ULN2803 is going to drive a relay with a 5 V coil and the top of the relay is connected to a 5 V rail. The Darlington can only develop about a 3.5 V drop across the relay coil. The remaining 1.5 V will be across the Darlington.

Many 5 V relays have a pick-up voltage (the voltage at which the relay is guaranteed to operate) that is higher than 3.5 V. This means the ULN2803 may not be capable of driving the 5 V relay with only a 5 V rail.

Another place prepackaged Darlington drivers fall short is when people use them to drive digital inputs. The Darlington's voltage drop can eat up *all* of a CMOS device's noise margin, especially when the CMOS device is being driven from a 3.3 V or 2.7 V rail.

9.3.2 MOSFETs

Metal oxide semiconductor field effect transistors (MOSFETs) have many advantages over BJTs. Their low $R_{DS(ON)}$ allows MOSFETs to switch much higher currents than BJTs. They can be paralleled to share current. MOSFETs have a theoretically zero sustained drive current.

Disadvantages include maximum V_{GS} restrictions—usually around ± 20 V. MOSFETs are notoriously sensitive to ESD damage. Through careful design, these issues can be managed.

There are hundreds of MOSFETs on the market. Table 9.2 highlights several inexpensive devices that cover a range of performance. The devices in Table 9.2 can be driven with a | VGS | of 5 V, allowing 5 V logic to turn them on and off.

The last two lines in Table 9.2 indicate that very high-drain currents are possible. The conditions under which these currents could be obtained are highly unlikely to be realized in a practical design. The case temperatures must be maintained at 25°C to get the -74 or $+75$ amp drain currents listed by the manufacturer. Perhaps with some sort of spray cooling or chilled heat-sink technology this might be possible. Most designs are lucky to have circulating air in a vented box. Under these conditions, using the IRF4905S or IRF1010NS to switch a few amps is reasonable, provided that sufficient V_{GS} is available.

The circuit shown in Figure 9.11 uses an IRF4905S to switch a 30 V source into a load. A zener diode is used to clamp the maximum V_{GS} to less than the 20 volt maximum allowed by the data sheet.

Unlike BJTs, there are no gate resistors associated with MOSFETs that carry a significant current. This means less power is dissipated driving a MOSFET than driving a BJT; refer back to Figure 9.9b.

A neat trick that can be used to save a few pennies in a design, such as that shown in Figure 9.11, is to use a reverse-biased base-emitter junction in place of the zener diode. The Zetex

Table 9.2: These readily available SMT MOSFETs are good choices for new designs.

Manufacturer	P/N	Polarity	Package	VDS(MAX) (Volts)	ID(MAX) (Amps)	VGS(MAX) (Volts)
Fairchild Semiconductor	FDN5618P	P-Channel	SOT-23	−60	−1.25	±20
Fairchild Semiconductor	FDN5630	N-Channel	SOT-23	60	1.7	±20
Fairchild Semiconductor	NDS332P	P-Channel	SOT-23	−20	−1.0	±8
Fairchild Semiconductor	NDS335N	N-Channel	SOT-23	20	1.7	20
Fairchild Semiconductor	NDT2955	P-Channel	SOT-223	−60	−2.5	±20
International Rectifier	IRFL014	N-Channel	SOT-223	60	2.7	±20
International Rectifier	IRF4905S	P-Channel	D_2PAK	−55	−74.0	±20
International Rectifier	IRF1010NS	N-Channel	D_2PAK	55	75.0	±20

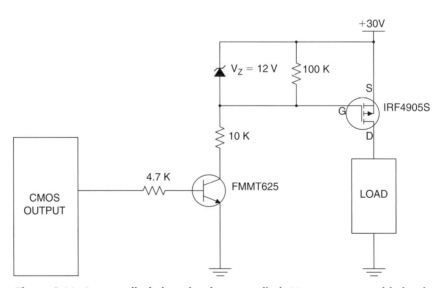

Figure 9.11: A zener diode is a simple way to limit V_{GS} to a reasonable level.

FMMT491 can be used for both the current sink NPN shown in Figure 9.11 and in place of the zener diode. The base-emitter junction breaks down around 10 V. Since the IRF4905S turns on pretty hard with a V_{GS} of −10 V, this technique will allow the elimination of a line item from the BOM without compromising the circuit's performance.

Some caution is advised with this sort of hack. The Zetex FMMT491 indicates the emitter base junction will breakdown at 5 V with 100 uA. Through experimentation, one can see that with 2 mA, the FMMT491's VEB is about 10 V. Hitting the FMMT491 with a can of freeze-mist and a blast from a hot-air gun will show the FMMT491 will work pretty well over a broad range of temperatures.

As a "general rule," engineers frown upon relying on a device's undocumented characteristics. Engineers are also the ones responsible for making trade-offs. Sometimes the careful application of common sense coupled with a little experimentation will yield an equitable trade-off that runs counter to the "general rules." Sometimes engineering is as much art as science.

As with BJT circuits, IC manufactures offer devices that have multiple MOSFET drivers in a single package. One such family of devices is the Power Logic available from Texas Instruments and STMicroelectronics.

The Power Logic devices are available in common logic functions, such as latches (TPIC6259, STPIC6A259, and TPIC6B273) and shift registers (TPIC6595). The output stage is an open drain N-channel MOSFET. There are integrated fly-back suppression diodes for driving inductive loads.

Depending on the number of Power Logic channels sinking current in a device and the ambient temperature, each channel can continuously sink 150 mA to 500 mA. Some devices allow peak currents up to 1.5 amperes. The outputs are usually rated for 45 to 50 V.

Unlike the Darlington high-current drivers, the Power Logic devices will pull their outputs very close to ground. Additionally, FETs share current well. Multiple channels of Power Logic outputs can be paralleled to obtain higher current drive capacity than is available from only one channel.

Texas Instruments offers the devices in DIP packages. STMicroelectronics has datasheets for SMT versions of some devices.

With the advent of inexpensive, robust and beefy MOSFETs, a designer has options that didn't exist with BJTs. In particular, higher load currents can be switched with lower drive currents.

9.3.3 Electromechanical Relays

When an application calls for a switch with low contact resistance, an electromechanical relay is sometimes the best choice. Like all electronic components, relays have evolved rapidly in the last few decades.

Relay manufacturers measure reliability in minimum expected operations. Today it's common to see tens of millions of expected operations before contact failure.

Relays are available in a SMT and PTH packages for circuits that must switch milliamps to a few amps. Milk carton-sized relays, called *contactors*, are also available for applications that need to switch hundreds of amps.

Small relays are often used to drive larger contactors. In these situations, the small relay is called a *pilot relay*. Since driving a contactor is simply a matter of driving a pilot relay, we'll examine smaller relays in a bit more detail.

The Omron G6B series PTH relay is notable for having contacts rated at 5 amps up to 250 VAC. The Omron contacts can switch a maximum 150-watt (1250 VA) resistive load. This is only impressive when one considers the device's volume is only about 2.4 cubic centimeters.

SMT relays such as the TQ-SMD series from NAIS are also available. These small DPDT relays occupy less than a cubic centimeter. The contacts are rated to switch a 60-watt (or 62.5 VA) resistive load with a maximum current of 2 amps.

Relays, like all switches, have a finite contact resistance. This is best measured under maximum current load.

The practical implication of contact resistance is heat. When the contacts carry current, the contact resistance causes Joule heating proportional to I^2R.

The Omron G6B has a 30 milliohm contact resistance. The smaller TQ-SMD contacts have more than twice the contact resistance of the Omron (75 milliohm).

The Omron G6B relays achieve such outstanding contact ratings by having BIG contacts. Opening a G6B with a hacksaw will reveal the disproportionate large contacts for the volume of the relay. To get the relay to perform with large contacts and a relatively small coil, Omron has incorporated a "helper" magnet in the G6B series.

The helper magnet effectively polarizes the relay. When the coil is energized with the proper polarity, the helper magnet is attracted to the coil and the relay's armature is actuated. When the coil is de-energized, the contacts move back into their original position. A reverse energized coil will actually repulse the armature and the contacts will not actuate.

Regardless whether the relay has a helper magnet or not, the driver circuit must be capable of handling the coil's fly-back voltage.

Placing a diode across the relay coil is the most common method of clamping the fly-back voltage to a level that will be tolerated by transistor drivers. Packaged drivers, such as the ULN2803 and UDN2985, shown in Figure 9.10, contain a fly-back suppression diode. Figure 9.12 shows an example of a discrete low-side BJT driver sinking current for a relay with an accompanying fly-back suppression diode.

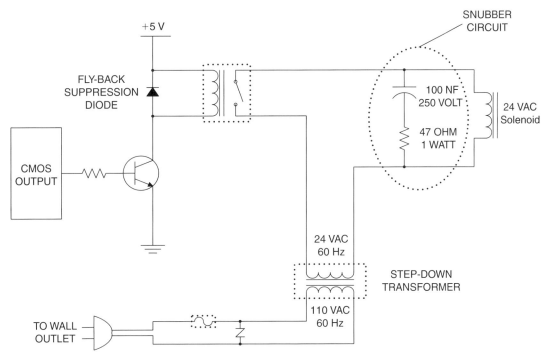

Figure 9.12: Fly-back suppression diodes and snubbers are both tools to manage the energy stored in inductive elements.

Relays are often used to drive motors, contactors, solenoids and other magnetic devices. When driving inductive loads, a designer must take into account the energy that will be stored in the load's magnetic field. If the load is driven with a DC source, a fly-back suppression diode can be used.

If AC is used to drive the load, a simple diode will not work. A simple RC network, called a snubber network, can be used quite effectively to dissipate stored energy. Figure 9.12 shows an example of a snubber network.

When the relay is opened, the snubber and load form a damped LCR circuit. Since both the inductor and capacitor are theoretically purely reactive devices, they will not dissipate any energy. The energy stored in the load will be dissipated as heat in the snubber's resistor.

In practice, all parasitic resistances will dissipate energy as heat. The effective series resistance of the capacitor will limit the (ripple) currents the capacitor can handle. The DC resistance of the inductor will generate heating (copper losses) in the inductor, as will core heating (iron) loses.

The consequences for not using a mechanism, like a snubber, to burn off undesired energy can be as innocuous as a little arcing on the relay contacts or as serious as arcing that reaches back into relay's coil driver. Small relays can offer several thousand volts of isolation between the contacts and coil circuit. Inductive fly-back can often exceed this level. In either case, undesired radiated emissions from the arcing will occur.

Figure 9.12 shows a snubber placed across an inductive load. This arrangement can reduce noise significantly. The snubber provides a conductive path for the current that the inductor's stored energy will produce. Inductive fly-back voltages will be significantly reduced when the relay opens.

In practical systems, the length of wire between the switch and the inductive load is often sufficient to produce a bit of back EMF across the relay. This will contribute to arcing in the switch contacts. Arcing generates unwanted EMI and if severe will cause deterioration of the relay contacts.

A common practice is to place the snubber directly across the switch contacts as shown in Figure 9.13. This minimizes arcing and the unwanted associated emissions. Another advantage to this configuration is that the relay is often located on a PCB. This means it is fairly easy to add a snubber as the resistor and capacitor are intended to be mounted on a PCB.

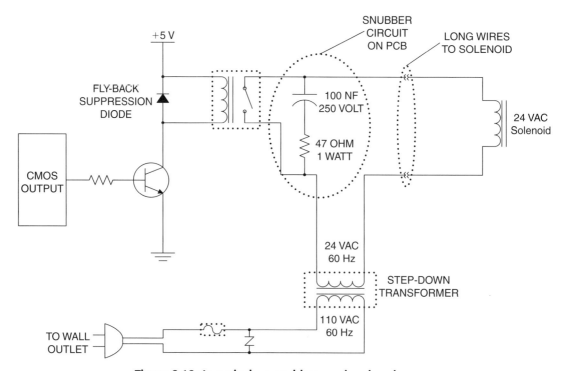

Figure 9.13: In a pinch, a snubber can be placed across the relay contacts instead of the inductive load.

A snubber configuration as shown in Figure 9.12 is sometimes overly costly because the solenoid may not have any PCB to which to mount a snubber. This means either the device must be "fly wired" and glued or a printed circuit board must be added. Either way, labor and material costs increase.

As a practical matter, if the wire lengths between the switch and the inductive load are long, the area enclosed by the wire loop can become quite large. Depending on the frequencies involved, this can make for a pretty good radiator. In some cases, a snubber may need to be placed across both the switch and the load to bring emission levels down to an acceptable level.

The decision to place a snubber across the switch depends upon the geometry of the system, the currents being switched and the magnitude of the inductive load. One other factor is the construction of the switch.

Mechanical contacts arc. Semiconductor based switches do not. A snubber that is used to reduce arcing and the associated broadband radiated emissions may not be required if the mechanical relay is replaced with a solid-state relay.

On the other hand, if the size of the inductive load is unknown or the length of wire between the switch and the load is long, a snubber near the switch is good insurance against high flyback voltages damaging the switch.

Whenever there is a snubber placed across a switch, leakage must be considered. The impedance of the snubber's capacitor is given by:

$$C_{\text{IMPEDANCE}}(\omega) = \frac{1}{C \cdot s} = \frac{1}{C \cdot j\omega} \Rightarrow ( )C_{\text{IMPEDANCE}}(f) = \frac{1}{C \cdot 2\pi f j}$$

$$\Rightarrow |C_{\text{IMPEDANCE}}(f)| = \frac{1}{C \cdot 2\pi f}$$

ω is radian frequency; $s = j\omega$; f is frequency in Hz; $j = \sqrt{-1}$

In Figure 9.13, when the relay contacts are open, there is still a conductive path for the 60 Hz AC across the contacts. The impedance is:

$$\text{Contact}_{\text{(OFF)IMPEDANCE}}(60\ Hz) = \frac{1}{100 \cdot 10^{-9} \cdot 2 \cdot \pi \cdot 60} \angle -90° + 47\angle 0° \approx 26.5\ k\Omega \angle -90°$$

In the 24 VAC system of Figure 9.13, 26.5 k ohms allows about a milliamp of leakage current. This is not sufficient to keep the solenoid energized.

In some cases, a snubber across a relay's contacts can be problematic. If the relay is switching a high voltage, the snubber can allow enough leakage current to give an incautious human a good jolt. Depending on the system this can be handled with safety interlocks and strict lockout/tag-out procedures. In other systems, it may simply be unacceptable to allow any leakage current across the relay contacts. These are trade-offs the designer must make.

Another situation to look out for is when a snubber allows enough leakage current to keep the magnetic load actuated when it should be off. The case of a pilot relay driving a contactor is a classic case.

Contactors (and relays) have a pick-up current and a hold current. The pick-up current is the amount of current required to develop a sufficient magnetic field to actuate the contactor. The hold current is required to hold the contactor in an actuated position once the contactor has been actuated.

The pick-up current is generally much higher than the hold current. Some relay manufacturers specify a pick-up voltage and a hold voltage. By using the impedance of the relay or contactor's coil, one can calculate the currents in question.

The hold current will often be specified as a maximum value. This means that even less current will be required to hold the device in an actuated state than the specified hold current.

An undesirable situation will occur if a pilot relay's snubber allows a leakage current higher than the *minimum* hold current for the contactor. In essence, the pilot relay will turn on the contactor, but the leakage current will never allow the contactor to fully de-energize.

If the system designer is suffering particularly bad karma, the snubber's leakage current will not be enough to keep most contactors energized. Symptoms of an inappropriately designed snubber may not show up until systems are deployed in the field. Besides being somewhat embarrassing, this type of bug can be hard to find and expensive to fix.

Snubbers are invaluable tools for managing energy stored in inductive loads. The designer will have to trade off component values to achieve the responsiveness, costs, size and leakage currents required for each specific system.

For small inductive loads (fractional horse power loads), values for the capacitor are often found between 10 nF and 100 nF. It's common to over specify the working voltage for the capacitor. In systems like those shown in Figure 9.13, high-voltage transients from the AC mains will be stepped down and coupled to the capacitor.

Another condition to be aware of is the failure of the galvanic isolation between the primary and secondary of the step-down transformer. If this occurs, the ideal situation would be for the fuse to open (see Figure 9.13). In another case, the solenoid may be damaged. If the snubber's capacitor is over specified, as is the case in Figure 9.13, under no circumstances should the controller board be damaged. Worse case, a technician may have to replace a fuse, the step-down transformer, and the solenoid. None of this requires repairing a PCB.

Resistor values tend to range from 10 ohms to 100 ohms. As a matter of good practice, the resistor should be specified to be flameproof and, if size and cost permit, over specified by a watt or two.

When making these trade-offs, the best advice is "test early and test often." Better to spend a little more money early on in a design to find out that a design trade-off will push emissions above allowable limits than to get a design set in concrete only to find the same information out later. A good test lab will be willing to work with a designer throughout the design process—not just at the end of a design.

Another useful tool is a circuit simulator, like SPICE. Simulators are notorious for being so strict and finicky that getting a simulation to complete is nearly impossible. Some other simulators are so over-simplified that getting data the designer can trust is difficult. Like all technologies, circuit simulators have improved over time.

Linear Technologies (www.linear.com) offers a free SPICE implementation called Switcher-CAD III. This tool has an easy to use graphical front end, while allowing access to the SPICE netlist. SwitcherCAD III is a third-generation SPICE engine and is optimized for simulating switching power supplies. The tool works wonderfully for the general-purpose simulation.

On the CD-ROM packaged with this book, we have provided simulations for the circuits shown in Figures 9.12 and 9.13. A quick visit to www.linear.com will let the reader download and install the most current version of SwitcherCAD III.

9.3.4 Solid-State Relays

Solid-state relays (SSRs) are simply optoisolated silicon switches. These devices are available in a wide range of form factors and power ratings. Currently, the "contact" side of SSRs can be found with BJTs, SCRs, TRIACs, and MOSFETs. Some devices even contain circuitry to delay the "contact" from turning on until the AC voltage across the "contact" is at a zero crossing.

SSRs, when correctly sized, offer an advantage in mean time between failure (MTFB) over electromechanical devices. This comes from the fact that SSRs have no moving parts. SSRs, like all electrical components, have their own trade-offs.

SSRs are more expensive than electromechanical devices, and tend to dissipate more heat than their electromechanical counterparts. When specifying an SSR, the designer must pay careful attention to the device's derating curve.

For example, the SSRs in the Crydom MCX family are sold as 5 amp relays. The derating curve shows that at ambient temperatures above 30°C the device is to be derated by 120 mA/°C. This means if the SSR is intended to operate in a 60°C ambient environment, the device is only capable of carrying about 3 amps.

This heavy temperature derating can cause a designer to have to specify larger SSRs and thereby incur even more cost. Electromechanical relays also have temperature derating curves, but SSRs generally have the more onerous restriction.

In addition to MTBF, SSRs offer an advantage over electromechanical devices on the drive side. Driving an SSR is simply driving an LED. SSRs often have a fairly wide input range. The Crydom MCX series has input options spanning 3–15 VDC to 90–140 VAC (rms).

SSR input currents are usually a few milliamps to a few tens of milliamps. The MCX datasheet lists typical input currents from 5 mA to 15 mA depending on input configuration.

An SSR's driver circuit does not have to contend with inductive fly-back. However, switching inductive loads is still something to which the system designer must pay careful attention. When switching DC into inductive loads, designers can use a fly-back suppression diode to manage back-EMF. Circuits switching AC into inductive loads will benefit from the same simple snubber circuit shown in Figures 9.12 and 9.13.

SSRs, when properly applied, have outstanding life expectancies. They may require substantial heat sinking or air circulation. SSRs are usually more expensive than their electromechanical cousins. SSRs are wonderful innovations, but they are not the best solution for every situation.

9.4 CPLDs and FPGAs

Complex programmable logic devices (CPLDs) and field programmable gate arrays (FPGAs) have been a boon for system designers. The high gate density available allows complex functions to be implemented in a configurable device. Programmable logic devices are available in ever-faster speed grades.

FPGAs are available as SRAM-based devices that must be programmed on boot. An 8-pin boot EEPROM is the usual method. There are some antifuse-based FPGAs that are the equivalent of OTP devices.

CPLDs are generally Flash-based devices. Once programmed, they retain their configuration even during power-down.

These devices can be used to best effect by placing them in the path of high-speed data to do filtering or other high-speed manipulation. Figure 9.14 shows how FPGAs can be placed in a video path to do image filtering and other video-related tasks.

CPLDs and FPGAs are architecturally different creatures, but for our purposes the distinction is academic. Generally we can assume a CPLD will be smaller, cheaper, and faster than FPGAs.

Figure 9.14 shows a block diagram of a reasonably sophisticated single CCD camera. The system provides a processed analog output as well as a Web connection. The Web connection

Interfacing to Sensors and Actuators 419

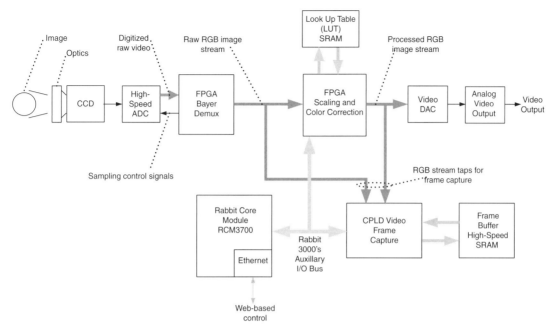

Figure 9.14: A digital video camera is an application that benefits from the versatility and high speed of FPGAs.

in this system is primarily used to configure the image processing. The system also allows a frame to be captured from either the raw RGB image path or from the processed RGB image path. The captured image can be pulled into the Rabbit-based RCM3700 for analysis or sent over the Ethernet connection.

Color cameras that use a single CCD have a color filter (or mask) over the CCD that makes some pixels sensitive to red light, some to green, and some to blue. The most common of these masks is called a *Bayer color filter*. When reading a Bayer encoded image, a system must decode (or demultiplex) the image and form single digital pixels each with a red, green, and blue component by weighting several adjacent color selective pixels from the sensor. In Figure 9.14, the first FPGA in the video path is dedicated to this function.

Once the raw video image has been converted to a digital steam of RGB information, the camera places a second FPGA in the signal path to apply additional processing to the image. This may include scaling the image, digitally zooming on a portion of the image, or doing color correction.

The RCM3700 can communicate with the image processing FPGA via the Rabbit's Auxiliary I/O bus. This is an 8-bit bus and is comparatively slow. The Rabbit isn't moving real time video data across the bus, just configuration data for the FPGA.

The look-up-table (LUT) connected to the second FPGA is a pretty common feature found in image processors. The RCM3700 can load the LUT through the image processing FPGA. Once the LUT is loaded, the FPGA can use the data in the LUT as part of the filtering algorithm.

Figure 9.14 shows a CPLD acting as a high-speed data acquisition controller and memory interface. The CPU can command the CPLD to start a high-speed capture of a single frame from either the raw or processed video streams. Once an image frame is collected in the frame buffer SRAM, the CPU uses the CPLD as glue logic to access the contents of the frame buffer.

The Rabbit 3000A has two new block move instructions (LSDDR and LSIDR) that allow the CPU to move data from repeated reads from a single IO port to multiple RAM locations. These block move instructions could be used to move a video image efficiently from the frame buffer into the RCM3700's internal SRAM.

If a system has need of a CPLD or FPGA, extra I/O pins can often be used to expand digital I/O lines. Since the expensive programmable logic already exists in the system, the extra I/O pins are essentially free.

For example, if the camera needed PTZ (pan/tilt/zoom) controls or a local man-machine interface (MMI), extra pins from the CPLD or image processing FPGA can serve these functions.

9.5 Analog Interfacing: An Overview

So far, we have looked at expanding a controller's capacity for turning devices on and off and sensing whether a device is on or off. Most of the world is analog. Temperature, color, strain, sound intensity, velocity, pressure, and innumerable other environmental quantities are analog. Our discussion will not be complete unless we look at pulling analog signals into a digital environment.

9.5.1 ADCs

An analog-to-digital converter (ADC) is a device that maps an analog signal, usually a voltage, to a digital code. These devices can be implemented in a variety of ways to optimize for certain characteristics. For example, delta-sigma ($\Delta\Sigma$) converters are generally slow but have very high resolution and moderate cost. The $\Delta\Sigma$ converters are found in many biophysical sensing instruments.

Flash converters are the fastest converters available but are expensive and have relatively low resolutions (6 to 12 bits). These types of converters are found in video applications, communications systems, and digitizing oscilloscopes.

Successive approximation (SAR) converters use a binary search algorithm and a single comparator to accomplish a conversion. They strike a middle ground between $\Delta\Sigma$ converters and

Flash converters. SAR converters are available between 8 and 16 bits with reasonably high sample rates at moderate prices. SAR converters are the workhorses found in many embedded systems.

ADCs have nonidealities. Most ADC datasheets do a fairly good job of characterizing the device performance. Sometimes a datasheet may give "typical" performance parameters or only characteristics at 25°C. In these situations, further testing and a few phone calls to the manufacturer will often yield a more concrete characterization.

An ADC is just one component in a chain of components that comprise a data acquisition channel (DAQ channel). Unless the ADC selected is a terrible device, the errors associated with the ADC will probably be swamped by errors associated with other components in the DAQ channel.

9.5.2 Project 1: Characterizing an Analog Channel

The ADC's reference voltage (or current), signal conditioning op-amps, gain resistors, filter capacitors, noisy power supplies, parasitic noise coupling from poor PCB layout, and even raw sensor errors will introduce noise and errors into any DAQ channel. Characterizing an ADC is less important than characterizing an overall DAQ channel.

There are a few types of analysis that can be used to characterize a DAQ channel. These roughly divide into two groups. DC characterization is concerned with identifying offset errors, gain errors and noise when a DC signal is applied to the DAQ channel. AC characterization is concerned with looking at these errors as various frequencies are applied to the DAQ.

There are many papers that discuss academically rigorous methods for characterizing DAQ channels. In some systems an exhaustive characterization may be justified. However, most practicing engineers will get a long way with simple DC characterization. In this section we work through a DC noise analysis of one of the Rabbit Semiconductor's RCM3400 DAQ channels.

Linux-based tools will be used to do the data collection and analysis. These tools are well developed and, best of all, are free. Linux is an ideal platform for laboratory PCs collecting, logging, and analyzing data.

Some level of noise is present in any analog circuit, dependent on the physical properties of the ICs, noise coupled in through the physical layout of the printed circuit board, ripple present on the DC power supplies, and so forth. By applying a known DC signal to our DAQ channel, we can sample our ADC readings and learn a great deal about the quality of the system.

For this section, we will use equipment set up as shown in Figure 9.15.

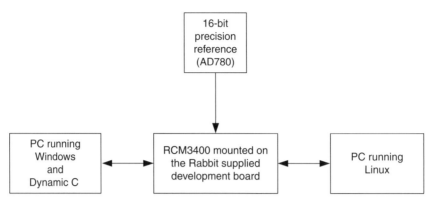

Figure 9.15: A simple setup allows a quick characterization of the RCM3400's analog channels.

Our analysis will yield a quantitative set of numbers characterizing the offset and noise performance of the DAQ channel. From the acquired dataset, we will also generate a histogram as a qualitative aid for the engineer to see how the noise is distributed.

The characterization will proceed as follows:

1. The RCM3400 will acquire measurements of the precision reference.
2. The RCM3400 will transmit the sampled data to the Linux PC.
3. Linux-based software will analyze the data statistically and display it graphically.

The data collection phase will complete entirely before the sampled data is transmitted to the Linux PC. This reduces the chance that any activity on the serial ports will be coupled as noise into the ADC channels. Furthermore, sampling the ADC channel as quickly as possible minimizes the likelihood of thermal drift and offset changes smearing the data in our histogram. We are trying to characterize the RCM3400's best-case performance.

9.5.2.1 Sampling the Precision Reference with Dynamic C

We must decide how many ADC samples we need in order for our findings to be statistically significant. Arbitrarily, we pick 32,768 (32 K) samples. This number is not so large that it will take too long to acquire the data, and it doesn't seem so small as to be insignificant. Picking a "statistically meaningful" sample size for a project like this is a bit of an art.

For our example, we are using a battery-powered, precision 16-bit voltage reference that produces 2.4995 ± 0.0005 V with noise less than 1 part in 65,536 (16-bits). The RCM3400 ADC will accept voltages up to 20 V, so we are well within the physical limits of the Rabbit's hardware.

Let's consider the storage required for our sampled data. The RCM3400 API includes a function named `anaIn()` that returns an int when it reads from the ADC. The ADC itself returns a value between 0 and 2047, and the driver expands that to include two special (negative) values indicating abnormal conditions. While the valid ADC values fit into 11 bits, it is convenient to treat them as 16-bit integers for our purpose. The RCM3400 has ample RAM available to store the extra information.

If we lacked sufficient RAM, we could use another, more compact format to store our samples. With no loss in fidelity, we could pack eight 11-bit samples into 11 bytes, a savings of 5 bytes over using 16-bit integers. Alternatively, if we could accept some loss in fidelity, we could store the difference between successive samples instead of the samples themselves. If we assume that successive samples are likely to be nearly equal, the difference between them should be close to zero. Storing an 8-bit difference between samples would require half of the storage space used to store 16-bit integers. A limitation of this approach is that if the sampled signal changes too much between successive samples, the difference will not fit into 8 bits and thus we lose fidelity.

Although the RCM 3400 has sufficient RAM available to store 32 K integers, declaring an array like this:

```
int storage[32768];
```

yields the compiler error message: "Array dimension is too large; the size of a dimension cannot exceed 32768." Even if we decide that 32767 samples would be sufficient, we cannot proceed naively because (a) if the storage array is declared as a global variable, we get the compiler error "Out of variable data space." and (b) if the storage array is declared as an "auto" variable and thus stored on the runtime stack, our program fails at runtime because we've corrupted the stack with our data array.

Fortunately, other methods to use memory on the Rabbit exist and are easy. Dynamic C allows us to allocate large memory blocks from extended memory at runtime using the `xalloc()` function call. Additionally, Dynamic C lets us read and write to integers stored in extended memory using the `xgetint()` and `xsetint()` functions. Now that we have considered the storage requirements, we can press ahead with understanding how to acquire the samples.

The RCM3400 library includes functions to configure the DAQ channel's gain and read the ADC. Sample code included with Dynamic C shows how these functions are used. We started with the sample code's function `sample_ad()` and then modified it to suit our purposes.

The enclosed CD contains the complete program, NoiseCheck.c, that was used to sample the RCM3400's ADC, store the samples, and report them via the serial port.

The code begins by defining several macros: `GAINSET` is used to configure the programmable ADC gain in the function `sample_ad()`; `NUMSAMPLES` is the number of ADC samples we will take; `DINBUFSIZE` and `DOUTBUFSIZE` configure the size of the receive and transmit buffers used by the serial drivers.

The function `sample_ad()` takes one parameter—an integer that represents which ADC channel we wish to sample from. The ADC has eight channels, numbered 0 through 7. The original `sample_ad()` function provided with Dynamic C had a second parameter— the number of samples to take and returned the average of the samples. Using the average of a number of samples is a common data acquisition technique to help mitigate the impact noise on the DAQ channel. Since our goal is to determine the noise that might be present, we do not want to average ADC samples. Thus, we removed the second parameter for our program.

9.5.2.2 Transmitting the Sampled Data to the Linux PC

The collected samples will be transferred from the RCM3400 to the Linux PC via RS-232. The RCM3400 has several serial channels available, and most PCs are equipped with at least one RS-232 port.

Z-World provides a convenient library of functions in Dynamic C for handling the serial ports. The functions that we will use are `serDopen()`, `serDwrFlush()`, `serDrdFlush()`, `serDclose()`, `serDputs()`, and `serDwrUsed()`. The prefix "serD" is used in the function names to indicate that the function affects serial channel D. Similar functions exist for all serial channels A through F. The open and close functions are housekeeping functions to configure the serial port. `wrFlush()` and `rdFlush()` are used to ensure that the transmit and receive buffers are cleared before we use them. `puts()` is used to queue a string in the output buffer for asynchronous transmission. `wrUsed()` returns the number of bytes that are currently queued for transmission. Behind the scenes, the serial library handles the asynchronous transmission and reception of serial data by installing interrupt handlers to deal with the various buffers and UART registers involved. This library makes using Rabbit UARTs for RS-232 convenient.

Under Linux we have even fewer functions to worry about. Historically, UNIX and its descendents (like Linux) treat hardware devices as files. To refer to a hardware device, the programmer need only refer to a special file name that corresponds to that device, using the same software functions as one would for any other file (e.g., `open()`, `close()`, `read()`, and `write()`). Additionally, we can use the usual shell commands for file manipulation to send and receive data from the serial ports under Unix/Linux.

It's quite convenient to write small programs that each solve a small piece of the complete a task. In this case, there will be one small program (GetData.sh on the enclosed CD) that reads data from the serial port and saves it to a data file. There will be a second program (Histogram.pl on the enclosed CD) that reads the data file and produces a statistical analysis of the sample data and saves a histogram of the data to a second file. A third program (gnuplot) will read the histogram data to produce a graph of the data.

We must choose a data format that both the RCM3400 and the Linux PC can support. Since there's no need for high-speed data transfer, we will use ASCII text to send the sample data. Because the data analysis will happen on the Linux PC, we might want to send each sample as a hexadecimal number, in a format easily handled by the usual UNIX tools. This leads to the simple format below:

```
<start of data><carriage return / linefeed>
<sample 1 value><comma><carriage return / linefeed>
...
<sample 32768 value><comma><carriage return / linefeed>
<end of data><carriage return / linefeed>
```

Each <sample> will look like: 0xXXXX where X is a digit 0..9 or letter a..f and the start and end of data messages are ASCII text.

Strictly speaking, the <comma> characters in the message stream are redundant since each line will be ended by a carriage return and line feed. However, they could be useful if the data file ends up being imported into a spreadsheet program. The start and end of message strings would normally be well defined, except that the data extraction tools on UNIX are generally very forgiving about such things.

In NoiseCheck.c, we used the string "Start of data transfer" to indicate the start of the message and the string "End of data transfer" for the end of message. Both humans and Linux can easily parse the resulting data file.

9.5.2.3 Linux Data Capture Program Listing

The enclosed CD contains the file getdata.sh, also shown in Listing 9.1. This is the bash shell script that was used to record the received ADC sample data on the Linux PC.

Linux is generally configured with online manuals for their system programs. For more information regarding the programs mentioned, please refer to the online "man" pages. For example, to get more information on stty at a Linux command prompt, just type man stty.

The script begins by defining two variables, COMPORT and FILEBASE, which hold the filename of the serial device and the base filename of the received data files, respectively. Next, the script uses the STTY program to set the receive and transmit baud rates for the serial device. The default settings for number of data bits, stop bits, and parity mode are acceptable and are not changed here.

> **Note:** UNIX serial ports were historically used to connect the computer to teletypewriters, and STTY reflects this heritage by having a plethora of options that would be used only for that purpose. In general, when you're using the serial ports on a UNIX machine for data communication, these options should be disabled.

The script continues by generating a unique filename where the received data will be stored. The DATE command is called to return the Julian date (the day of the year) followed by the time in 24-hour format. This data string is appended to the value of the FILEBASE variable to generate a filename.

Finally, a terse line explanation is given to the user and the heart of the script commences. The CAT program is used to concatenate the contents of a list of files and print the resulting data stream on the STDOUT device, which is by default the user's console. We invoke CAT giving the name of the serial device as the file that we want to send to STDOUT. We then pipe (|) that data stream to the TEE command, which copies its input to a file as well to the STDOUT data stream. The result of this command is that the contents of the serial port are both copied to the file named by FILENAME as well as echoed to the screen for the user to watch. This program ends when the user presses Control-C.

If the user didn't want to see the received serial data, the user could change the last line to read:

`cat $COMPORT > $FILE`

Also, if the user didn't want to write to a new file each time this script was run but instead append new data to the end of the current file, the last line could be changed to:

`cat $COMPORT | tee -a $FILEBASE`

Both of the previous changes would be written as:

`cat $COMPORT >> $FILEBASE`

If the user didn't want to have the current shell process to wait for a Control-C, the last command should be placed in the background via the & shell operator as follows:

`cat $COMPORT >> $FILEBASE &`

Note that even if the & is used to "background" the script, the script can continue to echo data to the console. This feature is not very useful, and probably demonstrates more flexibility than ease of use on the part of Unix and derivative operating systems.

Since power outages are common, it's reasonable to want to make a Linux PC begin logging data upon boot without user intervention. Unix provides many ways for the user to have programs invoked automatically, some based on a schedule via the AT and CRON programs, some based on the current state of the operating system via the INIT program.

> **Note:** The simplest of the mechanisms is a shell script that the system runs after all other boot-up functions have completed. This script is located in different places on different flavors of UNIX, but on many systems the script is named /etc/rc.local. One need only

> add the commands in Listing 9.1 to the end of the /etc/rc.local script, or the rc.local script could invoke a separate shell script that contains the commands from Listing 9.2. In this case, the user should definitely background the CAT command and not echo data to the STDOUT data stream.

```sh
#!/bin/sh
#
# acquire a batch of data from the serial port
#
# Kelly Hall, 2003

COMPORT=/dev/ttyS1
FILEBASE=data.log

# set up the com port
stty -F $COMPORT ispeed 19200 ospeed 19200

# get the time/date
DATE=`date +%j-%T`
FILE=$FILEBASE"-"$DATE

# tell the user what's happening
echo "logging data from $COMPORT to $FILE"
echo "press control-C to exit"

cat $COMPORT | tee $FILE
```

Listing 9.1: Bash shell script used to receive and record the sample data.

9.5.2.4 Analyzing the Data Graphically and Numerically

Once the ADC sample data has been transferred to the Linux PC, we want to analyze the data for trends. One form of qualitative graphical analysis is a histogram. This is a plot of each data value read versus the number of times that value occurred in the data set. Visual inspection of the histogram will quickly tell us if the noise "looks" right. For example, multiple humps, also called *modes*, are indicative of nonrandom (coupled) noise. If the histogram is a single bell-shaped curve, then we can look to a more quantitative analysis that assumes the noise is gaussian.

Other analyses we will perform include calculating the mean of the data set, the standard deviation of the data, and the calculated noise in the ADC channel represented as both voltages and bits of error. This type of analysis is quantitative.

Since the RCM3400 ADC accepts input voltages in the maximum range of 0 V through 20 V and returns a code based on the input voltage that ranges from 0 to 2047 (11 bits), each ADC code represents approximately (20 V–0 V) / 2048 = 9.8 mV.

Additionally, the ADC has a gain control that rescales the input signal: with a Gain of 2, the input signal ranges from 0 V to 10 V, so each ADC code would represent 4.9 mV. Gains of 4, 5, 8, 10, 16, and 20 are available, as well as a gain of 1, which corresponds to the 0 V–20 V range, discussed above. As the gain increases, the range of input voltage decreases accordingly.

Our experimental precision reference based on the AD780 generates a constant 2.500 V signal. We can use gains of 1, 2, 4, or 5 to measure the signal. If the gain was set any higher, the maximum input voltage would be lower than 2.500 V and thus our input signal would be out of range.

Once we know the gain setting of the ADC, we can convert between input voltages and ideal ADC codes by multiplying or dividing the size of each code by the appropriate amount. For example, with a gain of 1, a 3.3 V input should produce the ADC code of (3.3 V / 9.8 mV per code) = 337. Similarly, an ADC code of 1250 should correspond to an input voltage of (1250 * 9.8 mV per code) = 12.25 V.

The first part of our data analysis will compute the average returned ADC code for the input voltage. This will likely differ somewhat from the expected ADC code that we can calculate above using the gain and resolution of the converter. The difference between the expected value and the mean is called an "offset." We can express offset in ADC codes or in volts.

The standard deviation (σ) of the data is a measure of how spread out the data is from the mean. σ is conveniently mathematically equal to the RMS noise of the ADC channel. We can measure this noise in ADC codes, or in Volts. The base-2 logarithm of the standard deviation is the RMS noise in bits, and if we subtract that from the advertised resolution of the ADC we can obtain the effective resolution of the channel in bits.

The overall peak-to-peak noise in the sample set is taken to be $\pm 3 * \sigma$ (in ADC codes). Some texts take $\pm 3.3\sigma$ to be the peak-to-peak noise. We can convert this into bits of peak-to-peak noise via the base-2 logarithm of $3 * \sigma$, and thus we compute the "noise-free resolution" of the ADC channel by subtracting the bits of noise from the advertised resolution of the ADC.

The quantitative characterization techniques presented here are useful in comparing one system to another as long as the engineer consistently applies the same functions to data from different systems.

9.5.2.5 Linux Data Reduction Program Listing

Listing 9.2 shows the Perl script used to calculate the histogram of the ADC sample data, as well as the various statistical data we are looking for to characterize the noise on the ADC channel.

Perl comes standard with most Linux distributions. It is also available for free download for both Linux and Windows® from www.ActiveState.com.

The script begins by loading the Perl Module 'Statistics::Lite'. This module is available from the Comprehensive Perl Archive Network at www.cpan.org. This module provides simple access to standard deviation, mean, and variance functions over a data set.

The script continues by declaring a hash table to store a count of how many times we have seen each ADC code, and by declaring a list to hold the complete data set. Next, two constants are declared to hold the desired input voltage and the number of bits of resolution of the ADC.

Next, the script pulls the name of the data file and the maximum input range (in volts) from the command line. This allows us to reuse the script on different data files and with different gain settings.

Next, the script gets to work by resetting a counter of the number of data lines read, and proceeds to open the input file. The script loops through each line of the input file using the <> operator. For each line of input, we try to match the input line to a regular expression of the form:

$0xXXXX,^

where:

$ means the beginning of the line
0x are literal characters
X means a nonspace character (letter or digit)
, is a literal character
^ means the end of the line

The exact regular expression in the if statement is slightly different—it includes optional whitespace following the comma, and if successful, it saves off the four *X* characters into a special variable. If the if statement succeeds, then the matched part of the expression is used as the ADC code, and we save that into the variable $new_data. $new_data is a string that begins with "0x…." and while that's a legal hash table key for Perl, it's more convenient to convert that string to an integer before we use it as a key. We use the Perl function oct() to do this and save the integer into the variable named $num. Then $num is used as a key into the hash table, and we update the count stored at that location in the table. Then we insert the $new_data into our complete list of all data read stored in @data. Finally, we increment the number of lines read in the variable $count. When we've read all the data in the input file, the while loop terminates and we close the input file.

The script generates the histogram and writes it to a new file. The histogram is placed in a file with a name generated from the name of the file passed on the command line with

"-histogram" appended. We initialize a new variable that will hold the sum of all the data values we read. Then we loop through the hash table, in order of the keys. For each key, we print out the key itself, some whitespace, and the count from the hash table. We also update our running sum. When all the keys have been processed, we close the output file and print out the $sum we've calculated. It should be the same as the number of lines of input data we read above.

Next the script computes some statistics. The Statistics::Lite module generates all of the data that it can, and it saves its results in the hash named %results. The last thing the script does is print out the results.

```perl
#!/usr/bin/perl

# some handy stats functions
use Statistics::Lite qw(:all);

# we need a hash to store the codes we see
my %codes;
my @data;

# constants for this analysis
$inputvoltage = 2.500;
$ADCbits = 11;

# get the file name from the command line
$fname = $ARGV[0] or die "usage: $0 fname maxRange";
$maxRange = $ARGV[1] or die "usage: $0 fname maxRange";

open DATA, $fname or die "can't open $fname";
$count = 0;
while(<DATA>) {
  # we just want lines of the form $0xXXXX,^
  if( m/^(0x\w\w\w\w),\s*$/ ) { # if the regexp matches, save off the data
    $new_data = $1;              # update the bin
    $codes{$new_data} += 1;
    $num = oct $new_data;        # convert to an integer
    push @data, $num;            # add to our raw data array
    $count++;                    # update our count
  }
}
close DATA;

# open the output file
$fname = $fname . "-histogram";
```

Listing 9.2: Perl script used for data analysis.

```
open OFILE, „>$fname" or die "can't open output file $fname";
$sum = 0;
foreach $code (sort keys %codes) {
  print OFILE oct($code) . " " . $codes{$code} . "\n";
  $sum += $codes{$code};
}
close OFILE;

# now do the analysis
%results = statshash @data;

printf "\nAssuming an %3d bit converter with 0 to %3dV input range\
n", $ADCbits, $maxRange;
printf "Assuming a %6.3f volt precision reference\n\n",
$inputvoltage;
$voltspercode = $maxRange / (1<<$ADCbits);
printf "1 Code = %7.3g Volts\n\n", $voltspercode;

printf "Measured Mean (codes) = %9.4f\n", $results{mean};
printf "Expected Mean (codes) = %9.4f\n", ($inputvoltage /
$voltspercode);
$offset = $inputvoltage - $voltspercode * $results{mean};
printf "offset (volts)        = %7.3f Volts\n\n", $offset;

$noiseRMSvolts = $results{stddev}*$voltspercode;
$noiseRMScodes = $results{stddev};
printf "RMSnoise(ADC codes)       = %7.3f\n", $noiseRMScodes;
printf "RMSnoise(volts)           = %8.4f\n\n", $noiseRMSvolts;
printf "Noise(pk-to-pk)(ADC codes) = %7.3f\n", 3.3*$noiseRMScodes;
printf "Noise(pk-to-pk)(volts)     = %8.4f\n", 3.3*$noiseRMSvolts;
```

Listing 9.2: Continued

9.5.2.6 Linux Histogram Visualization

Gnuplot was used to plot and save the histogram graphics. Gnuplot is highly adept at plotting data from files such as these. Invoke gnuplot as follows:

```
$ gnuplot
```

This will return a report similar to:

```
        G N U P L O T
        Version 3.7 patchlevel 3
        last modified Thu Dec 12 13:00:00 GMT 2002
        System: Linux 2.4.20-20.9smp

        Copyright(C) 1986-1993, 1998-2002
        Thomas Williams, Colin Kelley and many others
```

```
      Type `help` to access the on-line reference manual
      The gnuplot FAQ is available from
      http://www.gnuplot.info/gnuplot-faq.html

  Send comments and requests for help to <info-
  gnuplot@dartmouth.edu>
     Send bugs, suggestions and mods to <bug-gnuplot@dartmouth.edu>

Terminal type set to 'unknown'
gnuplot>
```

At the gnuplot> prompt, just type the command

```
Plot "HISTOGRAMFILE" with boxes
```

where HISTOGRAMFILE is the name of the file containing the histogram data. For example,

```
plot "data.log-352-15:44:06-histogram" with boxes
```

Gnuplot can generate encapsulated postscript files (EPS) suitable for importing into word processors:

```
gnuplot> set terminal postscript eps
gnuplot> set output "histogram.eps"
gnuplot> plot "data.log-352-15:44:06-histogram" with boxes
gnuplot> set output
gnuplot> exit
```

For more information on the versatile gnuplot program, refer to the gnuplot man page or the gnuplot homepage at www.gnuplot.org.

9.5.2.7 Sample ADC Noise Quantification and Visualization

Here we put together the work in the preceding sections and characterize the first DAQ channel on the RCM3400 development board.

Since our voltage reference was 2.5 V we had the option of using ADC gain settings of 1, 2, 4, and 5. We ran the analysis for all gain settings.

The histograms showed that the data reported from the ADC was tightly clumped around the mean value. In each histogram, only two or three bins had data in them. An ideal ADC in a noiseless system would only report data in one bin. In real life systems, a histogram with only two or three bins is excellent. Figure 9.16 shows a representative histogram from our experiments. The reported ADC codes are on the horizontal axis. The number of occurrences is shown on the vertical axis.

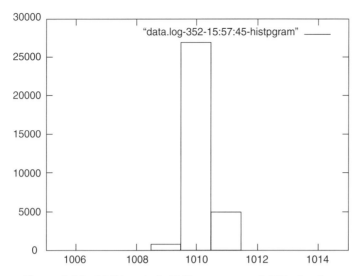

Figure 9.16: GAIN = 4, 0–5 V input range, 2.5 V stimulus.

An example of the textual output from our Perl script is shown here:

```
Assuming an 11 bit converter with 0 to 20V input range
Assuming a 2.500 volt precision reference

1 Code = 0.00977 Volts

Measured Mean (codes) = 251.9695
Expected Mean (codes) = 256.0000
offset (volts)        = 0.039 Volts

RMSnoise(ADC codes)           = 0.172
RMSnoise(volts)               = 0.0017

Noise(pk-to-pk)(ADC codes)    = 0.568
Noise(pk-to-pk)(volts)        = 0.0055v
```

The results of our four experiments are summarized in Table 9.3.

If the RMS noise on a channel is greater than 1 bit, we can compute the effective resolution of the channel from:

$$\text{RESOLUTION}_{\text{EFFECTIVE}} = \text{ADC bits} - \text{LOG}_2 (\text{Noise}_{\text{RMS}} \text{ in codes})$$

The RCM3400 has RMS noise levels so low as to render the computation of effective resolution meaningless. The logarithm becomes negative for RMS noise levels less than one code.

Table 9.3: The RCM3400 performed admirably over all the ranges measured.

	Gain of 1 0–20 V Range	Gain of 2 0–19 V Range	Gain of 4 0–5 V Range	Gain of 5 0–4 V Range
Volts per code	9.77 mV	4.88 mV	2.44 mV	1.95 mV
Measured offset	39 mV	36 mV	34 mV	32 mV
Measured RMS noise in codes	0.172	0.409	0.403	0.675
Measured RMS noise in volts	1.7 mV	2.4 mV	1.0 mV	1.3 mV
Measured pk-pk noise in codes	0.568	1.617	1.329	2.227
Measured pk-pk noise in volts	5.5 mV	7.9 mV	3.2 mV	4.3 mV

The conclusion we can draw from our experiments is that the effective resolution of the RCM3400 is a full 11 bits for gains of 1, 2, 4, and 5.

If the peak-to-peak noise on a channel is greater than 1 bit, we can compute the noise free resolution of the channel from:

$$\text{RESOLUTION}_{\text{NOISE FREE}} = \text{ADC bits} - \text{LOG}_2 (\text{Noise}_{\text{PK-PK}} \text{ in codes})$$

In our experiment, we see that the worst case peak to peak noise on the RCM3400 channel 1 occurs with a gain of 5 and is 2.227 codes. From the equation above, we compute:

$$\text{RESOLUTION}_{\text{NOISE FREE Gain=5}} = 11 - \text{LOG}_2 (2.227) = 11 - \frac{\ln(2.227)}{\ln(2)} = 9.84 \text{ bits}$$

Our analysis of the RCM3400's DAQ channel showed admirable performance. For this example, we only examined one of the eight single-ended (or four differential) DAQ channels on the RCM3400. A careful engineer will characterize all DAQ channels used in a system.

The DC analysis techniques presented here will allow an engineer to get a good feel for how accurately the ADC is reporting sensor data under ideal conditions. This is the best performance that the engineer can expect from the system without calibration or averaging data samples.

9.6 Conclusion

Companies are rolling more and more features into silicon, making the system designer's job easier. However, the designer must still exercise caution when devising an interface between real world sensors and a processor. Issues of ESD, bus loading and power consumption still exist and must be handled by the system engineer.

The Rabbit 3000 has simplified the bus-loading issue by providing an auxiliary I/O bus allowing peripheral devices to be added without unduly loading the high-speed memory bus. The core module designs address issues of memory interfacing, battery backup, system reset, power supervision, and, on some cores, analog interfacing.

Some core modules from Rabbit Semiconductor have provided low-noise DAQ channels. For example, the RCM3400 will drop right into many applications needing one to eight solid 11-bit resolution DAQ channels.

Endnote

Tinder, Richard, *Digital Engineering Design: A Modern Approach*, Prentice Hall, 1991.

CHAPTER 10
Other Useful Hardware Design Tips and Techniques

Jack Ganssle
Ken Arnold

10.1 Introduction

Embedded systems are a blend of hardware and software. Each must complement the other. Hardware people can make the firmware easier to implement.

Many of the suggestions in this chapter will make a system easier to debug. Remember that a good design works; a *great* design is also one that's easy to debug.

10.2 Diagnostics

In the nonembedded world, a favorite debugging trick is to seed print statements into the code. These tell the programmer whether the execution stream ever got to the point of the print. But firmware people rarely have this option. So, add a handful of unassigned parallel I/O bits. The firmware people desperately need these as a cheap way to instrument their code. Seeding I/O instructions into the code that drives these outputs is a simple and fast way to see what the program is doing.

Developers can assert a bit when entering a routine or ISR, then drive it low when exiting. A scope or logic analyzer then immediately shows the code snippet's execution time.

Another trick is to cycle an output bit high when the system is busy and low when idle. Connect a voltmeter to the pin, one of the old-fashioned units with an analog needle. The meter will integrate the binary pulse stream, so the displayed voltage will be proportional to system loading.

If space and costs permit, include an entire 8-bit register connected to a row of 0.1 inch spaced vias or headers. Software state machines can output their current "state" to this port. A logic analyzer captures the data and shows all the sequencing, with nearly zero impact on the code's execution time.

At least one LED is needed to signal the developer—and perhaps even customers—that the system is alive and working. It's a confidence indicator driven by a low-priority task or idle loop, which shows the system is alive and not stuck somewhere in an infinite loop. A lot of embedded systems have no user interface; a blinking LED can be a simple "system OK" indication.

Highly integrated CPUs now offer a lot of on-chip peripherals, sometimes more than we need in a particular system. If there's an extra UART, connect the pins to an RS-232 level shifting chip (e.g., MAX232A or similar). There's no need to actually load the chip onto the board except for prototyping. The firmware developers may find themselves in a corner where their tools just aren't adequate and will then want to add a software monitor (see www.simtel.com) to the code. The RS-232 port makes this possible and easy.

If PCB real estate is so limited that there's no room for the level shifter, then at least bring Tx, Rx, and ground to accessible vias so it's possible to suspend a MAX232 on green wires above the circuit board.

> **Note:** Attention, developers: if you do use this port, don't be in such a panic to implement the monitor that you implement the RS-232 drivers with polled I/O. Take the time to create decent interrupt-driven code. In my experience, polled I/O on a monitor leads to missed characters, an unreliable tool, and massive frustration.

Bring the reset line to a switch or jumper, so engineers can assert the signal independently of the normal power-up reset. Power-up problems can sometimes be isolated by connecting reset to a pulse generator, creating a repeatable scenario that's easy to study with an oscilloscope.

10.3 Connecting Tools

Orient the CPU chip so that it's possible to connect an emulator, if you're using one. Sometimes the target board is so buried inside of a cabinet that access is limited at best. Most emulator pods have form factors that favor a particular direction of insertion.

Watch out for vertical clearance, too! A pod stacked atop a large SMT adaptor might need 4 to 6 inches of space above the board. Be sure there's nothing over the top of the board that will interfere with the pod.

Don't use a "clip-on" adaptor on a SMT package. They are simply not reliable (the one exception is PLCC packages, which have a large lead pitch). A butterfly waving its wings in Brazil creates enough air movement to topple the thing over. Better, remove the CPU and install a soldered-down adaptor. The PCB will be a prototype forever, but at least it will be a reliable prototype.

Leave margin in the system's timing. If every nanosecond is accounted for, no emulator will work reliably. An extra 5 nsec or so in the read and write cycle—and especially in wait state circuits—does not impact most designs.

If your processor has a BDM or JTAG debug port, be sure to add the appropriate connector on the PCB. Even if you're planning to use a full-blown emulator or some other development tool, at least add PCB pads and wiring for the BDM connector. The connector's cost approaches zero and may save a project suffering from tool woes.

A logic analyzer is a fantastic debugging tool yet is always a source of tremendous frustration. By the time you've finished connecting 100 clip leads, the first 50 have popped off. There's a better solution: Surround your CPU with AMP's Mictor connectors. These are high-density, controlled impedance parts that can propagate the system's address, data, and control buses off-board. Both Tektronix and Agilent support the Mictor. Both companies sell cables that lead directly from the logic analyzer to a Mictor. No clip leads, no need to make custom cables, and a guaranteed reliable connection in just seconds. Remove the connectors from production versions of the board or just leave the PCB pads without loading the parts.

Some signals are especially prone to distortion when we connect tools. Address latch enable (ALE), also known as address strobe (AS) on Motorola parts, distinguishes address from data on multiplexed buses. The tiniest bit of noise induced from an emulator or even a probe on this signal will cause the system to crash. Ditto for any edge-triggered interrupt input (like NMI on many CPUs). Terminate these signals with a twin-resistor network. Though your design may be perfect without the terminations, connecting tools and probing signals may corrupt the signals.

Add test points! Unless its ground connection is very short, a scope cannot accurately display the high-speed signals endemic to our modern designs. In the good old days it was easy to solder a bit of wire to a logic device's pins to create an instant ground connection. With SMT this is either difficult or impossible, so distribute plenty of accessible ground points around the board.

Other signals we'll probe a lot and that must be accessible include clock, read, write, and all interrupt inputs. Make sure these each have either test points or a via of sufficient size that a developer can solder a wire (usually a resistor lead) to the signal.

Do add a Vcc test point. Logic probes are old but still very useful tools. Most need a power connection.

10.4 Other Thoughts

Make all output ports readable. This is especially true for control registers in ASICs because there's no way to probe these.

Be careful with bit ordering. If reading from an A/D, for instance, a bad design that flips bit 7 to input bit 0, 6 to 1, etc. is a nightmare. Sure, the firmware folks can write code to fix the mixup, but most processors aren't good at this. The code will be slow and ugly.

Use many narrow I/O ports rather than a few wide ones. When a single port controls three LEDs, two interrupt masks, and a stepper motor, changing any output means managing every output. The code becomes a convoluted mess of ANDs/ORs. Any small hardware change requires a lot of software tuning. Wide ports do minimize part counts when implemented using discrete logic, but inside a PLD or FPGA there's no cost advantage.

Avoid tying unused digital inputs directly to Vcc. In the olden days this practice was verboten, since 74LS inputs were more susceptible to transients than the Vcc pin. All unused inputs went to Vcc via resistor pull-ups. That's no longer needed with logic devices, but it is still a good practice. It's much easier to probe and change a node that's not hardwired to power.

However, if you must connect power directly to these unused inputs, be very careful with the PCB layout. Don't run power through a pin; that is, don't use the pin as a convenient way to get the supply to the other pins or to the other side of the board. It's much better to carefully run all power and ground connections to input signals as tracks on the PCB's outside layers, so they are visible when the IC is soldered in place. Then developers can easily cut the tracks with an X-Acto knife and make changes.

Pull-up resistors bring their own challenges. Many debugging tools have their own pull-ups, which can bias nodes oddly. It's best to use lower values rather than the high ones permitted by CMOS (say 10k instead of 100k).

PCB silkscreens are oft-neglected debugging aids. Label switches and jumpers. Always denote pin 1 because there's no standard pin 1 position in the SMT world. And add tick-marks every 5 or 10 pins around big SMT packages, and indicate whether pin numbers increase in a CW or CCW direction. Otherwise, finding pin 139 is a nightmare, especially for bifocal-wearing developers suffering from caffeine-induced tremors.

Key connectors so that there's no guessing about which way the cable is supposed to go.

Please add comments to your schematic diagrams! For all off-page routes, indicate the page the route goes to. Don't hide the pin numbers associated with power and ground—explicitly label these.

When the design is complete, check every input to every device and make absolutely sure that each is connected to something—even if it's not used. I have seen hundreds of systems fail in the field because an unused input drifted to an asserted state. You may expect the software folks to mask these off in the code, but that's not always possible, and even when it is, it's often forgotten.

Try to avoid hardware state machines. They're hard to debug and are often quite closely coupled to the firmware, making that, too, debug-unfriendly. It's easier to implement these completely in the code. Tools (e.g., VisualState from IAR) can automatically generate the state machine code.

10.5 Construction Methods

Embedded controllers can be constructed using any one of several techniques, but the most common method is a *printed circuit board* (PCB). The PCB is constructed of insulating material, such as epoxy impregnated glass cloth, laminated with a thin sheet of copper. Multiple layers of copper and insulating material can be laminated into a multilayer PCB. By drilling and plating holes in the material, it is possible to interconnect the layers and provide mounting locations for through-hole components.

In designing the layout, or interconnecting pattern of the PCB, there are many conflicting requirements that must be addressed to make a reliable, cost-effective, and producible device. For low-speed circuits, the parasitic effects can be ignored and are often assumed to be ideal connections. Unfortunately, real circuits are not ideal, and the wires and insulating material have an effect on the circuit, especially for signals with fast signal rise/fall times. The traces, or wires, on the PCB have stray resistance, capacitance, and inductance. At high speeds, these stray effects delay and distort the signals. Special care must be taken when designing a PC board to avoid problems with transmission line effects, noise, and unwanted electromagnetic emissions.

10.5.1 Power and Ground Planes

When possible, it is a good idea to use two layers of a four-or-more-layer PCB dedicated to the Vcc and ground signals. These are referred to as *power* and *ground planes*. One advantage is that there is a beneficial high-frequency parasitic power supply decoupling capacitance, which reduces the power supply noise to the ICs. Power planes also reduce the undesirable emission of electromagnetic radiation that can cause interference and reduce the circuit's susceptibility to externally induced noise. The power planes tend to act as a shield to reduce the susceptibility to external noise and radiation of noise from the system.

10.5.2 Ground Problems

Although the concept of an ideal circuit ground may seem relatively simple, a great many system problems can be directly traced to ground problems in actual applications. At the least, this can cause undesirable noise or erroneous operation; at the worst, it can result in safety problems, including possibly even death by electrocution. Lest you dismiss the importance of this possibility too quickly, the author has narrowly missed electrocution while testing a device in which the grounding was improperly implemented!

These problems are most often caused by one of the following problems:

- Excessive inductance or resistance in the ground circuit, resulting in "ground loops"
- Lack of or insufficient isolation between the different grounds in a system: earth, safety, digital, and analog grounds
- Nonideal grounding paths, resulting in the currents flowing in one circuit inducing a voltage in another circuit

The solutions to these problems vary, depending on the type of problem and the frequency range in which they occur. Usually they can be simplified to reducing the currents flowing in common impedances of circuits that need to remain isolated using a single point ground and the prudent application of shields and insulation to prevent unwanted parasitic signal coupling.

10.6 Electromagnetic Compatibility

Electromagnetic compatibility (EMC) issues have become much more significant now that there are a large number of electronic devices which unintentionally radiate electromagnetic energy in the same frequency ranges used for communication, navigation, and instrumentation. Regulatory agencies—such as the Federal Communications Commission (FCC) in the United States, the Department of Communications (DOC) in Canada, and similar organizations in Europe—have defined limits to the amount of energy such electronic devices are allowed to emit at various frequencies. Even more stringent requirements are placed on life-critical equipment, such as aircraft navigation and life support equipment, because of the sensitive nature of the applications. Among other things, these devices are required to provide a minimum level of immunity to externally induced noise (radiated and conducted susceptibility).

In solving an EMC problem, the first step is to identify the source of the noise, the path to the problem area, and the destination at which the problem manifests itself. Once these three characteristics of an EMC problem are identified, the engineer can evaluate the relative merits of eliminating the noise at its source, breaking the path using shielding and similar techniques, and reducing the sensitivity of the affected circuit. There are several useful resources, including publications, seminars, test labs, and consultants who specialize in solving EMC problems. The best solution is usually to begin testing a new design at the earliest possible point in the prototype phase to determine the potential problem areas so that they can be addressed with the least cost and schedule impact.

10.7 Electrostatic Discharge Effects

Electrostatic discharge (ESD) is an important design consideration in embedded applications because of the potential for failure and erroneous operation in the presence of external electric fields. ESD voltages are commonly impressed on embedded interfaces—on the order of tens of thousands of volts—when someone walks across a floor in a low-humidity environment before touching an electronic device. One of the most common places where this becomes an issue is in the keyboard or user input device, which comes in direct contact with the outside world. This effect can cause immediate damage or upset or may cause latent failures that show up months after the ESD event. Designers most often use shielding and grounding techniques similar to those used for safety and emission-reduction techniques to minimize the effects of

ESD. The same resources that are available for EMC problems are also generally of use for ESD problems.

10.7.1 Fault Tolerance

Increasingly, fault tolerance has become a requirement in embedded systems as they find their way into applications where failure is simply unacceptable. Many hardware and software solutions have been developed to address this need.

To understand how to deal with these faults, we must first identify and understand the types and nature of each type of fault. Every fault can be categorized as a "hard" or a "soft" fault. Hard faults cause an error that does not go away—for example, pushing reset or powering down does not result in recovery from the fault condition. Soft faults are due to transient events or, in some cases, program errors.

Self-test and diagnostic programs may be able to identify and diagnose the failure if it is not too severe. Depending on the type of fault that occurs and which device(s) are affected, it may be possible to design a system to detect the fault, possibly even isolating the location of the fault to some degree. In the event of a soft failure, it may be possible for the designer to make the system recover from the fault automatically.

A built-in self-test program can be written for an embedded processor that will be able to detect faults in the following types of devices:

- Processor (if the fault is not too severe)
- Memory
- ROM
- RAM
- E/EEPROM
- Peripheral devices

Note that it is difficult, if not impossible, to detect faults in the control circuits or "glue logic" in a system. Other devices, such as memories, lend themselves to diagnostic methods.

The data contents of ROM devices can be tested for errors using one or more of the following techniques:

- Parity
- Checksum
- Cyclic redundancy check (CRC)

RAM memories and the integrity of information stored in RAM by the processor can be tested for proper operation using one of the following techniques:

- Hardware error detection and correction
- Data/address pattern tests
- Data structure integrity by checking stack limits and address range validity

Additionally, the integrity of the program and proper execution sequence by the CPU can be checked using one or more of the following techniques:

- Hardware parity error detection
- Duplicate, redundant hardware and cross checking or voting
- "Watch dog" timer that operates the CPU chip's reset line
- Diagnostics that run constantly, when the CPU has nothing else to do

10.8 Hardware Development Tools

There are two general classes of hardware development tools available to the embedded developer: *passive* analysis tools which allow looking at the operation of the system, and *active* tools which allow the designer to intrude on the operation of the system while it's running (even making changes to the system's configuration and software while it is under test). The system under test is usually referred to as the "target" system, and the computer that is used to develop, edit, compile, assemble, and download the code to the target system is called the "host" system.

Passive tools include:

- Logic probes to look at static logic levels and detect pulses
- Oscilloscopes to look at signal waveforms
- Logic analyzers, with processor specific probes
- Software to assist hardware development, scope loops

Active tools include:

- In-circuit emulators (ICE) for HW/SW integration are plugged into the application circuit (the "target" system) in place of the CPU, allowing the designer to "see inside" the microcontroller, download, and execute programs selectively.
- ROM emulators (ROM ICE) allow the designer to reduce the time it takes to edit-compile-load-debug programs by replacing the program EPROM with a RAM that can be loaded quickly and easily from the host computer.

10.8.1 Instrumentation Issues

One of the most significant, but often ignored, problems designers must address is the proper selection and use of test instrumentation. Improper selection and application of these tools are frequently the source of much wasted time and confusion for the designer. Two common usage problems relate to the use of oscilloscope and logic analyzer probes.

A typical scope or logic analyzer is supplied with probes that might not be expected to have an effect on the observed signal or distort the data gathered. With input impedances in the meg-ohm range and parasitic capacitances of tens of picofarads, it might seem that the test equipment would have little or no effect on the measurement, but this is definitely not the case.

There are two common causes for measurement problems: excessive ground lead inductance and excessive capacitive loading. These things cause at the least a potential for erroneous measurements or, at worst, they can cause the circuit under test to behave differently. Two things can be done to mitigate these problems:

1. Use the shortest possible test leads, especially for the ground connection on fast logic.
2. Use high-impedance probes, especially designed for high-speed applications, such as high-speed FET input scope probes.

Other instrumentation problems can be caused by misinterpretation of the sampling effects in digital scopes, the lack of glitch detection in logic analyzers, and other obscure but potentially painful "learning experiences." These can only be avoided with a good understanding of the operation of the equipment in use and some practical experience.

10.9 Software Development Tools

Most of the software development tools available to the embedded system designer fall into one of three categories: language translators, debuggers, and utility programs that generally run on the host computer. Most of the available tools have been designed to run on the x86 architecture PC, and many are available as freeware, shareware, or low-cost commercial products for the more common target processor architecture. Translators include:

- Assembler
- Compiler
- Linker
- Interpreter

Debuggers include:

- Software/firmware monitors

- Processor In-Circuit Emulator (ICE)
- ROM ICE

Utilities include:

- PROM Programming
- Performance measurement
- Execution frequency histograms

10.10 Other Specialized Design Considerations

There are several other characteristics that the embedded system designer should become at least somewhat familiar with. These include the thermal characteristics of a system and the concept of thermal resistance, power dissipation, and the effects on device temperature and reliability. Another issue of importance in portable, handheld, and remotely located systems is the application of battery power storage.

10.10.1 Thermal Analysis and Design

The temperature of a semiconductor device, such as a voltage regulator or even a CPU chip, is a critical system operating parameter. The reliability of these devices is also closely related to temperature, so much so because the device's reliability drops *exponentially* with increasing temperature. Fortunately, calculating the operating temperature of a device is not too difficult, since there is a simple electrical circuit analogy that is most often used to compute temperature of a device. The temperature is analogous to voltage, the power dissipated is equivalent to current, and the thermal resistance is equivalent to electrical resistance. In other words:

$$\text{Temperature rise (°C)} = \text{power (watts)} * \text{thermal resistance (°C/watt)}$$

The thermal resistance of multiple mechanical components stacked one upon the other add, just as series resistors are equivalent to a single resistor equal to the sum of the individual values.

For example: Given a 5 V linear voltage regulator with a 9 V input providing 1 ampere of load current, the regulator will dissipate:

$$P = V * I = (9 - 5) \text{ volts} * 1 \text{ amp, or 4 watts, of power}$$

If the regulator is specified with a thermal resistance between the semiconductor junction and case of 1°C/watt (signified as Θjc), and the heat sink the regulator is mounted to has a thermal resistance from the regulator mounting surface to still ambient air of 10°C/watt (signified as Θca), then the total thermal resistance between the semiconductor junction and ambient air is:

$$\Theta ja = \Theta jc + \Theta ca = 1 + 10 = 11°C/watt$$

The temperature rise of the junction above that of the air surrounding the regulator will then be given by:

$$T = P * \Theta ja = 4 \text{ watts} * 11°C/watt = 44°C \text{ above ambient}$$

If the regulator was specified to operate at a maximum junction temperature of 85°C, the device should not be operated in ambient air of temperature higher than 85 − 44 = 41°C or the regulator will fail prematurely. If this is not acceptable, the designer must reduce the input voltage to reduce the power dissipated, reduce the thermal resistance by forced air flow, or change the design to another type (e.g., a switch mode regulator) so as to keep the regulator junction within operating constraints.

10.10.2 Battery-Powered System Design Considerations

The rapid increase in the use of portable, battery-operated electronic devices has spurred the development of new battery technologies for these applications. The older single-use and rechargeable battery chemistries have been supplanted by newer ones, providing improved power densities, operating life, and other enhancements. Unfortunately, these new energy storage devices come with new and different characteristics and limitations.

Batteries are generally divided into two common groups: primary (one-time discharge and discard) and secondary (rechargeable) batteries. Primary memories include the nonrechargeable alkaline and lithium cells sold commercially; secondary cells include the older lead-acid and nickel-cadmium (NiCd) chemistries as well as the newer nickel metal hydride (NiMH) and rechargeable alkaline and lithium ion chemistry products. There is also a wide range of special-purpose batteries that are optimized for some specific characteristic, such as the zinc-air primary cell, which uses atmospheric air as an "electrode" to provide very high energy density at low operating current.

Primary batteries, such as alkalines and lithium coin cells, are relatively simple to use but are often limited to one to three years of operation. This is primarily due to the shelf-life limit imposed by internal leakage current that discharges the battery slowly over time, especially at high temperatures.

The secondary, rechargeable battery types each have slightly differing charge/discharge requirements and limitations that must be considered for effective application in a battery-powered system. There are special algorithms to optimize the performance and service life of the batteries, and there are even chips designed specifically to manage the charge and discharge of common secondary battery types.

Many embedded devices must be designed to operate for long periods of time with very little power obtained from solar cells, batteries, and other limited power sources. As a result, there are CMOS processors and memories which have been designed with very low power consumption operating modes, frequently referred to as "sleep," "power down," or "idle" modes, that consume current in the μA range.

10.11 Processor Performance Metrics

In an effort to compare different types of computers, manufacturers have come up with a host of metrics to quantify processor performance.

The successful application of these devices in an embedded system usually hinges on the following characteristics:

- IPS (instructions per second)
- OPS (operations per second)
- FLOPS (floating-point OPS)
- Benchmarks (standardized and proprietary "sample programs") that are short samples indicative of processor performance in small application programs

10.11.1 IPS

The term IPS, or the more common forms, MIPS (millions of IPS) and BIPS (billions of IPS), is commonly thrown about but are essentially worthless marketing hype because they only describe the rate at which the fastest instruction executes on a machine. Often that instruction is the NOP instruction, so 500 MIPS may mean that the processor can do nothing 500 million times per second!

10.11.2 OPS

In response to the weakness in the IPS measurement, OPS (as well as MOPS and BOPS, which sound fun at least) are instruction execution times based on a mix of different instructions. The intent is to use a standard execution frequency weighted instruction mix that more accurately represents the "nominal" instruction execution time. FLOPS (megaFLOPS, giga-FLOPS, etc.) are similar except that they weight floating-point instructions heavily to represent heavy computational applications, such as continuous simulations and finite element analysis. The problem with the OPS metric is that the resulting number is heavily dependent on the instruction mix that is used to compute it, which may not accurately represent the intended application instruction execution frequency.

10.11.3 Benchmarks

Benchmarks are short, self-contained programs that perform a critical part of an application—such as a sorting algorithm—and are used to compare functionally equivalent code on different machines. The programs are run for some number of iterations, and the time is measured and compared with that of other CPUs. The weakness here is that the benchmark is not only a measure of the processor but also of the programmer and the tools used to implement the program. As a result, the best benchmark is the one you write yourself, since it allows you to discover how efficiently the code you write will execute on a given processor with the tools available. That's as close to the real application performance as you're likely to get, short of fully implementing the application on each processor under evaluation.

APPENDIX A
Schematic Symbols

Tammy Noergaard

These symbols are a subset of industry-accepted schematic symbols representing electronic elements on schematic diagrams. Note that symbols for the same electronic device can differ internationally as well as depending on the standards being adhered to by a particular organization (NEMA, IEEE, JEDEC, ANSI, IEC, DoD, etc.). If there are any unfamiliar symbols within a schematic, it is always best to ask the engineer responsible for drafting the schematic.

AC Voltage Source		Voltage source that generates alternating current (AC). Because an AC voltage source can come from a variety of components (outlet, oscillator, signal generator, etc.), the type of AC source is typically stated somewhere on the schematic.
Antenna Balanced General Loop (Shielded) Loop (Unshielded) Unbalanced		A transducer made up of conductive material (i.e., wires, metal rod, etc.) used to transmit and receive wireless signals (i.e., radio waves, IR, etc.).
Attenuator Fixed Variable		Commonly used for a variety of purposes, including to extend the dynamic range of certain devices (i.e., power meters, amplifiers, etc.), reduce signal levels, match circuits, and balance out unequal signal levels in transmission lines, just to name a few.
Battery/DC Cell		Voltage source that creates voltage through a chemical reaction in a battery.

Buffer {Amplifier}	▷	An electrical device that is used to provide compatibility between two signals (i.e., interfacing the output of a CMOS to the input of a TTL).
Capacitors Non-Polarized General Feedthrough Non-Polarized/Bipolar Fixed Polarized Fixed (Electrolytic) Variable Single Split-Stator	─┤├─ ─┴─ ─┤(─ ─┤(─ ─┤(─ ─┤(┤(─	A passive electrical element that stores electric charge in a circuit. The feedthrough capacitor is uniquely constructed to provide lower parallel inductance, better decoupling capability for all high di/dt environments, significant noise reduction in digital circuits, EMI suppression, broadband I/O filtering, Vcc power line conditioning in comparison to other types of capacitors. A non-polar/bipolar fixed capacitor has no "implicit" polarity, thus can be connected in any way into a circuit. A fixed polarized has an "explicit" polarity, thus there is only one way to connect it into a circuit. A variable capacitor has capacitance that can be varied on the fly. The split-stator capacitor is a variable capacitor used to preserve balance in a circuit.
Cathode Cold Directly Heated Indirectly Heated	ǀ ⋀ ⋀ ⌐	[1] The positively charged pole (terminal) of a voltage source. [2] The negatively charged electrode of a device (i.e., diode) that acts as an electron source.
Cavity Resonator	─[oo]─	A component that contains and maintains an oscillating electromagnetic field.
Circuit Breaker (single pole)	─⌒─	An electrical component that ensures that a current load doesn't get too large by shutting down the circuit when its overheat sensor senses there is too much current.
Coaxial Cable	⌀ ═══	A type of cabling made up of two layers of physical wire, one center wire and one grounded wire shielding. Coaxial cables also include two layers of insulation, one between the wire shielding and center wire

			and one layer above the wire shielding. The shielding allows for a decrease in interference (electrical, RF, etc.).
Connector Female Male			An electrical component that interconnects different types of subsystems.
Crystal			An electrical component that determines an oscillator's frequency. A crystal is typically made up of two metal plates separated by quartz, with two terminals attached to each plate. The quartz within a crystal vibrates when current is applied to the terminals, and it is this frequency that impacts the frequency at which the oscillator operates.
Delay Line			An electrical component that delays the transmission of a signal.
Diode Diode Light-Emitting Diode [LED] Photodiode/Photosensitive Zener			Two-terminal semiconductor device that allows current flow in one direction and blocks current flow in the opposite direction. Diode is typically cheaper and more common, made of silicon or germanium. All diodes emit light. LEDs are made from special semiconductive material, which optimizes the light. The photodiode optimizes the fact that diodes are light sensitive, i.e., solar cells that convert light into electrical energy. The zener diode is designed with a specific reverse-breakdown voltage that causes a specific amount of resistance when blocking current flow.
Flip-Flop RS JK D			Flip-flops are sequential circuits that are called such because they function by alternating (flip-flopping) between two output states (0 and 1) depending on the input. The RS flip-flop alternates between the two output lines (Q and Q NOT) depending on the R and S inputs.

			The JK flip-flop alternates between the two output lines (Q and Q NOT) depending on the J and K inputs as well as the clock signal (C).
			The D flip-flop alternates between the two output lines (Q and Q NOT) depending on the D input as well as the clock signal (C).
Fuse			An electrical component that protects a circuit from too much current by breaking the circuit when a high enough current passes through it.
Gates AND OR NOT/inverter NAND NOR XOR	Standard	NEMA	ANSI	A more complex type of electronic switching circuit designed to perform logical binary operations. An AND gate's output is 1 when both inputs are 1. An OR gate's output is 1 if either of the inputs is 1. A NOT/inverter is an electrical device that inverts (i.e., a HIGH to a LOW or vice versa) a logical level input. A NAND gate's output is 0 when both inputs are 1. A NOR gate's output is 0 either of the inputs are 1. A XOR gate's output is 1 (or on, or high, etc.) if only one input (but not both) is 1.
Ground Circuit Earth Special			An arbitrary point for "0" potential voltage that a circuit is connected to.
Inductor (Coil) Air Core Iron Core Tapped Variable			An electrical component made up of coiled wire surrounding some type of core (air, iron, etc.). When a current is applied to a conductor, energy is stored in the magnetic field surrounding the coil, allowing for an energy storing and filtering effect.

Integrated Circuit (IC) Generic		An electrical device made up of several other discrete electrical active elements, passive elements, and devices (transistors, resistors, etc.), all fabricated and interconnected on a continuous substrate (chip).
Jack Coaxial 2 Conductor 3 Conductor Phono		An electrical device designed to accept a plug
Lamp Incandescent Neon Xenon Flash		An electrical device that produces light. An incandescent lamp produces light via heat. A neon lamp produces light via neon gas. Xenon flash lamps produces large flashes of bright white light via some combination that includes high voltage, electrodes, and gas.
Loudspeaker		A type of transducer that coverts variations of electrical current into sound waves.
Meter Ammeter Galvanometer Voltmeter Wattmeter		A measurement device that measures some from of electrical energy. An ammeter is a meter that measures in a current circuit. A galvanometer is a meter that measures smaller amounts of current in a circuit. A voltmeter is a meter that measures voltage. A wattmeter is a meter that measures power.

Microphone Condenser Microphone Dynamic Electret ECM Microphone		A type of transducer that converts sound waves into electrical current. A condenser microphone uses changes in capacitance in proportion to changes in sound waves to produce its conversions. A dynamic microphone uses a coil that vibrates to sound waves, and a magnetic field to generate a voltage that varies in proportion to sound variations. An electret microphone is dynamic and uses a small transistor amplifier.
Plug 2-Conductor 3-Conductor Phono/RCA		Electrical components used to connect one subsystem into the jack of another subsystem
Rectifier Semiconductor Silicon-Controlled (thyristor) Tube-Type		A four-layer PNPN (3 P-N junction) device that functions as a cross between a diode and transistor.
Relay DPDT DPST SPDT SPST		An electromagnetic switch. A Double Pole Double Throw (DPDT) relay contains two contacts that can be toggled both ways (on and off). A Double Pole Single Throw (DPST) relay contains two contacts that can only be switched on or off. A Single Pole Double Throw (SPDT) relay contains one contact that can be toggled both ways (on and off). A Single Pole Single Throw (SPST) relay contains one set of contacts and can only be switched one way (on or off).

Schematic Symbols

Resistor		Used to limit current in a circuit.
Fixed	USA/Japan, Europe	Fixed resistors have resistance value set at manufacturing.
		Variable resistors have a dial that allows a change in resistance values on the fly.
Variable/Potentiometers	USA/Japan, Europe	Potentiometer Variable-resistance control, similar to potentiometer but with three discrete areas of control. The part of the circuit connected off the arrow can be varied in resistance to the two circuit points connected to the other two leads.
Rheostat	Europe	
Photosensitive/Photoresistor		Photosensitive resistors have resistance that changes on the fly depending on the amount of light photo resistor are exposed to.
Thermally Sensitive/Thermistor		Thermistors have a resistance changes on the fly depending on the temperature the thermistor is exposed to (typically resistance decreases as temperature increases).
Switch		An electrical device is used to turn an electrical current flow on or off.
Single Pole Single Throw		SPST switch contains one set of contacts that can only be switched on or off (one way).
Single Pole Double Throw		SPDT switch contains one contact that can be toggled on and off (both ways).
Double Pole Single Throw		DPST switch contains two contacts that can only be switched one way (on or off).
Double Pole Double Throw		DPDT contains two contacts that can be toggled on and off (both ways).
Normally Closed Push Button		A normally closed push-button switch is a switch in the form of a button that is normally closed.
Normally Open Push Button		A normally open push-button switch is a switch in the form of a button that is normally open.
Thermocouple		An electronic circuit that relays temperature differences via current flowing through two wires joined at either end. Each wire is made of different materials, with one junction of the connected wires at the stable lower temperature while the other junction is connected at the temperature to measured.

Component	Symbol	Description
Transformer Air Core Iron Core Tapped Primary Tapped Secondary		A type of inductor that can increase or decrease the voltage of an AC signal.
Transistor		Three-terminal semiconductor device that provides current amplification as well as can acts as a switch.
Bipolar/BJT (Bipolar Junction Transistor)	NPN, PNP	A bipolar transistor is made of alternating P type and N type semiconductive material (meaning both positive and negative charges used to conduct, hence the name "bipolar").
Junction FET (Field Effect Transistor)	N Channel, P Channel	A junction FET is also made up of both N type and P type material, however unipolar, involving only positive or negative charges to conduct. Gate voltage applied across P-N Junction.
MOSFET (Metal Oxide Semiconductor FET)	N Channel Depletion, N Channel Enhancement, P Channel Depletion, P Channel Enhancement	A MOSFET is similar to Junction FET except gate voltage applied across insulator.
Photosensitive (phototransistor)		A photosensitive transistor is a bipolar transistor designed to leverage a transistor's sensitivity to light.
Wire Wire		Wires are conductors that carry signals between the other components on a board.
Wires Crossing and Connected		The wires crossing and connected symbol represents two connected wires.
Wires Crossing and Unconnected		The wires crossing and unconnected symbol represents two wires crossing on the board but not connected.

APPENDIX B
Acronyms and Abbreviations

Tammy Noergaard

A

AC	Alternating Current
ACK	Acknowledge
A/D	Analog-to-Digital
ADC	Analog-to-Digital Converter
ALU	Arithmetic Logic Unit
AM	Amplitude Modulation
AMP	Ampere
ANSI	American National Standards Institute
AOT	Ahead of Time
API	Application Programming Interface
ARIB-BML	Association of Radio Industries and Business of Japan
AS	Address Strobe
ASCII	American Standard Code for Information Interchange
ASIC	Application-Specific Integrated Circuit
ATM	Asynchronous Transfer Mode, Automated Teller Machine
ATSC	Advanced Television Standards Committee
ATVEF	Advanced Television Enhancement Forum

B

BDM	Background Debug Mode
BER	Bit Error Rate
BIOS	Basic Input/Output System
BML	Broadcast Markup Language
BOM	Bill of Materials
bps	Bits per Second
BSP	Board Support Package
BSS	Block Started by Symbol, Block Storage Segment, Blank Storage Space, ...

C

CAD	Computer-Aided Design
CAN	Controller Area Network
CAS	Column Address Select
CASE	Computer-Aided Software Engineering
CBIC	Cell-Based IC or Cell-Based ASIC
CDC	Connected Device Configuration
CEA	Consumer Electronics Association
CEN	European Committee for Standardization
CISC	Complex Instruction Set Computer
CLDC	Connected Limited Device Configuration
CMOS	Complementary Metal Oxide Silicon
CPU	Central Processing Unit
COFF	Common Object File Format
CPLD	Complex Programmable Logic Device
CRT	Cathode Ray Tube
CTS	Clear to Send

D

DAC	Digital-to-Analog Converter
DAG	Data Address Generator
DASE	Digital TV Applications Software Environment
DAVIC	Digital Audio Visual Council
dB	Decibel
DC	Direct Current
D-Cache	Data Cache
DCE	Data Communications Equipment
Demux	Demultiplexor
DHCP	Dynamic Host Configuration Protocol
DIMM	Dual Inline Memory Module
DIP	Dual Inline Package
DMA	Direct Memory Access
DNS	Domain Name Server, Domain Name System, Domain Name Service
DPRAM	Dual-Port RAM

DRAM	Dynamic Random Access Memory
DSL	Digital Subscriber Line
DSP	Digital Signal Processor
DTE	Data Terminal Equipment
DTVIA	Digital Television Industrial Alliance of China
DVB	Digital Video Broadcasting

E

EDA	Electronic Design Automation
EDF	Earliest Deadline First
EDO RAM	Extended Data Out Random Access Memory
EEMBC	Embedded Microprocessor Benchmarking Consortium
EEPROM	Electrically Erasable Programmable Read-Only Memory
EIA	Electronic Industries Alliance
ELF	Extensible Linker Format
EMI	Electromagnetic Interference
EPROM	Erasable Programmable Read-Only Memory
ESD	Electrostatic Discharge
EU	European Union

F

FAT	File Allocation Table
FCFS	First Come, First Served
FDA	Food and Drug Administration (USA)
FDMA	Frequency Division Multiple Access
FET	Field Effect Transistor
FIFO	First In, First Out
FFS	Flash File System
FM	Frequency Modulation
FPGA	Field Programmable Gate Array
FPU	Floating-Point Unit
FSM	Finite State Machine
FTP	File Transfer Protocol

G

GB	Gigabyte
GBit	Gigabit
GCC	GNU C Compiler
GDB	GNU Debugger
GHz	Gigahertz
GND	Ground
GPS	Global Positioning System
GUI	Graphical User Interface

H

HAVi	Home Audio/Video Interoperability
HDL	Hardware Description Language
HL7	Health Level Seven
HLDA	Hold Acknowledge
HLL	High-Level Language
HTML	HyperText Markup Language
HTTP	HyperText Transport Protocol
Hz	Hertz

I

IC	Integrated Circuit
I2C	Inter-Integrated Circuit Bus
I-Cache	Instruction Cache
ICE	In-Circuit Emulator
ICMP	Internet Control Message Protocol
IDE	Integrated Development Environment
IEC	International Engineering Consortium
IEEE	Institute of Electrical and Electronics Engineers
IETF	Internet Engineering Task Force
IGMP	Internet Group Management Protocol
INT	Interrupt
I/O	Input/Output
IP	Internet Protocol
IPC	Interprocess Communication

IR	Infrared
IRQ	Interrupt Request
ISA	Instruction Set Architecture
ISA Bus	Industry Standard Architecture Bus
ISO	International Standards Organization
ISP	In-System Programming
ISR	Interrupt Service Routine
ISS	Instruction Set Simulator
ITU	International Telecommunication Union

J

JIT	Just in Time
J2ME	Java 2 MicroEdition
JTAG	Joint Test Access Group
JVM	Java Virtual Machine

K

kB	Kilobyte
kbit	Kilobit
kbps	Kilo bits per second
kHz	Kilohertz
KVM	K Virtual Machine

L

LA	Logic Analyzer
LAN	Local Area Network
LCD	Liquid Crystal Display
LED	Light Emitting Diode
LIFO	Last In, First Out
LSb	Least Significant Bit
LSB	Least Significant Byte
LSI	Large-Scale Integration

M

mΩ	Milliohm
MΩ	Megaohm
MAN	Metropolitan Area Network
MCU	Microcontroller
MHP	Multimedia Home Platform
MIDP	Mobile Information Device Profile
MIPS	Millions of Instructions Per Second, Microprocessor without Interlocked Pipeline Stages
MMU	Memory Management Unit
MOSFET	Metal Oxide Silicon Field Effect Transistor
MPSD	Modular Port Scan Device
MPU	Microprocessor
MSb	Most Significant Bit
MSB	Most Significant Byte
MSI	Medium-Scale Integration
MTU	Maximum Transfer Unit
MUTEX	Mutual Exclusion

N

nSec	Nanosecond
NAK	Not Acknowledged
NAT	Network Address Translation
NCCLS	National Committee for Clinical Laboratory Standards
NFS	Network File System
NIST	National Institute of Standards and Technology
NMI	Nonmaskable Interrupt
NTSC	National Television Standards Committee
NVRAM	Nonvolatile Random Access Memory

O

OCAP	Open Cable Application Forum
OCD	On-Chip Debugging
OEM	Original Equipment Manufacturer
OO	Object Oriented
OOP	Object-Oriented Programming

OS	Operating System
OSGi	Open Systems Gateway Initiative
OSI	Open Systems Interconnection
OTP	One-Time Programmable

P

PAL	Programmable Array Logic, Phase Alternating Line
PAN	Personal Area Network
PC	Personal Computer
PCB	Printed Circuit Board
PCI	Peripheral Component Interconnect
PCP	Priority Ceiling Protocol
PDA	Personal Data Assistant
PDU	Protocol Data Unit
PE	Presentation Engine, Processing Element
PID	Proportional Integral Derivative
PIO	Parallel Input/Output
PIP	Priority Inheritance Protocol, Picture-in-Picture
PLC	Programmable Logic Controller, Program Location Counter
PLD	Programmable Logic Device
PLL	Phase Locked Loop
POSIX	Portable Operating System Interface X
POTS	Plain Old Telephone Service
PPC	PowerPC
PPM	Parts Per Million
PPP	Point-to-Point Protocol
PROM	Programmable Read-Only Memory
PSK	Phase Shift Keying
PSTN	Public Switched Telephone Network
PTE	Process Table Entry
PWM	Pulse Width Modulation

Q

QA	Quality Assurance

R

RAM	Random Access Memory
RARP	Reverse Address Resolution Protocol
RAS	Row Address Select
RF	Radio Frequency
RFC	Request for Comments
RFI	Radio Frequency Interference
RISC	Reduced Instruction Set Computer
RMA	Rate Monotonic Algorithm
RMS	Root Mean Square
ROM	Read-Only Memory
RPM	Revolutions Per Minute
RPU	Reconfigurable Processing Unit
RTC	Real-Time Clock
RTOS	Real-Time Operating System
RTS	Request to Send
RTSJ	Real-Time Specification for Java
R/W	Read/Write

S

SBC	Single-Board Computer
SCC	Serial Communications Controller
SECAM	Système Électronique pour Couleur avec Mémoire
SEI	Software Engineering Institute
SIMM	Single Inline Memory Module
SIO	Serial Input/Output
SLD	Source-Level Debugger
SLIP	Serial Line Internet Protocol
SMPTE	Society of Motion Picture and Television Engineers
SMT	Surface Mount
SNAP	Scalable Node Address Protocol
SNR	Signal-to-Noise Ratio
SoC	System-on-Chip
SOIC	Small Outline Integrated Circuit
SPDT	Single Pole Double Throw
SPI	Serial Peripheral Interface

SPST	Single Pole Single Throw
SRAM	Static Random Access Memory
SSB	Single Sideband Modulation
SSI	Small-Scale Integration

T

TC	Technical Committee
TCB	Task Control Block
TCP	Transmission Control Protocol
TDM	Time Division Multiplexing
TDMA	Time Division Multiple Access
TFTP	Trivial File Transfer Protocol
TLB	Translation Lookaside Buffer
TTL	Transistor-Transistor Logic

U

UART	Universal Asynchronous Receiver/Transmitter
UDM	Universal Design Methodology
UDP	User Datagram Protocol
ULSI	Ultra Large-Scale Integration
UML	Universal Modeling Language
UPS	Uninterruptible Power Supply
USA	United States of America
USART	Universal Synchronous-Asynchronous Receiver-Transmitter
USB	Universal Serial Bus
UTP	Untwisted Pair

V

VHDL	Very High-Speed Integrated Circuit Hardware Design Language
VLIW	Very Long Instruction Word
VLSI	Very Large-Scale Integration
VME	VersaModule Eurocard
VoIP	Voice-Over-Internet Protocol
VPN	Virtual Private Network

W

WAN	Wide Area Network
WAT	Way Ahead of Time
WDT	Watchdog Timer
WLAN	Wireless Local Area Network
WML	Wireless Markup Language
WOM	Write-Only Memoryv

X

XCVR	Transceiver
XHTML	Extensible HyperText Markup Language
XML	Extensible Markup Language

APPENDIX C
PC Board Design Issues

Walt Kester

C.1 Introduction

Printed circuit boards (PCBs) are by far the most common method of assembling modern electronic circuits. Composed of a sandwich of insulating layer (or layers) and one or more copper conductor patterns, they can introduce various forms of errors into a circuit, particularly if the circuit is operating at either high precision or high speed. PCBs, then, act as "unseen" components wherever they are used in precision circuit designs. Since designers don't always consider the PCB electrical characteristics as additional components of their circuit, overall performance can easily end up worse than predicted. This general topic, manifested in many forms, is the focus of this appendix.

PCB effects that are harmful to precision circuit performance include leakage resistances; spurious voltage drops in trace foils, vias, and ground planes; the influence of stray capacitance, dielectric absorption (DA), and the related "hook." In addition, the tendency of PCBs to absorb atmospheric moisture, hygroscopicity, means that changes in humidity often cause the contributions of some parasitic effects to vary from day to day.

In general, PCB effects can be divided into two broad categories: those that most noticeably affect the static or dc operation of the circuit and those that most noticeably affect dynamic or AC circuit operation.

Another very broad area of PCB design is the topic of grounding. Grounding is a problem area in itself for all analog designs, and it can be said that implementing a PCB-based circuit doesn't change that fact. Fortunately, certain principles of quality grounding, namely the use of ground planes, are intrinsic to the PCB environment. This factor is one of the more significant advantages to PCB-based analog designs, and an appreciable amount of this appendix is focused on this issue.

Some other aspects of grounding that must be managed include the control of spurious ground and signal return voltages that can degrade performance. These voltages can be due to external signal coupling, common currents, or simply excessive IR drops in ground conductors. Proper conductor routing and sizing as well as differential signal handling and ground isolation techniques enable control of such parasitic voltages.

One final area of grounding to be discussed is grounding appropriate for a mixed-signal, analog/digital environment. This topic is the subject of many application calls, and it is certainly true that interfacing with ADCs (or DACs) is a major part of the system design, and thus it shouldn't be overlooked. Indeed, the single issue of quality grounding can drive the entire layout philosophy of a high-performance mixed-signal PCB design—as well it should.

C.2 Resistance of Conductors

Every engineer is familiar with resistors, although perhaps fewer are aware of their idiosyncrasies. But too few engineers consider that all the wires and PCB traces with which their systems and circuits are assembled are also resistors. In higher-precision systems, even these trace resistances and simple wire interconnections can have degrading effects. Copper is not a superconductor—and too many engineers appear to think it is!

Figure C.1 illustrates a method of calculating the sheet resistance R of a copper square, given the length Z, the width X, and the thickness Y.

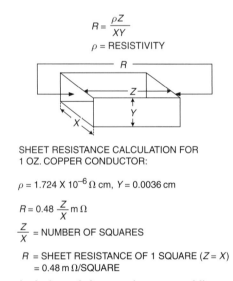

Figure C.1: Calculation of sheet resistance and linear resistance for standard copper PCB conductors.

At 25°C the resistivity of pure copper is $1.724 \times 10^{-6}\,\Omega\,cm$. The thickness of standard 1-ounce PCB copper foil is 0.036 mm (0.0014″). Using the relations shown, the resistance of such a standard copper element is therefore 0.48 Ω/square. One can readily calculate the resistance of a linear trace by effectively "stacking" a series of such squares end to end, to make up the line's length. The line length is Z and the width is X, so the line resistance R is simply a product of Z/X and the resistance of a single square, as noted in the figure.

For a given copper weight and trace width, a resistance/length calculation can be made. For example, the 0.25 mm (10 mil) wide traces frequently used in PCB designs equates to a resistance/length of about 19 mΩ/cm (48 mΩ/inch), which is quite large. Moreover, the temperature coefficient of resistance for copper is about 0.4%/°C around room temperature. This is a factor that shouldn't be ignored, in particular within low-impedance precision circuits, where the TC can shift the net impedance over temperature.

As shown in Figure C.2, PCB trace resistance can be a serious error when conditions aren't favorable. Consider a 16-bit ADC with a 5 kΩ input resistance, driven through 5 cm of 0.25 mm wide 1 oz PCB track between it and its signal source. The track resistance of nearly 0.1 Ω forms a divider with the 5 kΩ load, creating an error. The resulting voltage drop is a gain error of 0.1/5000 (~0.0019%), well over 1 LSB (0.0015% for 16 bits).

So, when dealing with precision circuits, the point is made that even simple design items such as PCB trace resistance cannot be dealt with casually. There are various solutions to address this issue, such as wider traces (which may take up excessive space), the use of heavier copper (which may be too expensive), or simply choosing a high-impedance converter. But the most important thing is to think it all through, avoiding any tendency to overlook items that appear innocuous on the surface.

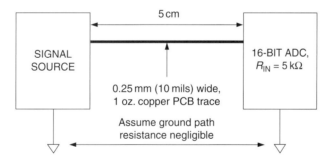

Figure C.2: Ohm's Law predicts >1 LSB of error due to drop in PCB conductor.

C.3 Voltage Drop in Signal Leads—"Kelvin" Feedback

The gain error resulting from resistive voltage drop in PCB signal leads is important only with high precision and/or at high resolutions (the Figure C.2 example) or where large signal currents flow. Where load impedance is constant and resistive, adjusting overall system gain can compensate for the error. In other circumstances, it may often be removed by the use of "Kelvin" or "voltage sensing" feedback, as shown in Figure C.3.

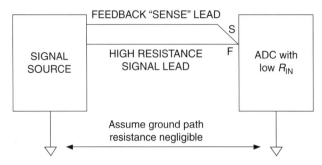

Figure C.3: Use of a sense connection moves accuracy to the load point.

In this modification to the case of Figure C.2, a long resistive PCB trace is still used to drive the input of a high-resolution ADC, with low input impedance. In this case, however, the voltage drop in the signal lead does *not* give rise to an error, because feedback is taken directly from the input pin of the ADC and returned to the driving source. This scheme allows full accuracy to be achieved in the signal presented to the ADC, despite any voltage drop across the signal trace.

The use of separate force (F) and sense (S) connections at the load removes any errors resulting from voltage drops in the force lead, but of course may only be used in systems where there is negative feedback. It is also impossible to use such an arrangement to drive two or more loads with equal accuracy, since feedback may only be taken from one point. Also, in this much-simplified system, errors in the common lead source/load path are ignored, the assumption being that ground path voltages are negligible. In many systems this may not necessarily be the case, and additional steps may be needed, as noted below.

C.4 Signal Return Currents

Kirchoff's Law tells us that at any point in a circuit the algebraic sum of the currents is zero. This tells us that all currents flow in circles and, particularly, that the return current must always be considered in analyzing a circuit, as is illustrated in Figure C.4 (see References 7 and 8).

In dealing with grounding issues, common human tendencies provide some insight into the way the correct thinking about the circuit can be helpful in analysis. Most engineers readily consider the ground return current, *I*, *when they are considering a fully differential circuit.*

However, in considering the more usual circuit case, where a single-ended signal is referred to "ground," it is common to assume that all the points on the circuit diagram where ground symbols are found are at the same potential. Unfortunately, this happy circumstance "just ain't necessarily so."

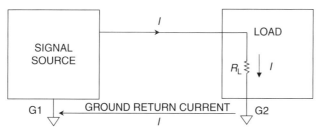

Figure C.4: Kirchoff's Law helps in analyzing voltage drops around a complete source/load coupled circuit.

This overly optimistic approach is illustrated in Figure C.5, where, if it really should exist, "infinite ground conductivity" would lead to zero ground voltage difference between source ground G1 and load ground G2. Unfortunately this approach isn't a wise practice, and when we're dealing with high precision circuits, it can lead to disasters.

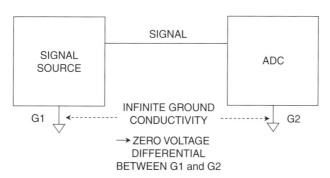

Figure C.5: Unlike this optimistic diagram, it is unrealistic to assume infinite conductivity between source/load grounds in a real-world system.

A more realistic approach to ground conductor integrity includes analysis of the impedance(s) involved and careful attention to minimizing spurious noise voltages. A more realistic model of a ground system is shown in Figure C.6. The signal return current flows in the complex impedance existing between ground points G1 and G2 as shown, giving rise to a voltage drop ΔV in this path. But it is important to note that additional external currents, such as I_{EXT}, may also flow in this same path. It is critical to understand that such currents may generate

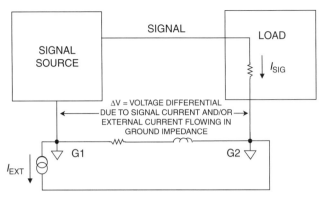

Figure C.6: A more realistic source-to-load grounding system view includes consideration of the impedance between G1 and G2, plus the effect of any nonsignal-related currents.

uncorrelated noise voltages between G1 and G2 (dependent upon the current magnitude and relative ground impedance).

Some portion of these undesired voltages may end up being seen at the signal's load end, and they can have the potential to corrupt the signal being transmitted.

C.5 Grounding in Mixed Analog/Digital Systems

Today's signal processing systems generally require mixed-signal devices such as analog-to-digital converters (ADCs) and digital-to-analog converters (DACs) as well as fast digital signal processors (DSPs). Requirements for processing analog signals having wide dynamic ranges increases the importance of high-performance ADCs and DACs. Maintaining wide dynamic range with low noise in hostile digital environments is dependent upon using good high-speed circuit design techniques, including proper signal routing, decoupling, and grounding.

In the past, "high-precision, low-speed" circuits have generally been viewed differently than so-called "high-speed" circuits. With respect to ADCs and DACs, the sampling (or update) frequency has generally been used as the distinguishing speed criteria. However, the following two examples show that in practice, most of today's signal processing ICs are really "high speed" and must therefore be treated as such in order to maintain high performance. This is certainly true of DSPs as well as ADCs and DACs.

All sampling ADCs (ADCs with an internal sample-and-hold circuit) suitable for signal processing applications operate with relatively high-speed clocks with fast rise and fall times (generally a few nanoseconds) and must be treated as high-speed devices, even though throughput rates may appear low. For example, a medium-speed 12-bit successive approximation (SAR) ADC may operate on a 10 MHz internal clock, although the sampling rate is only 500 kSPS.

Sigma-delta (Σ-Δ) ADCs also require high-speed clocks because of their high oversampling ratios. Even high-resolution, so-called "low-frequency" Σ-Δ industrial measurement ADCs (having throughputs of 10 Hz to 7.5 kHz) operate on 5 MHz or higher clocks and offer resolution to 24 bits (for example, the Analog Devices AD77xx series).

To further complicate the issue, mixed-signal ICs have both analog and digital ports, and because of this, much confusion has resulted with respect to proper grounding techniques. In addition, some mixed-signal ICs have relatively low digital currents, whereas others have high digital currents. In many cases, these two types must be treated differently with respect to optimum grounding.

Digital and analog design engineers tend to view mixed-signal devices from different perspectives, and the purpose of this section is to develop a general grounding philosophy that will work for most mixed signal devices, without having to know the specific details of their internal circuits.

C.6 Ground and Power Planes

The importance of maintaining a low-impedance, large-area ground plane is critical to all analog circuits today. The ground plane not only acts as a low-impedance return path for decoupling high-frequency currents (caused by fast digital logic) but also minimizes EMI/RFI emissions. Because of the shielding action of the ground plane, the circuit's susceptibility to external EMI/RFI is also reduced.

Ground planes also allow the transmission of high-speed digital or analog signals using transmission line techniques (microstrip or stripline) where controlled impedances are required.

The use of "buss wire" is totally unacceptable as a "ground" because of its impedance at the equivalent frequency of most logic transitions. For instance, #22 gauge wire has about 20 nH/inch inductance. A transient current having a slew rate of 10 mA/ns created by a logic signal would develop an unwanted voltage drop of 200 mV at this frequency flowing through 1 inch of this wire:

$$\Delta v = L \frac{\Delta i}{\Delta t} = 20 \text{ nH} \times \frac{10 \text{ mA}}{\text{ns}} = 200 \text{ mV}. \tag{C.1}$$

For a signal having a 2 V peak-to-peak range, this translates into an error of about 200 mV, or 10% (approximate 3.5-bit accuracy). Even in all-digital circuits, this error would result in considerable degradation of logic noise margins.

Figure C.7 shows an illustration of a situation where the digital return current modulates the analog return current (top figure). The ground return wire inductance and resistance are shared

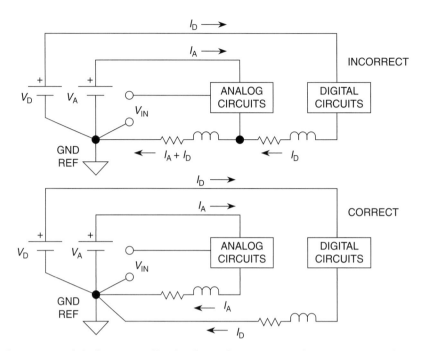

Figure C.7: Digital currents flowing in analog return path create error voltages

between the analog and digital circuits, and this is what causes the interaction and resulting error. A possible solution is to make the digital return current path flow directly to the GND REF, as shown in the bottom figure. This is the fundamental concept of a "star," or single-point ground system. Implementing the true single-point ground in a system which contains multiple high-frequency return paths is difficult because the physical length of the individual return current wires will introduce parasitic resistance and inductance, which can make obtaining a low-impedance, high-frequency ground difficult. In practice, the current returns must consist of large area ground planes for low impedance to high-frequency currents. Without a low-impedance ground plane, it is therefore almost impossible to avoid these shared impedances, especially at high frequencies.

All integrated circuit ground pins should be soldered directly to the low-impedance ground plane to minimize series inductance and resistance. The use of traditional IC sockets is not recommended with high-speed devices. The extra inductance and capacitance of even "low-profile" sockets may corrupt the device performance by introducing unwanted shared paths. If sockets must be used with DIP packages, as in prototyping, individual "pin sockets" or "cage jacks" may be acceptable. Both capped and uncapped versions of these pin sockets are available (AMP part numbers 5-330808-3 and 5-330808-6). They have spring-loaded gold contacts that make good electrical and mechanical connection to the IC pins. Multiple insertions, however, may degrade their performance.

Power supply pins should be decoupled directly to the ground plane using low-inductance ceramic surface-mount capacitors. If through-hole mounted ceramic capacitors must be used, their leads should be less than 1 mm. The ceramic capacitors should be located as close as possible to the IC power pins. Ferrite beads may be also required for additional decoupling.

C.7 Double-Sided versus Multilayer Printed Circuit Boards

Each PCB in the system should have at least one complete layer dedicated to the ground plane. Ideally, a double-sided board should have one side completely dedicated to ground and the other side for interconnections. In practice this is not possible, since some of the ground plane will certainly have to be removed to allow for signal and power crossovers, vias, and through-holes. Nevertheless, as much area as possible should be preserved, and at least 75% should remain. After completing an initial layout, the ground layer should be checked carefully to make sure there are no isolated ground "islands," because IC ground pins located in a ground "island" have no current return path to the ground plane. Also, the ground plane should be checked for "skinny" connections between adjacent large areas which may significantly reduce the effectiveness of the ground plane. Needless to say, autorouting board layout techniques will generally lead to a layout disaster on a mixed-signal board, so manual intervention is highly recommended.

Systems that are densely packed with surface-mount ICs will have a large number of interconnections; therefore multilayer boards are mandatory. This allows at least one complete layer to be dedicated to ground. A simple four-layer board would have internal ground and power plane layers, with the outer two layers used for interconnections between the surface mount components. Placing the power and ground planes adjacent to each other provides additional interplane capacitance which helps high-frequency decoupling of the power supply. In most systems, four layers are not enough, and additional layers are required for routing signals as well as power. Figure C.8 summarizes the key issues relating to ground planes.

- Use Large Area Ground (and Power) Planes for Low Impedance Current Return Paths (Must Use at Least a Double-Sided Board)
- Double-Sided Boards:
  - Avoid High-Density Interconnection Crossovers and Vias Which Reduce Ground Plane Area
  - Keep >75% BoardArea on One Side for Ground Plane
- Multilayer Boards: Mandatory for Dense Systems
  - Dedicate at Least One Layer for the Ground Plane
  - Dedicate at Least One Layer for the Power Plane
- Use at Least 30% to 40% of PCB Connector Pins for Ground
- Continue the Ground Plane on the Backplane Motherboard to Power Supply Return

Figure C.8: Ground planes are mandatory!

C.8 Multicard Mixed-Signal Systems

The best way of minimizing ground impedance in a multicard system is to use a "motherboard" PCB as a backplane for interconnections between cards, thus providing a continuous ground plane to the backplane. The PCB connector should have at least 30–40% of its pins devoted to ground, and these pins should be connected to the ground plane on the backplane mother card. To complete the overall system grounding scheme there are two possibilities:

1. The backplane ground plane can be connected to chassis ground at numerous points, thereby diffusing the various ground current return paths. This is commonly referred to as a "multipoint" grounding system and is shown in Figure C.9.

2. The ground plane can be connected to a single system "star ground" point (generally at the power supply).

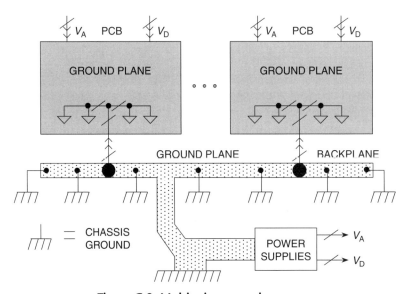

Figure C.9: Multipoint ground concept.

The first approach is most often used in all-digital systems but can be used in mixed-signal systems, provided that the ground currents due to digital circuits are sufficiently low and diffused over a large area. The low ground impedance is maintained all the way through the PC boards, the backplane, and ultimately the chassis. However, it is critical that good electrical contact be made where the grounds are connected to the sheet-metal chassis. This requires self-tapping sheet-metal screws or "biting" washers. Special care must be taken where anodized aluminum is used for the chassis material, since its surface acts as an insulator.

The second approach ("star ground") is often used in high-speed, mixed-signal systems having separate analog and digital ground systems and warrants further discussion.

C.9 Separating Analog and Digital Grounds

In mixed-signal systems with large amounts of digital circuitry, it is highly desirable to *physically* separate sensitive analog components from noisy digital components. It may also be beneficial to use separate ground planes for the analog and the digital circuitry. These planes should not overlap in order to minimize capacitive coupling between the two. The separate analog and digital ground planes are continued on the backplane using either motherboard ground planes or "ground screens," which are made up of a series of wired interconnections between the connector ground pins.

The arrangement shown in Figure C.10 illustrates that the two planes are kept separate all the way back to a common system "star" ground, generally located at the power supplies. The connections between the ground planes, the power supplies, and the "star" should be made up of multiple bus bars or wide copper braids for minimum resistance and inductance. The back-to-back Schottky diodes on each PCB are inserted to prevent accidental DC voltage from developing between the two ground systems when cards are plugged and unplugged. This voltage should be kept less than 300 mV to prevent damage to ICs that have connections to both the analog and digital ground planes. Schottky diodes are preferable because of their

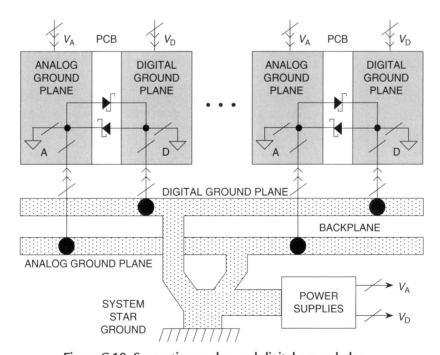

Figure C.10: Separating analog and digital ground planes.

low capacitance and low forward voltage drop. The low capacitance prevents AC coupling between the analog and digital ground planes. Schottky diodes begin to conduct at about 300 mV, and several parallel diodes in parallel may be required if high currents are expected. In some cases, ferrite beads can be used instead of Schottky diodes, but they introduce DC ground loops, which can be troublesome in precision systems.

It is mandatory that the impedance of the ground planes be kept as low as possible, all the way back to the system star ground. DC or AC voltages of more than 300 mV between the two ground planes not only can damage ICs, but they can cause false triggering of logic gates and possible latchup.

C.10 Grounding and Decoupling Mixed-Signal ICs with Low Digital Currents

Sensitive analog components such as amplifiers and voltage references are always referenced and de-coupled to the analog ground plane. *The ADCs and DACs (and other mixed-signal ICs) with low digital currents should generally be treated as analog components and also grounded and decoupled to the analog ground plane.* At first glance, this advice might seem somewhat contradictory, since a converter has an analog and digital interface and usually has pins designated as *analog ground* (AGND) and *digital ground* (DGND). The diagram shown in Figure C.11 will help explain this seeming dilemma.

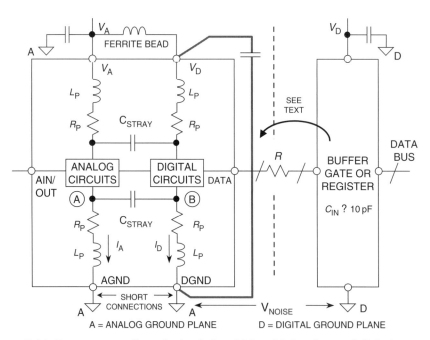

Figure C.11: Proper grounding of mixed-signal ICs with low internal digital currents.

Inside an IC that has both analog and digital circuits, such as an ADC or a DAC, the grounds are usually kept separate to avoid coupling digital signals into the analog circuits. Figure C.11 shows a simple model of a converter. There is nothing the IC designer can do about the wire-bond inductance and resistance associated with connecting the bond pads on the chip to the package pins except to realize it's there. The rapidly changing digital currents produce a voltage at point B which will inevitably couple into point A of the analog circuits through the stray capacitance, C_{STRAY}. In addition, there is approximately 0.2 pF unavoidable stray capacitance between every pin of the IC package. It's the IC designer's job to make the chip work in spite of this. However, to prevent further coupling, the AGND and DGND pins should be joined together externally to the *analog* ground plane with minimum lead lengths. Any extra impedance in the DGND connection will cause more digital noise to be developed at point B; it will, in turn, couple more digital noise into the analog circuit through the stray capacitance. Note that connecting DGND to the digital ground plane applies V_{NOISE} across the AGND and DGND pins and invites disaster.

> The name "DGND" on an IC tells us that this pin connects to the digital ground of the IC. This does not imply that this pin must be connected to the digital ground of the system.

It is true that this arrangement may inject a small amount of digital noise onto the analog ground plane. These currents should be quite small and can be minimized by ensuring that the converter output does not drive a large fanout (they normally can't, by design). Minimizing the fanout on the converter's digital port will also keep the converter logic transitions relatively free from ringing and minimize digital switching currents, thereby reducing any potential coupling into the analog port of the converter. The logic supply pin (V_D) can be further isolated from the analog supply by the insertion of a small lossy ferrite bead, as shown in Figure C.11. The internal transient digital currents of the converter will flow in the small loop from V_D through the decoupling capacitor and to DGND (this path is shown with a heavy line on the diagram). The transient digital currents will therefore not appear on the external analog ground plane, but are confined to the loop. The V_D pin decoupling capacitor should be mounted as close to the converter as possible to minimize parasitic inductance. These decoupling capacitors should be low inductance ceramic types, typically between 0.01 μF and 0.1 μF.

C.11 Treat the ADC Digital Outputs with Care

It is always a good idea (as shown in Figure C.11) to place a buffer register adjacent to the converter to isolate the converter's digital lines from noise on the data bus. The register also serves to minimize loading on the digital outputs of the converter and acts as a Faraday shield between the digital outputs and the data bus. Even though many converters have three-state outputs/inputs, this isolation register still represents good design practice. In some cases it may be desirable to add an additional buffer register on the analog ground plane next to the converter output to provide greater isolation.

The series resistors (labeled R in Figure C.11) between the ADC output and the buffer register input help to minimize the digital transient currents which may affect converter performance. The resistors isolate the digital output drivers from the capacitance of the buffer register inputs. In addition, the RC network formed by the series resistor and the buffer register input capacitance acts as a low-pass filter to slow down the fast edges.

A typical CMOS gate combined with PCB trace and a through-hole will create a load of approximately 10 pF. A logic output slew rate of 1 V/ns will produce 10 mA of dynamic current if there is no isolation resistor:

$$\Delta I = C \frac{\Delta v}{\Delta t} = 10\,\text{pF} \times 1\frac{\text{V}}{\text{ns}} = 10\,\text{mA}. \tag{C.2}$$

A 500 Ω series resistors will minimize this output current and result in a rise and fall time of approximately 11 ns when driving the 10 pF input capacitance of the register:

$$t_r = 2.2 \times \tau = 2.2 \times R \times C = 2.2 \times 500\,\Omega \times 10\,\text{pF} = 11\,\text{ns}. \tag{C.3}$$

TTL registers should be avoided because they can appreciably add to the dynamic switching currents due to their higher input capacitance.

The buffer register and other digital circuits should be grounded and decoupled to the *digital* ground plane of the PC board. Notice that any noise between the analog and digital ground plane reduces the noise margin at the converter digital interface. Since digital noise immunity is on the order of hundreds or thousands of millivolts, this is unlikely to matter. The analog ground plane will generally not be very noisy, but if the noise on the digital ground plane (relative to the analog ground plane) exceeds a few hundred millivolts, steps should be taken to reduce the digital ground plane impedance, thereby maintaining the digital noise margins at an acceptable level. Under no circumstances should the voltage between the two ground planes exceed 300 mV, or the ICs may be damaged.

Separate power supplies for analog and digital circuits are also highly desirable, even if the voltages are the same. The analog supply should be used to power the converter. If the converter has a pin designated as a digital supply pin (V_D), it should either be powered from a separate analog supply or filtered as shown in the diagram. All converter power pins should be decoupled to the analog ground plane, and all logic circuit power pins should be decoupled to the digital ground plane, as shown in Figure C.12.

In some cases it might not be possible to connect V_D to the analog supply. Some of the newer, high-speed ICs may have their analog circuits powered by 5 V, but the digital interface is

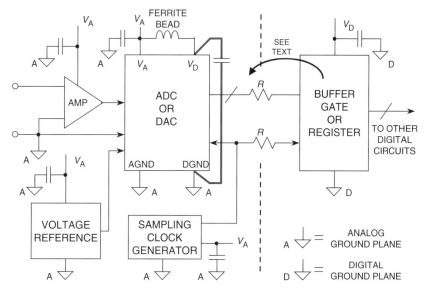

Figure C.12: Grounding and decoupling points.

powered by 3 V to interface to 3 V logic. In this case, the 3 V pin of the IC should be decoupled directly to the analog ground plane. It is also advisable to connect a ferrite bead in series with the power trace that connects the pin to the 3 V digital logic supply.

The sampling clock generation circuitry should be treated like analog circuitry and also be grounded and heavily decoupled to the analog ground plane. Phase noise on the sampling clock produces degradation in system SNR, as will be discussed shortly.

C.12 Sampling Clock Considerations

In a high-performance sampled data system, a low phase-noise crystal oscillator should be used to generate the ADC (or DAC) sampling clock because sampling clock jitter modulates the analog input/output signal and raises the noise and distortion floor. The sampling clock generator should be isolated from noisy digital circuits and grounded and decoupled to the analog ground plane, as is true for the op amp and the ADC.

The effect of sampling clock jitter on ADC signal-to-noise ratio (SNR) is given approximately by the equation:

$$\text{SNR} = 20 \log_{10} \left[\frac{1}{2\pi f t_j} \right] \tag{C.4}$$

where SNR is the SNR of a perfect ADC of infinite resolution where the only source of noise is that caused by the rms sampling clock jitter, t_j. Note that f in the equation is the analog input frequency. Just working through a simple example, if t_j = 50 ps rms, $f = 100$ kHz, then *SNR* = 90 dB, equivalent to about 15-bit dynamic range.

It should be noted that t_j in the above example is the root-sum-square (rss) value of the external clock jitter *and* the internal ADC clock jitter (called *aperture jitter*). However, in most high-performance ADCs, the internal aperture jitter is negligible compared to the jitter on the sampling clock.

Since degradation in SNR is primarily due to external clock jitter, steps must be taken to ensure that the sampling clock is as noise-free as possible and has the lowest possible phase jitter. This requires that a crystal oscillator be used. There are several manufacturers of small crystal oscillators with low jitter (less than 5 ps rms) CMOS compatible outputs. (For example, MF Electronics, 10 Commerce Dr., New Rochelle, NY 10801, Tel. 914-576-6570, and Wenzel Associates, Inc., 2215 Kramer Lane, Austin, Texas 78758, Tel. 512- 835-2038.)

Ideally, the sampling clock crystal oscillator should be referenced to the analog ground plane in a split-ground system. However, this is not always possible because of system constraints. In many cases, the sampling clock must be derived from a higher-frequency multipurpose system clock which is generated on the digital ground plane. It must then pass from its origin on the digital ground plane to the ADC on the analog ground plane. Ground noise between the two planes adds directly to the clock signal and will produce excess jitter. The jitter can cause degradation in the signal-to-noise ratio and also produce unwanted harmonics.

This can be somewhat remedied by transmitting the sampling clock signal as a differential signal using either a small RF transformer as shown in Figure C.13 or a high-speed differential

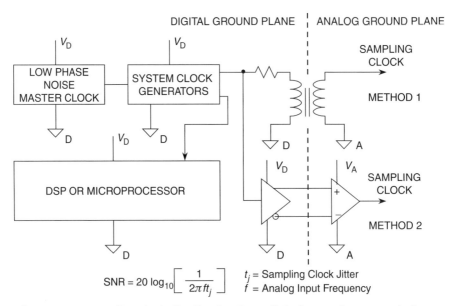

Figure C.13: Sampling clock distribution from digital to analog ground planes.

driver and receiver IC. If an active differential driver and receiver are used, they should be ECL to minimize phase jitter. In a single 5 V supply system, ECL logic can be connected between ground and 5 V (PECL), and the outputs AC coupled into the ADC sampling clock input. In either case, the original master system clock must be generated from a low-phase noise crystal oscillator, and not the clock output of a DSP, microprocessor, or microcontroller.

C.13 The Origins of the Confusion About Mixed-Signal Grounding: Applying Single-Card Grounding Concepts to Multicard Systems

Most ADC, DAC, and other mixed-signal device data sheets discuss grounding relative to a single PCB, usually the manufacturer's own evaluation board. This has been a source of confusion in trying to apply these principles to multicard or multi-ADC/DAC systems. The recommendation is usually to split the PCB ground plane into an analog plane and a digital plane. It is then further recommended that the AGND and DGND pins of a converter be tied together and that the analog ground plane and digital ground planes be connected at that same point as shown in Figure C.14. This essentially creates the system "star" ground at the mixed-signal device.

All noisy digital currents flow through the digital power supply to the digital ground plane and back to the digital supply; they are isolated from the sensitive analog portion of the board. The system star ground occurs where the analog and digital ground planes are joined together at

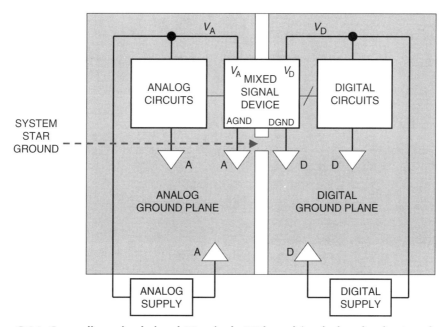

Figure C.14: Grounding mixed-signal ICs: single PC board (typical evaluation/test board).

the mixed-signal device. This approach will generally work in a simple system with a single PCB and single ADC/DAC, but it is not optimum for multicard mixed-signal systems. In systems having several ADCs or DACs on different PCBs (or on the same PCB, for that matter), the analog and digital ground planes become connected at several points, creating the possibility of ground loops and making a single-point "star" ground system impossible. For these reasons, this grounding approach is not recommended for multicard systems, and the approach previously discussed should be used for mixed-signal ICs with low digital currents.

C.14 Summary: Grounding Mixed-Signal Devices with Low Digital Currents in a Multicard System

Figure C.15 summarizes the approach previously described for grounding a mixed-signal device which has low digital currents. The analog ground plane is not corrupted because the small digital transient currents flow in the small loop between V_D, the decoupling capacitor, and DGND (shown as a heavy line). The mixed-signal device is for all intents and purposes treated as an analog component. The noise V_N between the ground planes reduces the noise margin at the digital interface but is generally not harmful if kept less than 300 mV by using a low-impedance digital ground plane all the way back to the system star ground.

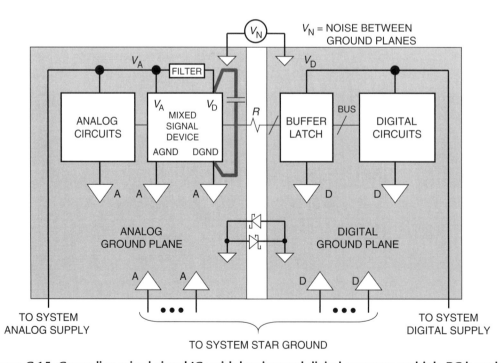

Figure C.15: Grounding mixed-signal ICs with low internal digital currents: multiple PC boards.

However, mixed-signal devices such as sigma-delta ADCs, codecs, and DSPs with on-chip analog functions are becoming more and more digitally intensive. Along with the additional digital circuitry come larger digital currents and noise. For example, a sigma-delta ADC or DAC contains a complex digital filter which adds considerably to the digital current in the device. The method previously discussed depends on the decoupling capacitor between V_D and DGND to keep the digital transient currents isolated in a small loop. However, if the digital currents are significant enough and have components at DC or low frequencies, the decoupling capacitor may have to be so large that it is impractical. Any digital current that flows outside the loop between V_D and DGND must flow through the analog ground plane. This may degrade performance, especially in high-resolution systems.

It is difficult to predict what level of digital current flowing into the analog ground plane will become unacceptable in a system. All we can do at this point is to suggest an alternative grounding method which may yield better performance.

C.15 Summary: Grounding Mixed-Signal Devices with High Digital Currents in a Multicard System

An alternative grounding method for a mixed-signal device with high levels of digital currents is shown in Figure C.16. The AGND of the mixed-signal device is connected to the analog ground plane, and the DGND of the device is connected to the digital ground plane. The digital currents are isolated from the analog ground plane, but the noise between the two ground planes is applied directly between the AGND and DGND pins of the device. For this method to be successful, the analog and digital circuits within the mixed signal device must be well isolated. The noise between AGND and DGND pins must not be large enough to reduce internal noise margins or cause corruption of the internal analog circuits.

Figure C.16 shows optional Schottky diodes (back-to-back) or a ferrite bead connecting the analog and digital ground planes. The Schottky diodes prevent large DC voltages or low-frequency voltage spikes from developing across the two planes. These voltages can potentially damage the mixed-signal IC if they exceed 300 mV because they appear directly between the AGND and DGND pins. As an alternative to the back-to-back Schottky diodes, a ferrite bead provides a DC connection between the two planes but isolates them at frequencies above a few MHz where the ferrite bead becomes resistive. This protects the IC from DC voltages between AGND and DGND, but the DC connection provided by the ferrite bead can introduce unwanted DC ground loops and might not be suitable for high-resolution systems.

C.16 Grounding DSPs with Internal Phase-Locked Loops

As if dealing with mixed-signal ICs with AGND and DGNDs wasn't enough, DSPs such as the ADSP21160 SHARC with internal phase-locked-loops (PLLs) raise issues with respect

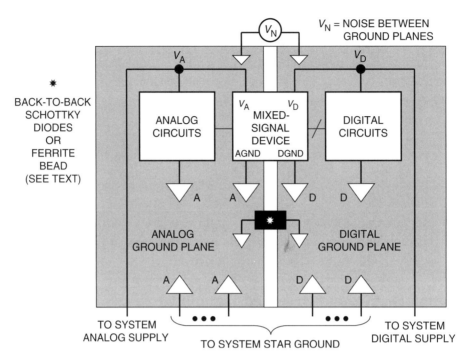

Figure C.16: Grounding alternative for mixed-signal ICs with high digital currents: multiple PC boards.

to proper grounding. The ADSP-21160 PLL allows the internal core clock (determines the instruction cycle time) to operate at a user-selectable ratio of 2, 3, or 4 times the external clock frequency, CLKIN. The CLKIN rate is the rate at which the synchronous external ports operate. Although this allows using a lower-frequency external clock, care must be taken with the power and ground connections to the internal PLL, as shown in Figure C.17.

To prevent internal coupling between digital currents and the PLL, the power and ground connections to the PLL are brought out separately on pins labeled AV_{DD} and AGND, respectively. The AV_{DD} 2.5 V supply should be derived from the $V_{DD\ INT}$ 2.5 V supply using the filter network as shown. This ensures a relatively noise-free supply for the internal PLL. The AGND pin of the PLL should be connected to the digital ground plane of the PC board using a short trace. The decoupling capacitors should be routed between the AV_{DD} pin and AGND pin using short traces.

C.17 Grounding Summary

No single grounding method will guarantee optimum performance 100% of the time. This section has presented a number of possible options, depending upon the characteristics of the particular mixed-signal devices in question. It is helpful, however, to provide for as many options as possible when laying out the initial PC board.

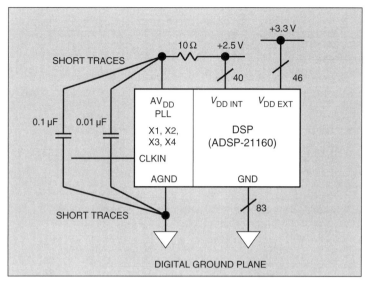

Figure C.17: Grounding DSPs with internal phase-locked loops (PLLs).

It is mandatory that at least one layer of the PC board be dedicated to ground plane. The initial board layout should provide for nonoverlapping analog and digital ground planes, but pads and vias should be provided at several locations for the installation of back-to-back Schottky diodes or ferrite beads, if required. Pads and vias should also be provided so that the analog and digital ground planes can be connected together with jumpers if required.

The AGND pins of mixed-signal devices should in general always be connected to the analog ground plane. An exception to this are DSPs such as the ADSP-21160 SHARC, which have internal phase-locked-loops (PLLs). The ground pin for the PLL is labeled AGND, but should be directly connected to the digital ground plane for the DSP. See Figure C.18 for a general summary of grounding philosophy.

C.16 Some General PC Board Layout Guidelines for Mixed-Signal Systems

It is evident that noise can be minimized by paying attention to the system layout and preventing different signals from interfering with each other. High-level analog signals should be separated from low-level analog signals, and both should be kept away from digital signals. We have seen elsewhere that in waveform sampling and reconstruction systems the sampling clock (which is a digital signal) is as vulnerable to noise as any analog signal but is as liable to cause noise as any digital signal and so must be kept isolated from both analog and digital systems. If clock driver packages are used in clock distribution, only one frequency clock should

- There is no single grounding method which is guaranteed to work 100% of the time
- Different methods may or may not give the same levels of performance
- At least one layer on each PC board MUST be dedicated to ground plane
- Do initial layout with split analog and digital ground planes
- Provide pads and vias on each PC board for back-to-back Schottky diodes and optional ferrite beads to connect the two planes
- Provide "jumpers" so that DGND pins of mixed-signal devices can be connected to AGND pins (analog ground plane) or to digital ground plane. (AGND of PLLs in DSPs should be connected to digital ground plane)
- Provide pads and vias for "jumpers" so that analog and digital ground planes can be joined together at several points on each PC board
- Follow recommendations on mixed signal device data sheet

Figure C.18: Grounding philosophy summary.

be passed through a single package. Sharing drivers between clocks of different frequencies in the same package will produce excess jitter and crosstalk and degrade performance.

The ground plane can act as a shield where sensitive signals cross. Figure C.19 shows a good layout for a data acquisition board where all sensitive areas are isolated from each other and

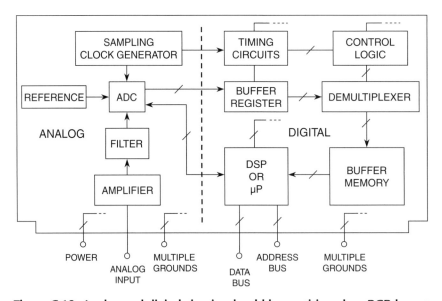

Figure C.19: Analog and digital circuits should be partitioned on PCB layout.

signal paths are kept as short as possible. While real life is rarely as tidy as this, the principle remains a valid one.

There are a number of important points to be considered when making signal and power connections. First, a connector is one of the few places in the system where all signal conductors must run in parallel; it is therefore imperative to separate them with ground pins (creating a Faraday shield) to reduce coupling between them.

Multiple ground pins are important for another reason: They keep down the ground impedance at the junction between the board and the backplane. The contact resistance of a single pin of a PCB connector is quite low (on the order of $10\,m\Omega$) when the board is new; as the board gets older, the contact resistance is likely to rise and the board's performance may be compromised. It is therefore well worthwhile to allocate extra PCB connector pins so that there are many ground connections (perhaps 30–40% of all the pins on the PCB connector should be ground pins). For similar reasons there should be several pins for each power connection, although there is no need to have as many as there are ground pins.

Analog Devices and other manufacturers of high-performance, mixed-signal ICs offer evaluation boards to assist customers in their initial evaluations and layout. ADC evaluation boards generally contain an on-board low-jitter sampling clock oscillator, output registers, and appropriate power and signal connectors. They also may have additional support circuitry such as the ADC input buffer amplifier and external reference.

The layout of the evaluation board is optimized in terms of grounding, decoupling, and signal routing and can be used as a model when laying out the ADC PC board in the system. The actual evaluation board layout is usually available from the ADC manufacturer in the form of computer CAD files (Gerber files). In many cases, the layout of the various layers appears on the data sheet for the device.

C.19 Skin Effect

At high frequencies, also consider *skin effect*, where inductive effects cause currents to flow only in the outer surface of conductors. Note that this is in contrast to the earlier discussions on DC resistance of conductors.

The skin effect has the consequence of increasing the resistance of a conductor at high frequencies. Note also that this effect is separate from the increase in impedance due to the effects of the self-inductance of conductors as frequency is increased.

Skin effect is quite a complex phenomenon, and detailed calculations are beyond the scope of this discussion. However, a good approximation for copper is that the skin depth in centimeters is $6.61/\sqrt{f}$ (f in Hz).

A summary of the skin effect within a typical PCB conductor foil is shown in Figure C.20. Note that this copper conductor cross-sectional view assumes looking into the side of the conducting trace.

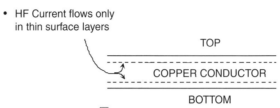

- Skin Depth: $6.61/\sqrt{f}$ cm, f in Hz
- Skin Resistance: $2.6 \times 10^{-7}\sqrt{f}$ ohms per square, f in Hz
- Since skin currents flow in both sides of a PC track, the value of skin resistance in PCBs must take account of this

Figure C.20: Skin depth in a PC conductor.

Assuming that skin effects become important when the skin depth is less than 50% of the thickness of the conductor, this tells us that for a typical PC foil, we must be concerned about skin effects at frequencies above approximately 12 MHz.

Where skin effect is important, the resistance for copper is $2.6 \times 10^{-7}\sqrt{f}$ ohms per square (f in Hz). This formula is invalid if the skin thickness is greater than the conductor thickness (i.e., at DC or low frequencies).

Figure C.21 illustrates a case of a PCB conductor with current flow, as separated from the ground plane underneath.

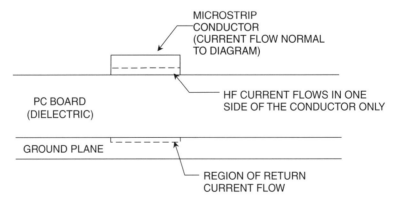

Figure C.21: Skin effect with PC conductor and ground plane.

In this diagram, note the (dotted) regions of high-frequency current flow, as reduced by the skin effect. When calculating skin effect in PCBs, it is important to remember that current generally flows in both sides of the PC foil (this is not necessarily the case in microstrip lines; see below), so the resistance per square of PC foil may be half the above value.

C.20 Transmission Lines

We earlier considered the benefits of outward and return signal paths being close together so that inductance is minimized. As shown previously in Figure C.22, when a high-frequency signal flows in a PC track running over a ground plane, the arrangement functions as a *microstrip* transmission line, and the majority of the return current flows in the ground plane underneath the line.

Figure C.22 shows the general parameters for a microstrip transmission line, given the conductor width w, dielectric thickness h, and the dielectric constant E_r.

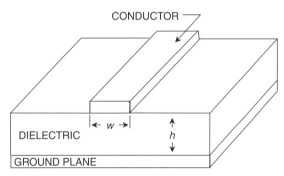

Figure C.22: A PCB microstrip transmission line is an example of a controlled impedance conductor pair.

The characteristic impedance of such a microstrip line will depend on the width of the track and the thickness and dielectric constant of the PCB material. Designs of microstrip lines are covered in more detail later in this chapter.

For most DC and lower-frequency applications, the characteristic impedance of PCB traces will be relatively unimportant. Even at frequencies where a track over a ground plane behaves as a transmission line, it is not necessary to worry about its characteristic impedance or proper termination if the free space wavelengths of the frequencies of interest are greater than 10 times the length of the line.

However, at VHF and higher frequencies, it is possible to use PCB tracks as microstrip lines within properly terminated transmission systems. Typically the microstrip will be designed to match standard coaxial cable impedances, such as $50\,\Omega$, $75\,\Omega$ or $100\,\Omega$, simplifying interfacing.

Note that if losses in such systems are to be minimized, the PCB material must be chosen for low high-frequency losses. This usually means the use of Teflon or some other comparably low-loss PCB material. Often, though, the losses in short lines on cheap glass-fiber board are small enough to be quite acceptable.

C.21 Be Careful with Ground Plane Breaks

Wherever there is a break in the ground plane beneath a conductor, the ground plane return current must by necessity flow *around* the break. As a result, both the inductance and the vulnerability of the circuit to external fields are increased. This situation is diagrammed in Figure C.23, where conductors A and B must cross one another.

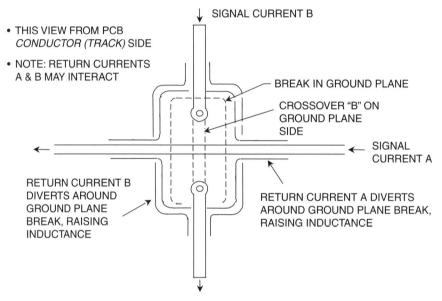

Figure C.23: A ground plane break raises circuit inductance and increases vulnerability to external fields.

Where such a break is made to allow a cross-over of two perpendicular conductors, it would be far better if the second signal were carried across both the first and the ground plane by means of a piece of wire or a resistor. The ground plane then acts as a shield between the two signal conductors, and the two ground return currents, flowing in opposite sides of the ground plane as a result of skin effects, do not interact.

With a multilayer board, both the crossover and the continuous ground plane can be accommodated without the need for a wire link. Multilayer PCBs are expensive and harder to troubleshoot than more simple double-sided boards, but do offer even better shielding and signal routing. The principles involved remain unchanged but the range of layout options is increased.

The use of double-sided or multilayer PCBs with at least one continuous ground plane is undoubtedly one of the most successful design approaches for high-performance, mixed-signal circuitry. Often the impedance of such a ground plane is sufficiently low to permit the use of a single ground plane for both analog and digital parts of the system. However, whether or not this is possible does depend on the resolution and bandwidth required and the amount of digital noise present in the system.

C.22 Ground Isolation Techniques

Although the use of ground planes does lower impedance and helps greatly in lowering ground noise, there may still be situations where a prohibitive level of noise exists. In such cases, the use of ground error minimization and isolation techniques can be helpful.

Another illustration of a common ground impedance coupling problem is shown in Figure C.24. In this circuit a precision gain-of-100 preamp amplifies a low-level signal V_{IN}, using an

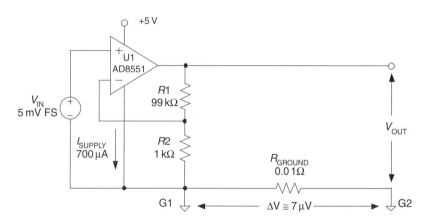

Figure C.24: Unless care is taken, even small common ground currents can degrade precision amplifier accuracy.

AD8551 chopper-stabilized amplifier for best DC accuracy. At the load end, the signal V_{OUT} is measured with respect to G2, the local ground. Because of the small 700 μA I_{SUPPLY} of the AD8551 flowing between G1 and G2, there is a 7 μV ground error—about seven times the typical input offset expected from the op amp.

This error can be avoided by routing the negative supply pin current of the op amp back to star ground G2 as opposed to ground G1 by using a separate trace. This step eliminates the G1-G2 path power supply current and so minimizes the ground leg voltage error. Note that there will be little error developed in the "hot" V_{OUT} lead so long as the current drain at the load end is small.

In some cases, there may be simply unavoidable ground voltage differences between a source signal and the load point where it is to be measured. Within the context of this "same-board" discussion, this might require rejecting ground error voltages of several tens of mV. Or, should the source signal originate from an "off-board" source, the magnitude of the common-mode voltages to be rejected can easily rise into a several volt range (or even tens of volts).

Fortunately, full signal transmission accuracy can still be accomplished in the face of such high-noise voltages by employing a principle discussed earlier. This is the use of a differential input, *ground isolation* amplifier. The ground isolation amplifier minimizes the effect of ground error voltages between stages by processing the signal in differential fashion, thereby rejecting common-mode voltages by a substantial margin (typically 60 dB or more). Note, however, that this approach is only effective for very low-frequency signals.

Two ground isolation amplifier solutions are shown in Figure C.25. This diagram can alternately employ either the AD629 to handle CM voltages up to ± 270 V or the AMP03, which is suitable for CM voltages up to ± 20 V.

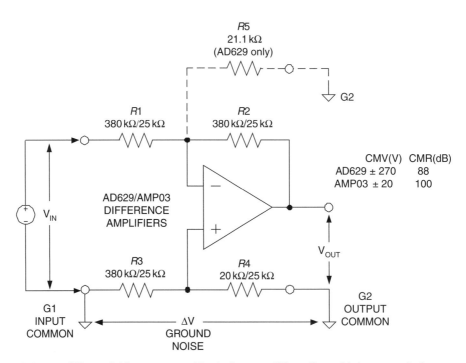

Figure C.25: A differential input ground isolating amplifier allows high transmission accuracy by rejecting ground noise voltage between source (G1) and measurement (G2) grounds.

In the circuit, input voltage V_{IN} is referred to G1 but must be measured with respect to G2. With the use of a high CMR unity-gain difference amplifier, the noise voltage ΔV existing between these two grounds is easily rejected. The AD629 offers a typical CMR of 88 dB, while the AMP03 typically achieves 100 dB. In the AD629, the high CMV rating is done by a combination of high CM attenuation, followed by differential gain, realizing a net differential gain of unity. The AD629 uses the first listed value resistors noted in the figure for $R1$–$R5$. The AMP03 operates as a precision four-resistor differential amplifier using the 25 kΩ value $R1$–$R4$ resistors noted. Both devices are complete, one package solutions to the ground-isolation amplifier.

This scheme allows relative freedom from tightly controlling ground drop voltages or running additional and/or larger PCB traces to minimize such error voltages. Note that it can be implemented with either the fixed gain difference amplifiers shown or with a standard in amp IC, configured for unity gain. The AD623, for example, also allows single-supply use. In any case, signal polarity is also controllable by simple reversal of the difference amplifier inputs.

In general terms, transmitting a signal from one point on a PCB to another for measurement or further processing can be optimized by two key interrelated techniques. These are the use of high impedance, differential signal handling techniques. The high impedance loading of an in amp minimizes voltage drops, and differential sensing of the remote voltage minimizes sensitivity to ground noise.

When the further signal processing is A/D conversion, these transmission criteria can be implemented *without* adding a differential ground isolation amplifier stage. Simply select an ADC that operates differentially. The high input impedance of the ADC minimizes load sensitivity to the PCB wiring resistance. In addition, the differential input feature allows the output of the source to be sensed directly at the source output terminals (even if single-ended). The CMR of the ADC then eliminates sensitivity to noise voltages between the ADC and source grounds.

An illustration of this concept using an ADC with high-impedance differential inputs is shown in Figure C.26. Note that the general concept can be extended to virtually any signal source, driving any load. All loads, even single-ended ones, become differential input by adding an appropriate differential input stage.

The differential input can be provided by either a fully developed high Z in amp or, in many cases, it can be a simple subtractor stage op amp, such as Figure C.25.

C.23 Static PCB Effects

Leakage resistance is the dominant static circuit board effect. Contamination of the PCB surface by flux residues, deposited salts, and other debris can create leakage paths between circuit nodes. Even on well-cleaned boards, it is not unusual to find 10 nA or more of leakage to nearby nodes from 15 V supply rails. Nanoamperes of leakage current into the wrong nodes

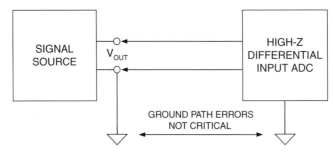

Figure C.26: A high-impedance differential input ADC also allows high transmission accuracy between source and load.

often cause volts of error at a circuit's output; for example, 10 nA into a 10 MΩ resistance causes a 0.1 V error. Unfortunately, the standard op amp pinout places the $-V_S$ supply pin next to the + input, which is often hoped to be at high impedance. To help identify nodes sensitive to the effects of leakage currents, ask the simple question: If a spurious current of a few nanoamperes or more were injected into this node, would it matter?

If the circuit is already built, it is possible to localize moisture sensitivity to a suspect node with a classic test. While observing circuit operation, blow on potential trouble spots through a simple soda straw. The straw focuses the breath's moisture, which, with the board's salt content in susceptible portions of the design, disrupts circuit operation upon contact. There are several means of eliminating simple surface leakage problems. Thorough washing of circuit boards to remove residues helps considerably. A simple procedure includes vigorously brushing the boards with isopropyl alcohol, followed by thorough washing with deionized water and an 85°C bakeout for a few hours. Be careful when selecting board-washing solvents, though. When cleaned with certain solvents, some water-soluble fluxes create salt deposits, exacerbating the leakage problem.

Unfortunately, if a circuit displays sensitivity to leakage, even the most rigorous cleaning can offer only a temporary solution. Problems soon return upon handling or exposure to foul atmospheres and high humidity. Some additional means must be sought to stabilize circuit behavior, such as conformal surface coating.

Fortunately, there is an answer to this problem, namely *guarding*, which offers a fairly reliable and permanent solution to the problem of surface leakage. Well-designed guards can eliminate leakage problems, even for circuits exposed to harsh industrial environments. Two schematics illustrate the basic guarding principle, as applied to typical inverting and noninverting op amp circuits.

Figure C.27 illustrates an inverting mode guard application. In this case, the op amp reference input is grounded, so the guard is a grounded ring surrounding all leads to the inverting input, as noted by the dotted line.

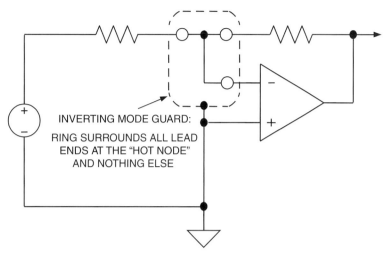

Figure C.27: Inverting mode guard encloses all op amp inverting input connections within a grounded guard ring.

Basic guarding principles are simple: *Completely* surround sensitive nodes with conductors that can readily sink stray currents, and maintain the guard conductors at the exact potential of the sensitive node (as otherwise the guard will serve as a leakage source rather than a leakage sink). For example, to keep leakage into a node below 1 pA (assuming 1000 MΩ leakage resistance) the guard and guarded node must be within 1 mV. Generally, the low offset of a modern op amp is sufficient to meet this criterion.

There are important caveats to be noted with implementing a true high-quality guard. For traditional through-hole PCB connections, to be most effective the guard pattern should appear on *both* sides of the circuit board. It should also be connected along its length by several vias. Finally, when either justified or required by the system design parameters, do make an effort to include guards in the PCB design process from the outset—there is little likelihood that a proper guard can be added as an afterthought.

Figure C.28 illustrates the case for a noninverting guard. In this instance the op amp reference input is directly driven by the source, which complicates matters considerably. Again, the guard ring completely surrounds all of the input nodal connections. In this instance however, the guard is driven from the low impedance feedback divider connected to the inverting input.

Usually the guard-to-divider junction will be a direct connection, but in some cases a unity gain buffer might be used at *X* to drive a cable shield or to maintain the lowest possible impedance at the guard ring.

In lieu of the buffer, another useful step is to use an additional, directly grounded screen ring, *Y*, which surrounds the inner guard and the feedback nodes as shown. This step costs nothing

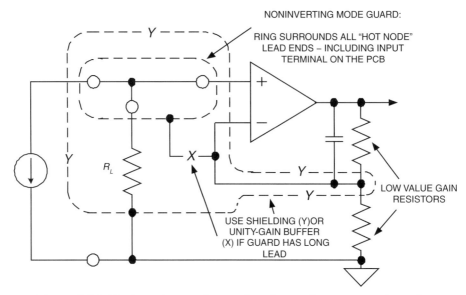

Figure C.28: Noninverting mode guard encloses all op amp noninverting input connections within a low impedance, driven guard ring.

except some added layout time and will greatly help buffer leakage effects into the higher-impedance inner guard ring.

Of course what hasn't been addressed to this point is just how the op amp itself is connected into these guarded islands without compromising performance. The traditional method using a TO-99 metal can package device was to employ double-sided PCB guard rings, with both op amp inputs terminated within the guarded ring.

Many high-impedance sensors use the above described method. The next section illustrates how more modern IC packages can be mounted to PCB patterns and take advantage of guarding and low leakage operation.

C.24 Sample MINIDIP and SOIC Op Amp PCB Guard Layouts

Modern assembly practices have favored smaller plastic packages such as 8-pin MINIDIP and SOIC types. Some suggested partial layouts for guard circuits using these packages is shown in the next two figures. While guard traces may also be possible with even more tiny op amp footprints, such as SOT-23, SC70, etc., the required trace separations become even more confining, challenging the layout designer as well as the manufacturing processes.

For the ADI "N" style MINIDIP package, Figure C.29 illustrates how guarding can be accomplished for inverting (left) and noninverting (right) operating modes. This setup would also be applicable to other op amp devices where relatively high voltages occur at pin 1 or 4. Using

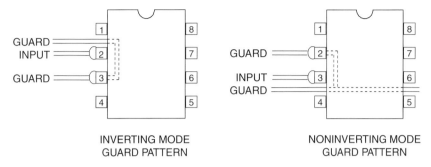

Figure C.29: PCB guard patterns for inverting and noninverting mode op amps using 8-pin MINIDIP (N) package.

a standard 8-pin DIP outline for a single op amp, it can be noted that this package's 0.1" pin spacing allows a PC trace (here, the guard trace) to pass between adjacent pins. This is the key to implementing effective DIP package guarding, as it can adequately prevent a leakage path from the $-V_S$ supply at pin 4 or from similar high potentials at pin 1.

For the left-side inverting mode, note that the grounded guard traces connected to pin 3 surround the op amp inverting input (pin 2), and run parallel to the input trace. This guard would be continued out to and around the source and feedback connections of Figure C.27 (or other similar circuit), including an input pad in the case of a cable. In the right-side noninverting mode, the guard voltage is the feedback divider voltage to pin 2. This corresponds to the inverting input node of the amplifier, from Figure C.28.

Note that in both of the cases of Figure C.29, the guard physical connections shown are only partial—an actual layout would include all sensitive nodes within the circuit. In both the inverting and the noninverting modes using the MINIDIP or other through-hole style package, the PCB guard traces should be located on both sides of the board, with top and bottom traces connected with several vias.

Things become slightly more complicated when using guarding techniques with the SOIC surface-mount ("R") package, as the 0.05" pin spacing doesn't easily allow routing of PCB traces between the pins. But there is still an effective guarding answer, at least for the inverting case. Figure C.30 shows guards for the ADI "R" style SOIC package.

Note that for many single op amp devices in this SOIC "R" package, pins 1, 5, and 8 are "No Connect" pins. For such instances, this means that these locations can be employed in the layout to route guard traces.

In the case of the inverting mode (left), the guarding is still completely effective, with the dummy pin 1 and pin 3 serving as the grounded guard trace. This is a fully effective guard without compromise. Also, with SOIC op amps, much of the circuitry around the device will

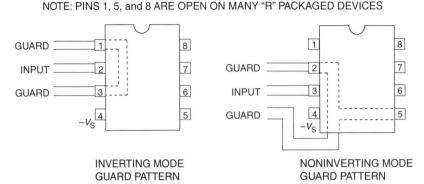

Figure C.30: PCB guard patterns for inverting and noninverting mode op amps using 8-pin SOIC (R) package.

not use through-hole components. So, the guard ring may only be necessary on the op amp PCB side.

In the case of the follower stage (right), the guard trace must be routed around the negative supply at pin 4, and thus pin 4 to pin 3 leakage isn't fully guarded. For this reason, a precision high-impedance follower stage using an SOIC package op amp isn't generally recommended, as guarding isn't as effective for dual supply connected devices.

However, an exception to this caveat does apply to the use of a *single-supply* op amp as a non-inverting stage. For example, if the AD8551 is used, pin 4 becomes ground, and some degree of intrinsic guarding is then established by default.

C.25 Dynamic PCB Effects

Although static PCB effects can come and go with changes in humidity or board contamination, problems that most noticeably affect the dynamic performance of a circuit usually remain relatively constant. Short of a new design, washing or any other simple fixes can't fix them. As such, they can permanently and adversely affect a design's specifications and performance. The problems of stray capacitance, linked to lead and component placement, are reasonably well known to most circuit designers. Since lead placement can be permanently dealt with by correct layout, any remaining difficulty is solved by training assembly personnel to orient components or bend leads optimally.

Dielectric absorption (DA), on the other hand, represents a more troublesome and still poorly understood circuit-board phenomenon. Like DA in discrete capacitors, DA in a printed-circuit board can be modeled by a series resistor and capacitor connecting two closely spaced nodes. Its effect is inverse with spacing and linear with length.

As shown in Figure C.31, the RC model for this effective capacitance ranges from 0.1 pF to 2.0 pF, with the resistance ranging from 50 MΩ to 500 MΩ. Values of 0.5 pF and 100 MΩ are most common. Consequently, circuit-board DA interacts most strongly with high-impedance circuits.

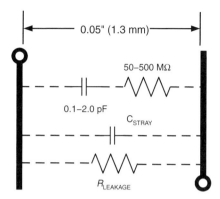

Figure C.31: DA plagues dynamic response of PCB-based circuits.

PCB DA most noticeably influences dynamic circuit response, for example, settling time. Unlike circuit leakage, the effects aren't usually linked to humidity or other environmental conditions, but rather, are a function of the board's dielectric properties. The chemistry involved in producing plated-through holes seems to exacerbate the problem. If circuits don't meet expected transient response specs, consider PCB DA as a possible cause. Fortunately, there are solutions. As in the case of capacitor DA, external components can be used to compensate for the effect. More importantly, surface guards that totally isolate sensitive nodes from parasitic coupling often eliminate the problem. (Note that these guards should be duplicated on both sides of the board, in cases of through-hole components.) As noted previously, low-loss PCB dielectrics are also available at higher costs.

PCB "hook," similar if not identical to DA, is characterized by variation in effective circuit-board capacitance with frequency (see Reference 1). In general, it affects high-impedance circuit transient response where board capacitance is an appreciable portion of the total in the circuit. Circuits operating at frequencies below 10 kHz are the most susceptible. As in circuit-board DA, the board's chemical makeup very much influences its effects.

C.26 Stray Capacitance

When two conductors aren't short-circuited together or totally screened from each other by a conducting (Faraday) screen, there is a capacitance between them. So, on any PCB, there will be a large number of capacitors associated with any circuit (which may or may not be considered in models of the circuit). Where high-frequency performance matters (and even DC and VLF circuits may use devices with high Ft and therefore be vulnerable to high-frequency instability), it is very important to consider the effects of this stray capacitance.

Any basic textbook will provide formulas for the capacitance of parallel wires and other geometric configurations (see References 9 and 10). The example we need consider in this discussion is the parallel plate capacitor, often formed by conductors on opposite sides of a PCB. The basic diagram describing this capacitance is shown in Figure C.32.

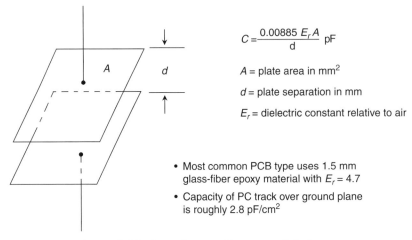

Figure C.32: Capacitance of two parallel plates.

$$C = \frac{0.00885\, E_r A}{d} \text{ pF}$$

A = plate area in mm²

d = plate separation in mm

E_r = dielectric constant relative to air

- Most common PCB type uses 1.5 mm glass-fiber epoxy material with E_r = 4.7
- Capacity of PC track over ground plane is roughly 2.8 pF/cm²

Neglecting edge effects, the capacitance of two parallel plates of area A mm² and separation d mm in a medium of dielectric constant E_r relative to air is 0.00885 E_r A/d pF.

From this formula, we can calculate that for general-purpose PCB material (E_r = 4.7, d = 1.5 mm), the capacitance between conductors on opposite sides of the board is just under 3 pF/cm². In general, such capacitance will be parasitic, and circuits must be designed so that it does not affect their performance.

While it is possible to use PCB capacitance in place of small discrete capacitors, the dielectric properties of common PCB substrate materials cause such capacitors to behave poorly. They have a rather high temperature coefficient and poor Q at high frequencies, which makes them unsuitable for many applications. Boards made with lower loss dielectrics such as Teflon are expensive exceptions to this rule.

C.27 Capacitive Noise and Faraday Shields

There is a capacitance between any two conductors separated by a dielectric (air or vacuum are dielectrics). If there is a change of voltage on one, there will be a movement of charge on the other. A basic model for this is shown in Figure C.33.

It is evident that the noise voltage $V_{COUPLED}$ appearing across Z_1 may be reduced by several means, all of which reduce noise current in Z_1. They are reduction of the signal voltage V_N,

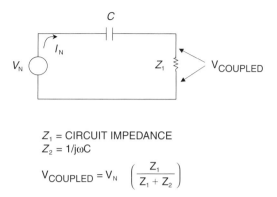

Figure C.33: Capacitive coupling equivalent circuit model.

reduction of the frequency involved, reduction of the capacitance, or reduction of Z_1 itself. Unfortunately, however, often none of these circuit parameters can be freely changed, and an alternate method is needed to minimize the interference. The best solution toward reducing the noise coupling effect of C is to insert a grounded conductor, also known as a *Faraday shield*, between the noise source and the affected circuit. This has the desirable effect of reducing Z_1 noise current, thus reducing $V_{COUPLED}$.

A Faraday shield model is shown by Figure C.34. In the left picture, the function of the shield is noted by the way it effectively divides the coupling capacitance, C. In the right picture, the net effect on the coupled voltage across Z_1 is shown. Although the noise current I_N still flows in the shield, most of it is now diverted away from Z_1. As a result, the coupled noise voltage $V_{COUPLED}$ across Z_1 is reduced.

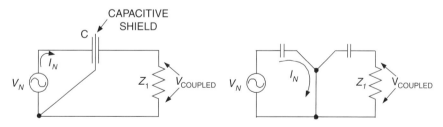

Figure C.34: An operational model of a Faraday shield.

A Faraday shield is easily implemented and almost always successful. Thus capacitively coupled noise is rarely an intractable problem. However, to be fully effective, a Faraday shield must completely block the electric field between the noise source and the shielded circuit. It must also be connected so that the displacement current returns to its source, without flowing in any part of the circuit where it can introduce conducted noise.

C.28 The Floating Shield Problem

It is quite important to note here that *a conductor that is intended to function as a Faraday shield must never be left floating, because this almost always increases capacity and exacerbates the noise problem.*

An example of this "floating shield" problem is seen in side-brazed ceramic IC packages. These DIP packages have a small square conducting Kovar lid soldered onto a metallized rim on the ceramic package top. Package manufacturers offer only two options: the metallized rim may be connected to one of the corner pins of the package, or it may be left unconnected.

Most logic circuits have a ground pin at one of the package corners, and therefore the lid is grounded. Alas, many analog circuits don't have a ground pin at a package corner, and the lid is left floating—acting as an antenna for noise. Such circuits turn out to be far more vulnerable to electric field noise than the same chip in a plastic DIP package, where the chip is completely unshielded.

Whenever practical, it is good practice for the user to ground the lid of any side-brazed ceramic IC where the lid is not grounded by the manufacturer, thus implementing an *effective* Faraday shield. This can be done with a wire soldered to the lid (this will not damage the device, as the chip is thermally and electrically isolated from the lid). If soldering to the lid is unacceptable, a grounded phosphor-bronze clip or conductive paint from the lid to the ground pin may be used to make the ground connection,.

A safety note is appropriate at this point. Never attempt to ground such a lid without first verifying that it is unconnected. Occasionally device types are found with the lid connected to a power supply rather than to ground.

A case where a Faraday shield is impractical is between IC chip bondwires. This can have important consequences, as the stray capacitance between chip bondwires and associated lead-frames is typically ≈ 0.2 pF, with observed values generally between 0.05 pF and 0.6 pF.

C.29 Buffering ADCs Against Logic Noise

If we have a high-resolution data converter (ADC or DAC) connected to a high-speed data bus that carries logic noise with a 2 V/ns–5 V/ns edge rate, this noise is easily connected to the converter analog port via stray capacitance across the device. Whenever the data bus is active, intolerable amounts of noise are capacitively coupled into the analog port, thus seriously degrading performance.

This particular effect is illustrated by the diagram of Figure C.35, where multiple package capacitors couple noisy edge signals from the data bus into the analog input of an ADC.

Present technology offers no cure for this problem, within the affected IC device itself. The problem also limits performance possible from other broadband monolithic mixed signal ICs

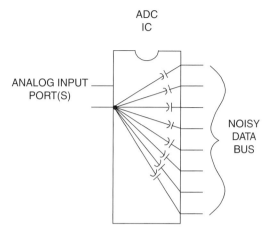

Figure C.35: A high-speed ADC IC sitting on a fast data bus couples digital noise into the analog port, thus limiting performance.

with single-chip analog and digital circuits. Fortunately, this coupled noise problem can simply be avoided by *not* connecting the data bus directly to the converter.

Instead, *use a CMOS latched buffer as a converter-to-bus interface*, as shown by Figure C.36. Now the CMOS buffer IC acts as a Faraday shield and dramatically reduces noise coupling from the digital bus. This solution costs money, occupies board area, reduces reliability (very

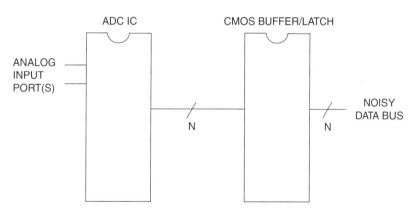

- THE OUTPUT BUFFER/LATCH ACTS AS A FARADAY SHIELD BETWEEN "N" LINES OF A FAST, NOISY DATA BUS AND A HIGH PERFORMANCE ADC

- THIS MEASURE ADDS COST, BOARD AREA, POWER CONSUMPTION, RELIABILITY REDUCTION, DESIGN COMPLEXITY AND, MOST IMPORTANTLY, IMPROVED PERFORMANCE

Figure C.36: A high-speed ADC IC using a CMOS buffer/latch at the output shows enhanced immunity of digital data bus noise.

slightly), consumes power, and it complicates the design—but it does improve the signal-to-noise ratio of the converter. The designer must decide whether it is worthwhile for individual cases, but in general it is highly recommended.

Bus switches can also be utilized to isolate data lines from buses.

Endnotes

[1] Doeling, W., Mark, W., Tadewald, T., and Reichenbacher, P., "Getting Rid of Hook: The Hidden PC-Board Capacitance," *Electronics*, October 12, 1978, p. 111–117.

[2] Rich, Alan, "Shielding and Guarding," *Analog Dialogue*, Vol. 17, No. 1, 1983, p. 8.

[3] Morrison, Ralph, *Grounding and Shielding Techniques*, 4th Edition, John Wiley, Inc., 1998, ISBN: 0471245186.

[4] Ott, Henry W., *Noise Reduction Techniques in Electronic Systems*, 2nd Edition, John Wiley, Inc., 1988, ISBN: 0-471-85068-3.

[5] Brokaw, Paul, "An IC Amplifier User's Guide to Decoupling, Grounding and Making Things Go Right for a Change," *Analog Devices AN202*.

[6] Brokaw, Paul, "Analog Signal-Handling for High Speed and Accuracy," *Analog Devices AN342*.

[7] Brokaw, Paul, and Barrow, Jeff, "Grounding for Low- and High-Frequency Circuits," *Analog Devices AN345*.

[8] Barrow, Jeff, "Avoiding Ground Problems in High Speed Circuits," *RF Design*, July 1989.

[9] Bleaney, B. I. & B., *Electricity & Magnetism*, Oxford at the Clarendon Press, 1957, pp. 23, 24, and 52.

[10] Dummer, G. W. A., and Nordenberg, H., *Fixed and Variable Capacitors,* McGraw-Hill, 1960, pp. 11–13.

[11] Rempfer, William C., "Get All the Fast ADC Bits You Pay For," *Electronic Design, Special Analog Issue*, June 24, 1996, p. 44.

[12] Sauerwald, Mark, "Keeping Analog Signals Pure in a Hostile Digital World," *Electronic Design, Special Analog Issue*, June 24, 1996, p. 57.

[13] Grame, Jerald, and Baker, Bonnie, "Design Equations Help Optimize Supply Bypassing for Op Amps," *Electronic Design, Special Analog Issue*, June 24, 1996, p. 9.

[14] Grame, Jerald, and Baker, Bonnie, "Fast Op Amps Demand More Than a Single-Capacitor Bypass," *Electronic Design, Special Analog Issue,* November 18, 1996, p. 9.

[15] Kester, Walt, and Bryant, James, "Grounding in High Speed Systems," *High Speed Design Techniques, Analog Devices,* 1996, Chapter 7, p. 7–27.

[16] Pattavina, Jeffrey S., "Bypassing PC Boards: Thumb Your Nose at Rules of Thumb," *EDN*, October 22, 1998, p. 149.

[17] Johnson, Howard W., and Graham, Martin, *High-Speed Digital Design,* PTR Prentice Hall, 1993, ISBN: 0133957241.

[18] Kester, Walt, "A Grounding Philosophy for Mixed-Signal Systems," *Electronic Design Analog Applications Issue*, June 23, 1997, p. 29.

[19] Morrison, Ralph, *Solving Interference Problems in Electronics,* John Wiley, 1995.

[20] Motchenbacher, C. D., and Connelly, J. A., *Low Noise Electronic System Design*, John Wiley, 1993.

[21] Crystal Oscillators: MF Electronics, 10 Commerce Drive, New Rochelle, NY, 10801, 914-576-6570.

[22] Crystal Oscillators: Wenzel Associates, Inc., 2215 Kramer Lane, Austin, Texas USA 78758, 512-835-2038, www.wenzel.com.

[23] Montrose, Mark, *EMC and the Printed Circuit Board,* IEEE Press, 1999 (IEEE Order Number PC5756).

Acknowledgments

Portions of this section were adapted from Grant, Doug, and Wurcer, Scott, "Avoiding Passive Component Pitfalls," originally published in *Analog Dialogue 17-2*, 1983.

Index

4K×8 SRAM logic circuit, 104
8×8 MOSFET Bipolar memory cells, 101
8×8 reading ROM circuit, 102
8×8 ROM logic circuit, 100
8-bit MOD-256 asynchronous counter, 94
8-bit MOD-256 synchronous counter, 96
8 bit register, 401, 437
 with 8 D flip-flops, 91
10Base-T cable, 157
16K×8 DRAM circuit, 106
16K×8 SRAM logic circuit, 105
802.11 hardware configuration
 with PCI card, 152
 with SoC, 152
 Standards, 148–150

A

AC catheterization, 421
AC circuits, 21–28
 Active devices, 28–32
 Capacitors, 23–27
 Inductors, 27–28
AC fan-out, 246–247
Acceptance filters, 386–387
Acknowledge bit (ACK), 347, 348, 362, 364, 385
Active components, 8
Active devices, 28–32
Active hardware development tools, 444
Adastra Neptune x86 board, 161
ADC clock jitter *see* Aperture jitter
Address latch enable (ALE), 439
Address strobe *see* Address latch enable
ADDRX bits, 351–352
Agilent, 439

Altera's Nios® device, 281
Alternate current, 21–28
AMD/National Semiconductor x86 reference board, 7, 59
Ampere, 12
Ampro MIPS reference board, 8, 60
Ampro PowerPC reference board, 8, 61
Ampro's Encore 400 board, 63
Analog and digital signals, 10–11
 Noise in, 11
 Separation, 479
Analog interface, 420–434
 ADCs, 420–421
  noise quantification and visualization, 432–434
 Analog channel, 421–422
 Graphical and numerical data, analyzing, 427–431
 Linux data capture program, 425–427
 Linux histogram visualization, 431–432
 Linux PC, sample data transmission, 424–425
 Precision reference sampling, with dynamic C, 422–424
Analog oscilloscope, 285
Analog return current, 475–476
Analog-to-digital converter (ADC), 420–421, 474, 480, 481–482
 Digital outputs, 481–483
 and Logic noise, buffering, 506–508
 Noise quantification and visualization, 432–434
Analog TV board, 174
 with Controller ISA implementation, 72

AND gate, 48, 49, 95
Aperture jitter, 484
Arbitration, 169, 344, 351, 381
 and Clock synchronization, 347
 I^2C, 350–351
 PCI, 177
Architectures, 64, 72
Arithmetic logic unit (ALU), 85–89
ARM architecture, 280–281
Assembler transmit, 330–331
Asymmetrical delay, 241
Asynchronous bus, 171, 204
Asynchronous counter, 93
 8-bit MOD-256, 94
Asynchronous logic, 240
Asynchronous memory, 203
Asynchronous memory controller (AMC), 204, 205
Asynchronous transfer, 119, 124, 141–143
ATA (AT Attachment), 209–210, 212
ATAPI (ATA Packet Interface), 210, 213
Attachment Unit Interface (AUI), 157–158
Autobuffer mode, 229
AVR, 348–350, 354
 Easy Ethernet, 375
 I^2C master-receiver mode code, 358–359
 Master I^2C code, 352–358, 368
AVR-to-PIC communications ball, 365
AVR-to-PIC grand I^2C ball, 362

B

Backplane bus, 168
Bash shell script, 427

BASIC
   C code, 341
   PicBasic Pro code, 336
   RS-232 instruction, 339
   Writing code in, 338
   Writing RS-232 Microcontroller routines in, 333–339
Battery-powered system, 447
Baud rate, 120, 143310, 328
Bayer color filter, 419
Benchmarks, 133, 448, 449
Billions of IPS (BIPS), 448
Binary Coded Decimal (BCD), 44, 46–47, 310
Binary logic, 43
   Cheat sheet, 46
Bipolar junction transistor (BJT), 405
Bipolar memory cells, 101
Bit rate, 120, 142–143
   I^2C SLOW
Bit S, 362
BJT-based drivers, 405–409
Blackfin cache organization, 189
Block diagrams, 1
   of Memory array, 70
   Net+ARM Ethernet, 160
   Serial components, 145
   von-Neumann-based I/O, 137
Blocked transferring scheme, 173
Board buses, 166
   Arbitration and timing, 168
   Integration, with other board components, 179–180
   Performance, 180–181
Board I/O, 137
   Component interfacing, 161–164
   Parallel I/O, 153–161
   Performance, 165–166
   Serial I/O, 140–152
Buffer Full (BF) bit, 362, 363
Burst transfer scheme *see* Blocked transferring scheme
Bus arbitration, 168, 169, 385–386
Bus arbitration and timing, 168
   I^2C (Inter IC) bus, 174–175
   Peripheral component interconnect (PCI) bus, 175–178
Bus contention, 243
Bus handshake, 171

Bus performance, 180–181
Byte ordering, 70, 167

C
Cache, 108–110, 187, 188, 191–193
   Architecture, 190
   Array, with tags, 194
   Concept, 189
   Contingent information, 191–193
   Data storage in, 110
   Definition, 188
   Direct-mapped cache, 190
   Fully associative cache, 190
   in Harvard models, 109
   in Memory hierarchy, 108
   N-way set-associative cache, 191
   in von Neumann model, 109
   Write-back data cache, 193–195
   Write-through data cache, 193–195
Cache hit, 109, 192
Cache miss, 109, 191, 192
Capacitive load, 246, 263
Capacitive noise, 504–505
Capacitive reactance, 23, 24
Capacitors, 23–27, 246, 247
   Ceramic capacitor, 477
   Decoupling capacitor, 481, 488
   Discrete capacitor, 502
   DRAM, 104, 105, 106
   Parasitic capacitor, energy storage in, 390
   Snubber's capacitor, impedance, 415
Carriage return/linefeed (CRLF), 340, 341
Carrier Sense Multiple Access/ Collision Detection (CSMA/CD), 381
Carrier sense system, 381
Cascaded adders, 88
Central Processing Unit (CPU), 5, 82–99, 109, 420
   Arithmetic logic unit, 85–89
   Components, 82
   Control unit, 97–98
   Counters, 93–97
   Execution time, 131, 132
   Fetch, decode, and execution cycle of, 83

   Flags, 92–93
   Internal buses, 84
   On-chip memory, 99
   MPC860 Processor, 83
   Registers, 89–92
   Requirements, 129–130
   and System (master) clock, 98–99
   x86, 278
Central-serialized arbitration, 170
Charge, 12
Cheat sheet, 45, 46
   Binary and hex, 46
Chip Select (CS) signal, 378
Circuits, 18–20, 40, 267, 498
   8×8 reading ROM circuit, 102
   AC circuits, 21–28
   Analog circuit, 26, 27, 421
   ANOE gate circuit, 30
   Clear to Send Circuit, 306
   Data Career Detect Circuit, 307
   Data Set Ready Circuit, 307
   Data Terminal Ready Circuit, 307
   Datacomm circuits, 28
   DC circuit, 12–21, 390
   Debugging, 285–286
   Digital circuit, 12, 53, 482, 490
   Diode OR circuit, 32
   Electrical path, 5
   Full address gate-level circuit, 87
   Gate-level circuit, 91
   High-speed circuit, 474
   I/O port circuit, 116
   Logic circuit, 34, 43, 85, 86, 100, 104, 105, 506
   Low-speed circuit, 441, 474
   Multifunction ALU gate-level circuit, 89
   Protective Ground Circuit, 306
   RC circuit, 25
   Received Data Circuit, 306
   Requested to Send Circuit, 306
   Ring Indicator Circuit, 308
   Signal Common Circuit, 307
   SR flip-flop gate-level circuit, 93
   Transmitted Data Circuit, 306
CISC vs. RISC, 75
Clear to Send Circuit (CTS), 302, 306
CLKIN, 488

Clock, 53–54
    Frequency, 244
    Signal, 98–99, 171, 244
Clock period, 131, 272
Clock stretching, 350, 359, 365
Clock synchronization, 347–351
Code snippet
CodeDesigner Lite, 334, 336
Coding system, 43–46
    BCD, 46–47
Collision, 381
Column Address Strobe (CAS), 106, 200
Combinatorial logic, 240
    AND gate, 48, 49
    Circuits, 50–53
    NAND gate, 48–49
    NOR gate, 49–50
    NOT gate, 47–48
    OR gate, 49
    Tristate devices, 53
    XOR gate, 50
Communication interface, 114, 127–128, 138, 164
Communication port, 114, 138, 163
Complementary logic MOS (CMOS), 255
    TTL compatible signal interfacing, 258, 259
Complex I/O subsystem, 115, 139
Complex instruction set computing (CISC) model, 74
    ISA implementation, 75
    vs. RISC model, 75
Complex programmable logic devices (CPLDs), 418–420
Conductor resistance, 470–471
Consultative committee on international telegraphy and telephony (CCITT), 303
Control unit (CU), 82, 97–98
    PowerPC Core and, 97
Controller Area Network (CAN)
    Architecture, 380–382
    Bus arbitration, 385–386
    Data formats, 382–385
    Message filtering, 386–387
Controller ISA model, 72
Coordinated protection, 393
Copper PCB conductors, 470–471
Current, 12–13

Custom Computer Services C Compiler, 319, 323, 324, 325, 329, 339, 340, 352, 354, 372
Custom peripherals building, with FPGAs, 281–282
Cycle-stealing DMA, 214
Cycle time see Clock period
Cycles per instruction (CPI), 131
Cyclic redundancy code, 385
Cyclical redundancy checking, 212
Cylinder, head, and sector (CHS) method, 212

D
D flip-flop, 54–55
Daisy-chain arbitration see Central-serialized arbitration
Darlington, 407–409, 411
Data acquisition channel (DAQ channel), 421
Data Carrier Detect Circuit CF, 307
Data Circuit-terminating Equipment see Data Terminal Equipment
Data Communications Equipment (DCE) device, 302, 303, 304
Data Length Code (DLC), 384
Data packet, RS-232, 311–312
Data Set Ready (DSR), 302, 307
Data Terminal Equipment (DTE), 144, 145, 302, 303, 304
Data Terminal Ready (DTR), 302, 307
Datapath ISA model, 72
DB9 connector, 145, 146
DB25 connector, 145, 146
DC characterization, 421
DC circuits, 12–21
    Circuits, 18–20
    Current, 12–13
    Power, 20–21
    Resistors, 14–17
    Voltage, 12–14
DDR2 SDRAM, 203
DEBUG functions, 335–336
Debuggers, 445–446
Debugging tricks, 437–438
DEBUGIN functions, 335, 336
De-rating delay, for excess CL, 266
Descriptor Array mode, 231

Descriptor-based DMA, 231
Descriptor List method, 231
Development hardware, selection, 282–285
Dielectric absorption (DA), 502–503
Digital inputs
    Expansion, 398–402
    Protection, 392–398
Digital interfacing, 389–404
    3.3 V and 5 V devices, mixing, 389–392
    Expanding digital inputs, 398–402
    Expanding digital outputs, 402–404
    Protecting digital inputs, 392–398
Digital oscilloscopes, 286
Digital outputs, 402–404
    Analog-to-digital converter, 482–483
    Expansion, 402–404
Digital return current, 475–476
Digital signal processor (DSP), 72, 474
    with Internal phase-locked loops, 487–488, 489
Digital system, 10, 474–475
Digital-to-analog converters (DACs), 474–475, 480
Diode, 31–32
    Schottky diode, 257
    Zener diode, 34, 394, 410
Direct current, 12
Direct memory access (DMA), 164, 214
    Classifications, 228
    Cycle-stealing DMA, 214
    Descriptor management, 231–234
    Direct-mapped cache, 190
    Descriptor-based DMA, 231
    DMA controller, 215–218
        Programming, 218
    External DMA, 235–236
    Register-based DMA, 228–231
    System performance tuning, 234
    Transfer configuration, 228
"Dirty" RS-232 circuitry, 318
Discrete cosine transform (DCT) engine, 281

Distributed arbitration scheme, 170, 171
Double buffering, 230, 360
Double data rate (DDR) SDRAM/ DDR1, 202–203
Double-sided vs. multilayer PCB, 477
Dynamic central parallel arbitration, 169–170
  PCI bus, 177
Dynamic RAM (DRAM), 103, 104, 106, 107
  (capacitor-based) memory cell, 105

E

Easy Ethernet AVR, 353, 375, 376
Easy Ethernet CS8900A, 343, 363, 375–376
eCos, 292
  Operating system, 292, 298–299
EEPROM (electrically erasable programmable ROM), 102–103, 206–207
Effective series inductance (ESL), 398
Effective series resistance (ESR), 398, 413
Electromagnetic compatibility (EMC) issues, 442
Electromechanical relays, 411–417, 418
Electromotive force (EMF), 12
Electronic Industries Association-232 (EIA-232), 144, 303
Electronics, 12
  AC circuits, 21–28
  Active devices, 28–32
  DC circuits, 12–21
Electrostatic discharge (ESD), 442
  Fault tolerance, 443–444
  Protection, 396–397, 398, 399
Embedded board, 5
  Hardware components, 5–6
  I/O device interfacing with, 162
  Port and device controllers, 115, 139
  and von Neumann model, 5–9
Embedded controller, of hardware design, 440–442
  Ground problems, 441–442
  Power and ground planes, 441
Embedded operating system, 287, 289–295
Embedded processors, 183, 443
  Internal processor design, 78–131
  ISA architecture model, 65–78
  Memory spaces, 183–187
  Performance, 131–133
Emitter-coupled logic (ECL), 259
EPROM (erasable programmable ROM), 101, 206
Error frames, 382, 385
ESD guns, 398
Ethernet cables, 156–157
Ethernet interface, 158, 160
Ethernet port, 158
Ethernet system model
  Adastra Neptune x86 board, 161
  Motorola/Freescale FADS board, 158–160
  Net Silicon ARM7 (6127001) development board, 160–161
Excalibur™ device, 281
Exclusive-OR *see* XOR
Expandable bus, 168, 173
External DMA, 235
External memory, 195
  Asynchronous memory, 203–206
  Nonvolatile memories, 206–207
  Synchronous memory, 195–203

F

Fall time, of signal, 241
Fan-out, 244
  CMOS drives LSTTL, 249–252
  Ground bounce, 253–255
  Transmission line-effect, 251–253
  Wiring capacitance calculation, 247–249
Faraday shields, 504–505, 506
Fault tolerance, in hardware designing, 443–444
Ferroelectric RAM (FRAM), 214
Field effect transistor (FET), 31, 257
Field-programmable gate arrays (FPGA), 418–420
  Custom peripherals, 281–282
Finite state machine with datapath (FSMD) model, 73
First in first out (FIFO), 169–170
Flags, 92–93, 353
Flash converters, 420, 421
Flip-flop, 54, 270
  Gate-level circuit, 91
  Metastability, 242, 243
  Timing specs, 271
  Worst-case timing analysis, 270–272
Floating-point OPS (FLOPS), 448
Floating shield problem, 506
Fly-back suppression diode, 413
Frame buffers, 106
Frames, 119, 141–142, 382
  Remote transfer, 385
Free software, consequences, 295–300
Full adder gate-level circuit, 87
Full adder logic equation, 87
Full adder logic symbol, 87
Full adder truth table, 87
Full duplex transmission scheme, 118, 119, 141
Fully associative cache, 190
Functional timing, 270, 271

G

Gas discharge tubes (GDTs), 392
Gate-level circuit
  of Flip-flop, 93
  Multifunction ALU, 89
  SR flip-flop, 93
Gate timing specs, 271
General Public License (GPL), 296–298
General-purpose register, 90
Geode, 132, 279
Geometric engine, 154
Getc function, 332, 333, 341
Gigabit Media Independent Interface (GMII), 160
Glow voltage *see* Holdover
GNU, 288
  Free software, consequence, 295–300
Gnuplot, 431, 432
GPIO, 328
Graphical design engines, 153
Ground and power planes, 475

Ground bounce, 253–255
Ground isolation techniques, 495–497
Ground plane breaks, carefulness with, 494–495
Ground problems, 441–442
Grounding and decoupling, 483
  Mixed signal ICs, with low digital currents, 479–480
Guarding, 498, 499

H

Half-adder logic circuits, 86
Half-adder logic symbol, 86
Half-duplex transmission scheme, 118, 141
Hard Disk storage
  AT Attachment (ATA), 209
  ATA Packet Interface (ATAPI), 209
  CHS method, 212
  Integrated Drive Electronics (IDE), 209
  Logical block addressing (LBA) mode, 212
Hard Drive Interfaces, 212
  Microdrive, 213
  SATA (Serial ATA), 212
  SCSI, 213
  USB/Firewire, 214
Hard Hat Linux, 290
Hardcopy graphics, 154
Hardware, 1–5
  components, 5–6
Hardware design, tips and techniques
  Battery-powered system, 447
  Connecting tools, 438–439
  Construction methods, 440–442
  Debugging tricks, 437–438
  Electromagnetic compatibility issues, 442
  Electrostatic discharge effects, 442–444
  Hardware development tools, 444–445
  Opinions, 439–440
  Processor performance metrics, 448–449
  Software development tools, 445–446

Thermal analysis, 446–447
Hardware development tools, 444
  Instrumentation issues, 445
Hardware design language (HDL), 281
Hardware drawings, 1–2
  Block diagrams, 1
  Logic diagrams/prints, 2
  Schematics, 1–2
  Timing diagrams, 2
  Wiring diagrams, 2
Harvard architecture model, 78, 183
  vs. Von Neumann 80
Heavy operating systems, 292
Hex, 43, 44
  Cheat sheet, 45
High-current outputs, 404–418
  BJT-based drivers, 405–409
  Electromechanical relays, 411–417
  MOSFETs, 409–411
  Solid-state relays, 417–418
High-speed signal transition, 215
Hold time, 242
Holdover, 392
Homegrown code, 330
"Host" system, 444
HyperTerminal software, 319, 320

I

I/O bus, 114, 138, 168
I/O components interfacing, 161
I/O controller, 114
  and master CPU interface, 164
  Requirements, 129–130
I/O device interfacing, with embedded board, 162–164
I/O hardware, 138
I/O performance, 165–166
I/O port sample circuit, 116
I/O subsystem, 114, 115, 116, 139, 140, 162
I^2C bus, 174, 175, 176, 342, 344–347
  ACKS and NAKS, 347
  Addressing, 351–352
  Arbitration and clock synchronization, 347–351
  AVR master-receiver mode, 358–359

AVR registers, 352–358
AVR-to-PIC communications ball, 365–378
Communication options, 378–387
Complete transfer session, 176
Construction, 344–347
Firmwares, 352
on MPC860, 179–180
PIC slave-transmitter mode, 359–365
Reasons for using, 343–344
and RS-232, comparison, 342
with SL clock, 172
START condition, 345
STOP condition, 345
Wired-AND function, 345
IC packages, 58
IDE (Integrated Drive Electronics), 209
Identifier Extension (IDE) flag, 384
Idle mode, 448
Idle RS-232 signal, 311
IEEE 802.11 wireless LAN
  Networking and communication, 148–153
In-circuit emulators (ICE), 444
Inductive load, 413–418
Inductors, 27–28
Instruction set architecture (ISA), 65
Integrated circuit (IC), 58–61
Integrated processor, 64
Intel x86, 288
Interface hardware, of RS-232, 314–319
Interfacing communication port, 163
Internal phase-locked loops, 487–488, 489
Internal processor design
  Central processing unit, 82–99
  On-chip memory, 99–113
  Processor buses, 130–131
  Processor input/output (I/O), 113–130
Interrupt driven I/O, 164
Interrupt request (IRQ) value, 168
Ions, 12

I

ISA architecture model
   Addressing modes, 71
   Application-specific, 72–74
   General-purpose, 74–75
   Instruction-level parallelism, 76–77
   Interrupts and exception handling, 72
   Operands, 68–69
   Operation
      formats, 67
      types, 65–66
   Storage, 69–71

J

Java virtual machine (JVM) model, 74
JK flip-flop, 55–56, 94
JTAG pod, 288–289

K

Karnaugh map, 389
"Kelvin" feedback
   Voltage drop, in signal lead, 471
Kirchoff's law, 472, 473

L

L1 data memory, 184, 187
L1 instruction memory, 184, 186
L1 memory architecture, 184
Least recently used (LRU), 193
Least significant bit (LSB), 351
Lesser GPL (LGPL), 297–298
Level-1 cache *see* Cache
Library GPL *see* Lesser GPL
Light emitting diode (LED), 437
Linux, 290–291
   Data capture program, 425–427
   Data reduction program, 428–431
   Histogram visualization, 431–432
Linux kernel, 296, 297, 298
Linux PC, Data transmission, 424–425
Load analysis, 264–265
Load-store architecture, 71
Loading analysis, 244, 246–247
   Ground bounce, 253–255
   Transmission line-effect, 251–253

Wiring capacitance calculation, 247–249
Logic analyzer, 439, 445
Logic circuit
   Coding system, 43–47
   Combinatorial logic, 47–53
   Integrated circuit, 58–61
   Sequential logic, 53–57
Logic diagrams/prints, 2
Logic family IC, 255–261
Logic high current sign, 246
Logic low current sign, 245
Logic noise, 506
Logic probes, 439
Logic threshold voltage, 255
Logic Wrap-up, 57
Logical block addressing (LBA), 212
Look-up-table (LUT), 420
LOOPBACK, 155
LP SDRAM *see* Mobile SDRAM
LSTTL, 249–251
   and CMOS processor, 268
   Gate DC parameters, 268
   Worst-case timing analysis, 270–272

M

Macraigor JTAG wiggler, 289
Magnetoresistive RAM (MRAM), 214
Main memory *see* RAM
Master CPU, 129
   I/O controller interfacing and, 164
Master processor communication, with I/O, 165, 168
Master Synchronous Serial Port (MSSP), 360
Maxim, 308–309
   MAX232CPE, 314, 318
Maximum load capacitance, 247
Media Access Control Component (MAC), 158
Media Independent Internet (MII), 160
Medium Attachment Unit (MAU), 157, 158, 159
Medium Dependent Interface (MDI), 157, 158
Memory, 5, 69–70

Memory array, 70
Memory cell, 101
Memory controller (MEMC), 106, 110
Memory hierarchy, 99
   Level 1 cache in, 108
Memory management units (MMUs), 110, 111
Memory map, 112
Memory organization, 112–113
Memory space, 183–187
   L1 data memory, 187
   L1 instruction memory, 186
Memory systems, 183
   Cache, 187–195
   Direct memory access (DMA), 214
   External memory, 195
   Memory spaces, 183–187
Message filtering, 386–387
Metal oxide semiconductor field effect transistors (MOSFETs), 31, 409–418
Metal oxide varistor (MOV), 393, 394, 395
Microchip 16-bit Peripheral Library, 379
Microcontroller and designs selection, 273
   Custom peripheral building, with FPGAs, 281–282
   Development hardware selection, 282–285
   Development toolchains, 286–289
   Free embedded operating systems, 289–295
   Free software, consequences, 295–300
   Laboratory equipment, 285–286
   Right core selection, 276–281
Microprocessor, 64
Mictor connectors, 439
Millions of instruction per second (MIPS), 133, 279, 448
MINIDIP, 500–502
MIPS32/MIPS I, 67
Mitsubishi analog TV reference board, 9, 61
Mixed analog/digital system grounding, 474

Mixed-signal devices, 485
  with High digital currents, in multicard system, 487
  with Low digital currents, in multicard system, 486–487
  Origins of confusion, 485
  PCB layout guidelines, 489–491
Mixed-signal ICs, with low digital currents
  Grounding and decoupling, 480–481
Mobile SDRAM, 201–202
MOD-256 counter
  Flip-flop CLK timing waveform, 94, 95, 96
Modem control signals, 302
Modes, 427
Monta Vista, 290
Motorola/Freescale MPC823 FADS board
  Ethernet system model, 158–160
  RS-232 model, 146–147
MPC823, 66, 67, 158, 159
MPC860, 79, 128
  CPU, powerPC core, 83
  Harvard architecture, 80
  I^2C on, 179
  Interfaced to Ethernet controller, 128
  Memory management and, 111
  Processor buses, 130
  Reference platform and I/O, 117
  Registers, within memory map, 112
  SCC, in UART mode, 122
  SMC, in UART mode, 124
    interfaced to RS-232, 128
  SPI, 125
    interfaced to ROM, 129
MPLAB IDE, 336
MROM (mask ROM), 101
MSI (medium-scale integration), 58
Multicard mixed-signal systems, 478–479, 485
Multifunction ALU, 89
Multilayer PCB, 477, 494
Multipoint grounding system, 478

N

N-bit register, with Flag and flip-flop, 93

N-channel metal oxide semiconductor (NMOS), 255, 257
N-way set-associative cache, 191
NAND flash memory, 207–209
NAND gate, 48–49
Negative acknowledge (NAK), 347, 348
Net silicon ARM7 (6127001) development board, 160
Net silicon ARM7 reference board, 7, 60
Net+ARM Ethernet, 160–161
NET+ARM50 embedded board 155–156
NetBSD, 291
Networking and communications
  Ethernet system, 156–158
  IEEE 802.11 Wireless LAN standards, 148–152
  RS-232, 144–146
Noise margin, 254
Nonexpandable bus, 168
  I^2C bus, 174–175
Nonvolatile memories, 99, 206
  Emerging technologies, 214
  IDE, ATA, and ATAPI, 209–212
  Microdrive, 212
  NAND flash memories, 207–209
  NOR flash memories, 207–209
  SATA, 212
  SCSI, 212
  USB/firewire, 214
NOR flash memory, 207–208
NOR gate, 47–48, 49–50
Novell®, 288
Null modem serial cables, 145
Number system, 43

O

Off-board I/O devices, 162
Off-chip memory, 185
Offset, 428
Ohm's Law, 14, 18, 20, 397, 471
Omron G6B, 412
On-chip memory, 99–113
  Cache, 108–110
  Management, 110–111
  Memory organization, 112–113
  Random access memory, 103–108
  Read-only memory, 99–103
Open collector outputs, 256
Open source license, 296, 298–299
OpenWatcom, 288
Operands, 68–69
Operation, 65–67
Operations per second (OPS), 448
Optical isolation, 260
OR gate, 49
OSCCAL value, 335
Oscilloscope, 35
OSI model
  Ethernet, 156
  IEEE 802.11 standard, 151
  RS-232, 144
Output Enable (OE), 53

P

Packets, 141–142
PalmOS® devices, 277, 293
Parallel circuits, 19
Parallel I/O, 127, 140
  Net+ARM50 embedded board, 155
  Networking and communication, Ethernet, 156–158
  Output and graphics I/O, 153–156
  vs. Serial I/Q, 118–121
Parallel interface, 127, 153
Parallel output and graphics I/O, 153–156
Parity bit, 311
Passive hardware development tool, 444
PC board (PCB) design issues, 469
  ADC digital outputs, 481–483
  ADCs and logic noise, buffering, 506–508
  Analog and digital grounds separation, 479–480
  Capacitive noise, 504–505
  Clock consideration sampling, 483–485
  Double layer versus multilayer PCBs, 477

PC board (PCB) design issues (*Continued*)
  DSPs, with internal phase-locked loops, 487–488
  Dynamic effects, 502–503
  Faraday shields, 504–505
  Floating shield problem, 506
  Ground and power planes, 475–477
  Ground isolation techniques, 495–497
  Ground plane breaks, 494–495
  Grounding summary, 488–489
  "Kelvin" feedback, 471–472
  MINIDIP and SOIC Op Amp guard layouts, 500–502
  Mixed analog/digital systems, grounding in, 474–475
  Mixed-signal devices
    with High digital currents, in multicard system, 487
    with Low digital currents, in multicard system, 486–487
  Mixed-signal grounding, origin of confusion, 485–486
  Mixed signal ICs, with low digital currents
    grounding and decoupling, 480–481
  Mixed-signal system guidelines, 489–491
  Multicard mixed-signal systems, 478–479
  Resistance of conductors, 470–471
  Signal return currents, 472–474
  Skin effect, 491–493
  Static effects, 497–500
  Stray capacitance, 503–504
  Transmission lines, 493–494
PCB effects, 497–500, 502–503
PCB "hook", 503
PCMCIA socket, 282
PDIR, 155
Peripheral component interconnect (PCI), 175–178
Perl Module, 429
Perl script, for data analysis, 430
Permittivity, 247
Personal computer, 313–314

Physical Coding Sub layer (PCS), 159, 160
Physical Layer Device (PHY), 159
Physical Layer Signaling (PLS), 158
Physical Medium Attachment (PMA), 157
Physical Medium Dependent (PMD), 159, 160
PIC I^{2}C slave-transmitter mode code, 359–365
PIC12F675, 310–311, 323, 324
PIC18F452, 342
PicBasic Pro compiler, 334, 337, 339–340
Pick-up current, 416
PICkit™ 1 FLASH Starter Kit, 310, 314, 316–318, 319
Pilot relay, 412
Pin-through-hole (PTH) device, 394, 412
Power, 20–21
  and ground planes, 441
Power supply, 10, 32
  Controls, 35–38
  Oscilloscope, 35
  Probes, 38–41
PowerPC core, 88, 91
  and ALU, 90
  and buses, 84
  and CU, 97
  MPC860 CPU, 83
  and register usage, 92
Print f function, 329, 341
Printed circuit board (PCB), 5, 378, 399
  Construction, 440–441
  Power and ground planes, 441
  Silkscreens, 440
Probes, 38–41
Processor buses, 130–131
Processor input/output, 113
  Master processor, with I/O controller, 127–130
  Parallel I/O, 127
  Processor serial I/O, 121–127
  Serial vs. parallel I/O, data management, 118–121
Processor performance metrics in Hardware designing, 448–449
Processor serial I/O, 121–127

Serial peripheral interface (SPI), 125–127
Universal asynchronous receiver-transmitter, 121–125
Processors, 63
PROM (programmable ROM), 101
Propagation delay, 241
Protective ground circuit, AA, 306
Pulse width, 244

R

RAM (random access memory), 103–108
Raster and display engine, 154
RC circuit, 25–26
Reactance, 23
READ (receive) transaction, 171
  PCI, 178
Ready-made operating system, 294
Real circuits, 19
Real hardware development, 294–295
Receive code, of RS-232, 331–333
Received data circuit (RD), 306
RedBoot, 292
RedHat
  eCos operating system, 292, 298–299
Reduced instruction set computing (RISC) model, 74–75
Register-based DMA, 228
Register-memory architecture, 71
Register set, 70–71
Relays, 413
Remote transfer frames, 382, 385
Remote Transmission Request (RTR) flag, 384
Rendering engine, 154
Request to send (RTS) signal, 302, 306
Resistors, 14–17
Ring Indicator Circuit (RI), 308
Ripple-carry adder, 87
Ripple counter, 56
Rise time, 241
RJ45 connector, 147
ROM (read-only memory), 99–103
  MPC860 SPI interface, 129
ROM emulators (ROM ICE), 444
ROW address strobe (RAS), 106
RS-232, 301

Basic hardware, 310–313
BASIC instruction, 339–341
BASIC writing, microcontroller routine in, 333–339
Code receiving, 331–333
Communication options, 378
Firmware writing, 319–326
History, 303–305
and I^{2}C bus, comparison, 342
Implementation, with microcontroller, 310
Interface, 145
Interface hardware, 314–319
MPC860 SMC interface, 128
Networking and communications, 144–146
Operating procedure, 305–308
Specifications list, 304
transmit code bit, 326–331
Transceiver building, 313–314
Voltage conversion considerations, 308–309
RTLinux, 290
RX_program_1, 333

S

SA-1100 instruction, 68
Safe operating area (SOA), 394
Sampling clock, 483
SATA (Serial ATA), 212
SBC (Single-board computer), 279
Schematic symbols, 451–458
Schematics, 1, 3–5
Schottky diodes, 479–480, 487
Schottky logic, 257
Scope see Oscilloscope
SCSI bus, 173
SDMA, 124, 127
Sensor and actuator interface
  Analog interface, 421–434
  CPLDs, 418–420
  Digital interfacing, 389–404
  FPGAs, 418–420
  High-current outputs, 404–418
Sequencer unit, 97–98
Sequential circuits, 53–57
  Logic wrap-up, 57
Sequential logic, 240
Serial clock line (SCL), 174, 175, 179, 342

Serial Clock (SCK) signal, 378, 379
Serial communication controller (SCC)
  Pinouts, 123
  in Receive mode, 122
  in Transmit mode, 123
Serial Data In (SDI), 378, 379
Serial Data Line (SDA), 174, 175, 179
Serial Data Out (SDO), 378, 379
Serial I/O, 140
  Ethernet, 156–158
  IEEE 802.11 wireless LAN, 148–152
  Motorola/Freescale FADS board, 146
  RS-232, 144–146
Serial interfaces, 121, 140
Serial management controller (SMC), 124
Serial peripheral interface (SPI), 125–127, 143, 378–380
Serial port, 144, 145, 334
Series circuits, 18–19
Set-reset (SR) flop, 54
Setup time, 242
Shell script, 426
Signal common circuit, 307
Signal return currents, 472–475
Signal-to-noise ratio (SNR), 483–484
Simple I/O subsystem, 116
Simple operand types, 68
Simple transistor drivers, 405, 406
Simplex transmission scheme, 118
Simplified UART
  for RS-232 device, 56–57
Single-bit addition circuits, 51
Single instruction, multiple data (SIMD) model, 76
Sinking driver see Simple transistor drivers
Sipex, 308–309
Six-transistor SRAM cell, 103
Skin effect, 491–493
Slave device, 168
  of I^{2}C bus, 344
Slave select (SS) signal, 378
SMC pins, 125
Snubber network, 413, 416
Softcopy graphics, 154

Software development tools, 445–446
SOIC, 500–502
Solid-state relays (SSR), 260, 417–418
Source memory, 190, 193, 194
Spark gas suppressors see Gas discharge tubes
Special-purpose register, 90
Specialized design considerations
  Battery-powered system, 447–448
  Thermal analysis, 446–447
SPI pins, 126
SR flip-flop
  Gate-level circuit, 93
SRAM, 204
SSI (small-scale integration), 58
SSPBUF, 359, 362, 363, 364–365
SSPCON, 359
SSPSTAT, 359, 362
Star ground, 478, 479, 485
Start bit, 142, 311, 335
Static RAM (SRAM), 103, 106, 107, 204
STOP bit, 142, 335
Stop mode, 230
Storage register, 89–90
Stray capacitance, 503–504
STROBE, 155
Substitute Remote Request (SRR) flag, 384
Successive approximation (SAR) converters, 420
Superconductors, 14
SuperH, 279
Superscalar machine model, 76
Surface mount technology (SMT)
  GDT, 393
  MOSFETs, 411
  Relay, 412
SurgX® technology, 395
SwitcherCAD III, 417
Sybase®, 288
Synchronous bus, 171
Synchronous Dynamic Random Access Memory (SDRAM), 185, 196, 201
  Commands, 197, 199
  Pin description, 197
  Refreshment, 201

Synchronous logic, 240
Synchronous memory, 195
   CAS latency, 200
   DDR SDRAM/DDR1, 202–203
   DDR2 SDRAM, 203
   Mobile SDRAM, 201–202
   SDRAM, 196–199
     refreshing, 201
Synchronous serial interface, 143, 378
Synchronous transfer, 119, 120, 141, 143
Synchronous transmission, 120
System bus, 167, 168
System performance tuning, 234

T

Tantalum, 24
"Target" system, 444
Tektronix, 40, 439
Tera Term Pro, 321, 322
Thermal analysis, 446–447
Three-stage pipeline, 82
Timing analysis, in embedded system
   Fan-out and loading analysis, 244
   Logic families and interfacing, 255–261
   Noise margin on design, 261–270
   Timing notation, 239–244
   Worst-case timing analysis, 270–272
Timing diagrams, 2
Timing notation, 239
   Clock frequency, 244
   Propagation delays, 241
   Pulse width, 244
   Rise and fall times, 241
   Setup and hold time, 241–243
   Tri-state bus interfacing, 243
Toolchain development, 286–289
Totem pole outputs, 256
Traister and Lisk method, 5
Transceiver, 157–158
Transferring mode schemes, 173–174

Transient voltage suppressor (TVS), 394, 395
Transistor–transistor logic (TTL), 309
   and CMOS, 256–257
   Gate DC parameters, 268
   Interfacing with CMOS, 258–259
   Logic voltages and noise margin, 255
   TTL-to-CMOS interface, 268
   Totem pole and open collector outputs, 256
Transmission-line effects, 251–253
Transmission lines, 493
Transmission medium, 114, 138
Transmitted data circuit, 306
Transorb, 395
Triode, 28
Tristate bus interfacing, 243
Tristate devices, 53
Truth table, 47
   Single-bit addition circuits, 51
TWEN bit, 352, 355, 357
TWI Enable Acknowledge (TWEA) bit, 359
TWINT bit, 355, 357
Two-Wire Interface (TWI), 352
Two-Wire Interface Bit Rate Register (TWBR), 352
Two-Wire Interface Control Register (TWCR), 352, 353
Two-Wire Interface Data Register (TWDR), 352
Two-Wire Interface Status Register (TWSR), 352, 360

U

UcLinux, 291
ULSI (ultra large-scale integration), 58
Ultra DMA, 212
Universal asynchronous receiver-transmitter (UART), 143
UNIX, 424
   Serial port, 425

USART, 369–372
   Transmit and receive line, 342
USB/Firewire, 214
Utility programs, 445, 446

V

Very long instruction word computing (VLIW) model, 77–78
VLSI (very large-scale integration), 58
Voltage, 12–14
Voltage conversions, of RS-232, 308–309
Voltage drop, in PCB signal leads, 471–472
Von Neumann model, 79, 81–82, 137
   and Embedded board, 5–9
   vs. Harvard architectures, 80
   Level-1 cache in, 109
   and Processor pins, 82

W

Watchdog timer, 324
Watcom C++, 288
Window of uncertainty, 242
Wireless transmission medium, 162
Wiring capacitance, calculation, 247–249
Wiring diagrams, 2
Worst case design, 239, 244
   Timing analysis, 270–272
WRITE (transmit) transaction, 171
   PCI, 178
Write-Back data cache, 193, 194
Write-through data cache, 193–194

X

x86, 277, 278, 279
   Ethernet, 161
   Von Neumann architecture, 81
XOR, 50

Z

Zener diode, 34, 394, 410